Membranes and Ion Transport

Volume 3

Date Due

		MAR 0 6 REC'D	

Membranes
and
Ion Transport

Volume 3

Edited by

E. Edward Bittar

Department of Physiology,
The University of Wisconsin,
Madison, Wisconsin, U.S.A.

WILEY-INTERSCIENCE

a division of John Wiley & Sons Ltd.

LONDON   NEW YORK   SYDNEY   TORONTO

Library of Congress Catalog card No. 71-110649

ISBN 0 471 07712 7

Printed in Great Britain by
Dawson & Goodall Ltd.
The Mendip Press, Bath.

This work is dedicated to
the late
EDWARD J. CONWAY
1894–1968

Preface

Thirty years have now elapsed since the publication of the celebrated paper by Boyle and Conway. During these years much has happened in the field of membrane metabolism and ion transport. The following work is an attempt to deal with the notable advances that have already been made and to treat the subject both systematically and critically. Membrane transport may be said to be an integral part of the life sciences but it does not yet seem to cover a comprehensive field. To undertake a general survey of it is therefore to invite criticism, in particular at a time when research is being conducted more intensively than ever. Written primarily for the novice and professed student of the field, the work aims at providing an intelligible view of the wide scope and importance of the subject, and of the different lines of research that may lead to a more unified discipline.

The work is divided into three books. Volume 1 contains a section on the structure, chemistry and behaviour of both artificial and natural membranes, and a section on the theoretical aspects of transport. A useful glossary has been appended at the beginning of this volume. Volume 2 deals with ion movements in symmetrical cells and sub-cellular organelles. And Volume 3 begins with an account of ion movements in asymmetrical cells, as well as water movements in cells in general, and ends with a collection of chapters dealing with ion regulatory mechanisms and cellular interaction and continuity.

The field of ion transport owes a great deal to the late Professor E. J. Conway. It therefore seemed more then fitting to dedicate this work to him.

Madison, Wisconsin
April, 1970

E. EDWARD BITTAR

Contributors to volume 3

E. Edward Bittar — Department of Physiology, University of Wisconsin, Madison, Wisconsin, U.S.A.

Maurice B. Burg — Laboratory of Kidney and Electrolyte Metabolism, National Heart Institute, National Institutes of Health, Bethesda, Maryland 20014, U.S.A.

D. A. T. Dick — Department of Anatomy, The University of Dundee, Dundee, Scotland.

C. J. Edmonds — Medical Research Council Department of Clinical Research, University College Hospital Medical School, University Street, London E.C.1, England.

John G. Forte — Department of Physiology-Anatomy, University of California, Berkeley, California, U.S.A.

D. Gingell — Department of Biology as Applied to Medicine, Middlesex Hospital Medical School, London W.1, England.

Jared J. Grantham — Laboratory of Kidney and Electrolyte Metabolism, National Heart Institute, National Institutes of Health, Bethesda, Maryland 20014, U.S.A.

Francisco C. Herrera — Departmento de Hidrobiologíca, Instituto Venezolano de Investigationes Científicas (I.V.I.C.), Caracas, Venezuela.

Brian M. Johnstone — Department of Physiology, The University of Western Australia, Nedlands, Western Australia.

V. Everett Kinsey — Institute of Biological Sciences, Oakland University, Rochester, Michigan 48063, U.S.A.

A. G. LOWE

Department of Biological Chemistry, The University, Manchester 13, *England.*

SIDNEY PESTKA

Roche Institute of Molecular Biology, Nutley, New Jersey, U.S.A.

Contents

IV. THE CELL SURFACE

I
Ion Movements in Asymmetrical Cells

CHAPTER 1

Frog Skin and Toad Bladder

Francisco C. Herrera

Departamento de Hidrobiología,
Instituto Venezolano de Investigaciones Científicas (I.V.I.C.),
Caracas, Venezuela

I. INTRODUCTION

The toad bladder and the amphibian skin belong to a group of epithelia whose main function is the secretion or absorption of different chemical species. These epithelia may cover the external surface of the animal, such as the skin, and they may form the external surface of specialized external

1

organs, such as the gills, or they may constitute the epithelium of the kidney tubules, the gut, and the many organs capable of secretion and absorption. These epithelia have in common the property of moving different substances in a preferred direction and come under the general heading of polar epithelia (Caparro, 1967). Those of the frog skin and toad bladder have served as fruitful experimental models, forming the basis of much work in the movement of different chemical species, especially sodium, chloride and water. Polar epithelia exhibit morphological, functional and, in many, electrical asymmetry between their apical and basal surfaces.

A morphological polarity or asymmetry is clearly exhibited by the cells composing many of these epithelia. The toad bladder epithelium, for example, has a free or luminal border covered by numerous microvilli from which a nap of fine filaments, the *antennulae microvillares*, extend into the bladder lumen. The basal surface consists only of simple cytoplasmic convolutions extending into irregular intercellular spaces. The frog skin epithelium shows a regular arrangement of cells forming several layers; the deepest layer is constituted by cylindrical cells which become more cubical and finally completely flattened towards the surface of the epithelium. The functional or physiological asymmetry is manifested by the differential permeability of the apical and basal surface of the epithelial cells towards different ions. Net transport across these epithelia would be impossible if they were completely symmetrical, thus their transport properties are a consequence of their functional polarity. The morphologic and functional polarity of these epithelia is associated with an electrical asymmetry; that is, when exposed to identical solutions on both sides, a potential difference is developed across them. The serosal or blood side is positive relative to the mucosal or outer surface. If the two membranes of the epithelium were perfectly symmetrical, no potential difference and no net transfer of material would be expected.

II. STRUCTURE OF THE SKIN AND BLADDER

A. The Frog Skin

The structure of frog skin is shown in figure 1. The skin is composed of an outermost stratified epithelium which rests on a basement membrane that separates the epithelial cells from the corium, a subjacent layer of loose connective tissue containing melanocytes, mucous glands and blood vessels. The epithelium consists of several layers of cells. The more basally located cells tend to be cylindrical and the cells become progressively more flattened towards the surface (Voute, 1963). The epithelial cells are connected with their neighbours by numerous desmosomes or false cell bridges (Ussing, 1965a). It is known that the cell membrane continues through the desmosomes but it may be postulated that the cell membranes at these points are

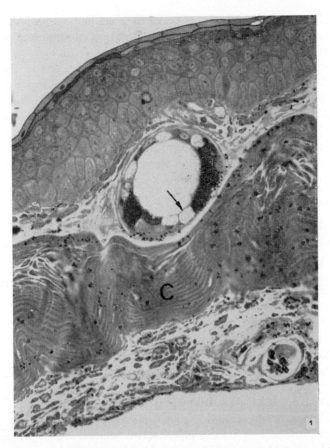

Figure 1. A light micrograph of a section of frog skin. Two layers of cornified cells are evident at the top, with focal areas where the underlying epidermal cells are beginning to separate. A single gland is evident between the base of the epidermis and the dense collagen (C) of the corium. (From Voute, 1967).

very permeable to sodium; thus the whole epithelium is in effect a continuous sodium compartment. This should be kept in mind when the origin of the skin potential is discussed below. Electron microscopic evidence indicates that the outermost surface of the epidermis is covered by a layer of one or two cornified epithelial cells. The structure of these cells indicates that originally they were the same as the underlying epithelial cells. The cells of the deeper epithelial layers are all similar in appearance. They interlock with neighbouring cells by multiple interdigitations and desmosomes, leaving intercellular spaces between the zones of attachment. The cytoplasm of the cells is filled with closely packed tonofilaments which converge towards the desmosomes. Mitochondria, Golgi membranes and pinocytotic vesicles are also prominent in the cells.

Occluding zonules, that is areas where the outer leaflets of the apposed cell membranes fuse into a single 35Å thick dense band, form a continuous belt around the cornified epithelial cells at the line of reflection from the superficial to the lateral aspects of the cells of this layer. The zone of fusion measures 0·1 to 0·3μ in depth. These junctional complexes have been found in epithelia known to maintain chemical and electrical potential gradients between the lumen and subepithelial space. (Farquhar and Palade, 1964).

B. The Toad Bladder

The toad bladder is a bilobed sack which opens into the cloaca. The bladder wall consists of an epithelium, submucosa and serosa. The luminal surface of the organ is lined by a single layer of epithelial cells. The epithelium rests on a thin basement membrane. Within the epithelium there are four different cell types (figure 2). By far the most numerous are the granular cells which contain abundant PAS-positive granules; they amount to 83 per cent of the epithelial cells. Next in order of abundance are the mitochondria-rich cells and the mucous cells, each of which accounts for 11 per cent of the cells (Keller, 1963). A fourth type, the basal cells, are rather inconspicuous dense cells, located close to the basement membrane (Choi, 1965). The luminal surface of the epithelium is populated by short microvilli. A terminal bar with typical desmosome structure is found regularly between opposing cell surfaces close to the lumen. The roles played by the various types of cells in the transport process taking place across the bladder epithelium is at present unknown. However, Gatzy and Clarkson (1965) have suggested that a small fraction of the epithelial cells have a high mucosal permeability, whereas the remaining cells show a rather low but specific permeability to sodium. It has not yet been possible to correlate these differential permeabilities of the mucosal surface of the epithelial cells with any specific cell types.

It is interesting to note that Keller (1963) found a dense selective staining for ATPase in the basal and lateral epithelial cell membranes up to the terminal bars. This may represent staining of components of the cellular sodium pump in agreement with the location of the pump suggested by tracer and pharmacological studies (see below).

The subepithelial layer consists of a loose feltwork of collagen bundles, capillaries, wandering cells, nerves fibres and smooth muscle fibres, while the serosal surface consists of a thin layer of squamous epithelium (Choi, 1965).

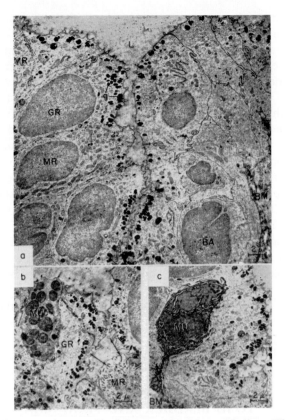

Figure 2. a: Electron micrograph of the toad bladder epithelium. The lumen, L, is at the top; BM, basement membrane; GR, granular cells; MR, mitochondria rich cells; BA basal cells. Microvilli may be seen projecting into the lumen from the epithelial surface.
b: Three types of cells are shown in this low power electron micrograph: a mucous cell, MU, with dense round mucous globules and a dense cytoplasmic matrix; a mitochondria rich cell, MR, with tubules or intervillous spaces (arrows) cut longitudinally; granular cells, GR, with dense granules are also shown.
c: Another type of mucous cell, MU, with a stalk which terminates in an arborization in the vicinity of the basement membrane, BM. (From Choi, 1963).

III. PHYSIOLOGICAL ROLES OF THE AMPHIBIAN SKIN AND BLADDER

The terrestrial habitat has imposed several physiological characteristics upon the amphibian, one of which is the ability to conserve water. This is achieved by means of several mechanisms: for example, they are able to restrict water loss from the kidney. They are also able to store and reabsorb water and sodium from the urinary bladder and to reabsorb sodium and water through the skin (Bentley, 1966). Thus, both the bladder and skin reabsorb sodium and water from what essentially represents the external environment of the animal (since the lumen of the bladder is in indirect communication with the exterior) and transport them towards the blood or vascular side of the epithelium. Not surprisingly, since they are polar epithelia, both exhibit a spontaneous electrical potential difference ranging between 10 and more than 100 mV between their luminal and blood surface. As will be discussed presently, this potential difference is associated with the ionic movements and ionic gradients established by the active reabsorption of sodium across the epithelium. Furthermore, both epithelia respond to antidiuretic hormone by showing an increase in sodium transport and water permeability. This response is similar to that of the collecting duct of the kidney to the hormone.

The water uptake through the skin of *Bufo marinus* was doubled by oxytocin and virtually all the water was retained by the toad with a functional kidney and bladder. By contrast, only about half of the water taken up through the skin was conserved if the bladder had been previously removed. Therefore, nearly 50 per cent of the water retained under the influence of oxytocin is reabsorbed from the bladder. The urinary bladder thus appears to play an important role in water conservation (Steen, 1929; Ewer, 1950; Ewer, 1952a; Ewer, 1952b; Bentley, 1967).

IV. EXPERIMENTAL PREPARATIONS OF THE TOAD BLADDER AND FROG SKIN

Before proceeding to the study of the processes involved in the movement of ions across the amphibian skin and bladder, the experimental preparation employed *in vitro* in the study of these processes should be reviewed.

A. The Transepithelial Potential Difference

Both the frog skin and the toad bladder remain viable *in vitro* for many hours. They are capable of generating a potential difference between the external or, in the case of the bladder, urinary surface and the blood surface of the epithelium; the blood side being positive relative to the opposite face. The frog skin was studied as far back as 1848 when Du Bois-Reymond (1848) first observed the potential difference across it. Galeotti (1904, 1907) found that sodium or lithium was required for the maintenance of the potential. Huf (1935) reported the net transport of chloride across the skin and assumed that the tissue actively transported sodium chloride from the outer face to the interior. Ussing (1949) found that isolated frog skin could transport sodium inwards against a steep concentration gradient. Moreover, the transport of chloride was shown to be a consequence of the electrical potential difference generated by the transporting skin. Calculations showed that the potential difference was sufficient to raise the electrochemical potential of the chloride ions of the external solution above that of the internal medium. For sodium, however, the electrochemical potential of the outside solution was much lower than that of the inner medium. Thus sodium had to overcome the combined effects of the concentration gradient and the electrical potential difference, both of which oppose its inward transport. According to Rosenberg (1954), 'active transport is a transport against the electrochemical potential'. The transport of sodium fulfills this criterion. However, this definition is too restrictive since active transport can take place down the electrochemical gradient, as has been suggested by Bricker and coworkers (1963).

However, the complexity of the frog skin with its stratified epithelium prompted Leaf and coworkers (1958) to investigate the toad urinary bladder which seemed to be simpler in structure. Although it is only 50 to 100 microns in thickness, the toad bladder is capable of maintaining a potential difference of up to 120 millivolts, the serosal side being positive relative to the urinary surface, and transporting sodium from the luminal side to the serosal side of the epithelium.

B. The Short-circuited Preparation

Ussing and his colleagues put forward the idea that the potential difference observed across the frog skin was very likely to be the consequence of the inward transport of positively charged sodium ions. Proof of the relationship between the electrical asymmetry of the skin and the active transport of sodium was obtained by the 'short-circuiting technique' (Ussing and Zerahn, 1951). An analogue of the skin may be imagined whereby the mechanism responsible for the active inward movement of positively charged sodium ions is represented by a battery (positive side inwards) with an internal resistance. This internal resistance represents the opposition offered by the transporting mechanism to the movement of sodium. In parallel with this 'active branch', one may imagine a low resistance shunt representing the pathway for the passive movement of ions represented mainly by a large anionic (chiefly chloride) flow and a small sodium and potassium leak. This analogue of the skin is shown in figure 3a. The potential difference is measured between the points A and B where the 'active branch' and the 'passive branch' of the circuit join. Therefore, the active inward flux of positively charged sodium ions may be represented by a left to right flow of positive current through the upper branch of the circuit. The circuit is

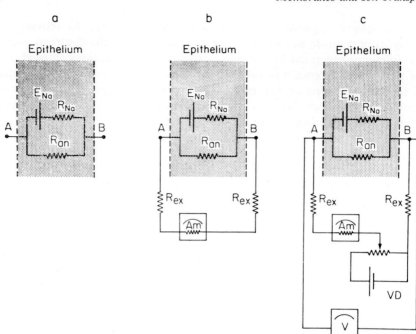

Figure 3. a: Schematic representation of the equivalent circuit for a polar epithelium. The active transport pathway is represented by E_{Na}, the electromotive force of the sodium transporting mechanism and R_{Na}, the internal resistance of the sodium pathway; positive current, represented by sodium ions, moves from left to right. The passive branch of the circuit is represented by R_{an}, the resistance opposed to the passive movement of ions, positive current (represented mainly by a left to right movement of anions) moves from right to left. The epithelial 'battery' is, therefore, partially shunted by the passive ionic flows. The resulting transepithelial potential difference is measured between the points A and B.

b: Equivalent circuit of a polar epithelium connected to an external current measuring circuit. R_{ex} represents the resistances of the reversible half cells and Ringer–Agar bridges required to connect the preparation to the ammeter, Am, which also contains an internal resistance. The sum of the external resistances is much greater than that of the passive pathway. Therefore, most of the current generated by the active transport branch will return through the passive shunt (mainly in the form of a left to right movement of anions) and only a small fraction of the current will flow through the external measuring circuit. Therefore, only a fraction of the current generated by the active transport mechanism will be recorded by the ammeter.

c: Diagram of a short-circuited epithelium. The potential difference between points A and B is abolished by applying at these points an external potential difference equal in magnitude but of opposite sign by means of a voltage divider, VD. No potential difference will be registered by the voltmeter, V, under these conditions. Hence no current will flow through the passive pathway, R_{an}, and all the current flowing through the active transport branch will also flow through the external measuring circuit composed of the ammeter, Am, and the external resistances. Therefore, the current registered in the ammeter is exactly equivalent to that generated by the active transport mechanism.

completed by a right to left flow of positive current through the passive pathway carried mainly by the passive left to right movement of negatively charged anions. If, as shown in figure 3b, an attempt is made to measure the ionic current flowing through the active branch by connecting an ammeter, via appropriate electrodes, in series with the preparation, only a very small fraction of the current will flow through the external measuring circuit (that is, the leads and the ammeter); most of the current will flow through the internal circuit represented by the passive shunt in parallel with both the 'active branch' of the analogue and the measuring circuit. This is because the resistance of the passive pathway is much lower than that of the external circuit employed for measuring the current. Therefore, the current indicated by the ammeter will be a small fraction of the sodium current flowing through the active transport branch of the circuit. Moreover, the ratio of the current flowing through the ammeter circuit to the current represented by the active transport of sodium through the 'active branch' will generally be unknown since it depends on the ratio of the resistance of the external circuit to that of the passive ionic pathway. This latter resistance cannot usually be directly determined. Ussing and Zerahn (1951) circumvented this difficulty by an ingenious experimental design. If, as shown in figure 3c, the potential drop across the preparation (mounted between identical physiological solutions) is reduced to zero by applying an external voltage across it, no current will flow through the passive ionic pathway, since there is no potential difference across it; in other words, the passive movement of ions will be equal in both directions. Therefore, all the current which flows through the active transport branch of the preparation will also flow through the external circuit. The current indicated on the ammeter included in the external circuit will there-fore be exactly equivalent to that carried by the sodium ions actively trans-ported across the preparation.

This method may be looked upon from a different point of view; since the preparation is bathed on both sides by identical solutions, and the potential difference across it has been abolished, there are no electrical or concentration gradients to drive ions across it. Moreover, a hydrostatic pressure difference capable of driving solvent which could carry solutes across the membrane can be easily ruled out by maintaining equal pressures on both sides of the membrane. Nevertheless, a current is registered by the ammeter. This current, therefore, must be driven by an active transport mechanism residing in the membrane.

In this fashion, concentration difference, potential difference and solvent drag, the three 'driving forces' which may account for the passive movement of ions across the membrane, have all been eliminated in the short-circuited preparation. Therefore, the persistence of a net ionic movement represented by the short-circuit current indicates that a mechanism residing in the epithelium moves ions, generally sodium, in a preferential direction. It should be noted that under certain circumstances the equality between net sodium transport and short-circuit current does not hold. Thus, it has been found that adrenaline stimulated the short-circuit current across the skin; however, the increase in current is represented by a net outward chloride flux (Koefoed-Johnsen and coworkers, 1953). Stimulation for 5 minutes at a frequency of 10 sec⁻¹ of the branch of the brachial plexus innervating the skin of the toad causes a transitory decrease, followed by a more prolonged increase in short-circuit current and potential difference. These effects seem to be mediated by the liberation of noradrenaline, due to stimulation of sympathetic nerve fibres. In *Rana temporaria*, the depolarization phase is accompanied by an increase of the outward fluxes of sodium and potassium. The hyperpolarization phase is attributed to the active transport of chloride towards the outer solution. In *Bufo bufo* the depolarization is associated with an active inward transport of chloride and the hyperpolarization with the active inward transport of potassium. (Schoffeniels and Salée, 1965; González and coworkers, 1966, 1967; Sanchez and coworkers, 1966); Salée, 1967; Salée and Vidrequin-Deliège, 1966, 1967a, 1967b.

V.   THE FORCES ACCOUNTING FOR THE PASSIVE MOVEMENT OF IONS

In the absence of water flow, ions move passively across the membrane under the influence of diffusion and the electric field. Thus, the ionic flux across an infinitely thin layer of thickness dx is described by the equation

$$\Phi = -\frac{u' \, C}{zF}\left(RT\frac{d \ln C}{dx} + zF\frac{dV}{dx}\right)$$

where Φ is expressed in moles/cm². sec, C represents the ionic concentration, u' the mobility of the ion in the membrane, z its charge, R,T and F, have their usual meaning, and V is the electrical potential (Whittembury and coworkers, 1961).

Two solutions for this equation, representing the inward and outward ionic fluxes across a membrane of finite thickness a, may be obtained by making the appropriate assumptions. By assuming that the electrical field across the membrane is constant, Goldman (1943) and Hodgkin and Katz (1949) integrated this equation, and the following solutions representing the inward, outward and net flux of ions were obtained

$$\Phi_{2-1} = \frac{PzFV}{RT}C_2\frac{\exp-zFV/RT}{1-\exp-zFV/RT}$$

$$\Phi_{1-2} = \frac{PzFV}{RT} \cdot \frac{C_1}{1-\exp-zFV/RT}$$

where P represents the permeability of the membrane to the ion moving through it and is given by RT u'/zFa.

The net passive ionic flux can be obtained from the difference between the unidirectional fluxes

$$\Phi_{net} = \frac{PzFV}{RT} \cdot \frac{C_1-C_2 \exp-zFV/RT}{1-\exp-zFV/RT}$$

Unfortunately these equations cannot be employed with certainty to calculate the fluxes across epithelial membranes, because the structure of the membrane is far from uniform, and the parameters pertaining to the membrane are, at best, difficult to define. However, an interesting relationship first derived by Ussing (1949) can be obtained from the ratio of the unidirectional fluxes. This relationship has the advantage that a knowledge of the value of P (which implies knowing the value of u' inside the membrane) is not required. In the absence of active transport and solvent drag, it can be shown that the flux ratio of passively moving ions,

$$\frac{\Phi_{1-2}}{\Phi_{2-1}} = \frac{C_1}{C_2}\exp\ zFV/RT$$

quantitatively describes their behaviour.

The flux ratio equation does not depend on any assumptions regarding the membrane and only parameters external to the epithelium, such as potential difference between the bathing solution and ionic concentration are involved. The flux ratio can, furthermore, be derived from thermodynamic considerations which relate the fluxes to the electrochemical gradients without involving any assumptions as to the interface across which the ionic movements take place. The flux ratio will indicate whether the ionic movements observed can be accounted for by the electrochemical gradients between the solution or if other mechanisms (such as solvent drag or active transport) are also involved.

Deviations of the experimentally determined flux ratios from the values predicted by the flux ratio equation imply that the mobilities of ions are not equal in both directions either because of interaction of the ions with other moving particles or because the apparent mobility in one direction has been increased because of the existence of an active transport mechanism for the ion. Thus the flux ratio for sodium across the frog skin and toad bladder can be more than a hundred times greater than that calculated from the concentration and potential differences. In the foregoing discussion, the possibility of ions moving through the membrane carried by a flow of solvent, i.e. water, has not been considered. If water flows across the membrane because of an applied hydrostatic pressure difference, it is conceivable that it will accelerate the movement of ions in its direction of flow, thus invalidating the flux ratio. Experimentally this possibility can be eliminated by maintaining equal pressures on both sides of the membrane. Other sources of deviation of the experimentally determined flux ratios from those predicted by the equation, such as single file diffusion, will not be considered here.

VI. NATURE OF THE TRANSEPITHELIAL POTENTIAL DIFFERENCE

Although the theoretical aspects of the generation of transmembrane potential differences and ion transport are reviewed elsewhere in this volume, specific models accounting for transepithelial transport and potential difference will be briefly discussed here.

A. The Frog Skin

On the basis of experiments in which the relationships between the potential difference across the skin and the sodium concentration of the outer solutions and the potassium concentration of the inner solutions were studied, Koefoed-Johnsen and Ussing (1958) proposed a model to explain the generation of the potential difference across the skin. According to these authors, the outer surface of the skin is selectively permeable to sodium and the inner surface to potassium. Therefore, the total skin potential difference results from the sum of two Nernst diffusion potentials generated at the sodium-permeable outer barrier and the potassium-permeable inner barrier of the epithelium. An active transport mechanism located at the inner border of the epithelial cells would maintain a low sodium and high potassium content inside the cell by effecting a 1:1 sodium-for-potassium exchange. Here it does not contribute to the membrane potential.

Several experimental findings favour this hypothesis. In the absence of penetrating anions, Koefoed-Johnsen and Ussing found that if the potassium concentration of the inner solutions were increased the transepithelial potential difference decreased with almost the correct slope for a potassium electrode, i.e., 60 mV per ten-fold increase in potassium concentration. Similarly, the transepithelial potential difference fell as the sodium concentration of the outer solution was decreased, and the slope of the line relating the fall in potential difference to the decreasing sodium concentration was, in the experiments of Koefoed-Johnsen and Ussing, close to the expected theoretical slope for a sodium electrode. Moreover, this model predicts that two positive potential steps relative to the outer solutions should be encountered in proceeding from the outer to the inner surface of the epithelium. Using glass microelectrodes, Engbaek and Hoshiko (1957) and Cereijido and Curran (1965) observed two positive steps relative to the outer side. However, other patterns were also observed. In some experiments, three or more potential jumps were seen and in a few experiments, only one step was found (Cereijido and Curran, 1965). Figure 4 is a diagrammatic correlation between skin structure and the potential profile (Hoshiko, 1961).

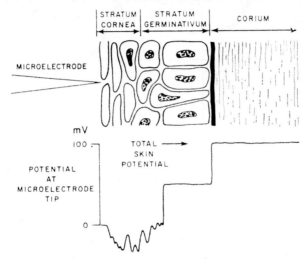

Figure 4. Diagrammatic correlation of the potential profile through the frog skin with skin structure. (From Hoshiko, 1961).

The observation of different patterns of the potential difference is not surprising since the epithelium of the skin is a complex structure containing several cell layers. Ussing and Windhager (1964) have suggested that sodium diffuses across a sodium-selective barrier located just underneath the cornified layer of cells, into the cells of the *stratum spinosum*, and across the desmosomes into the *stratum germinativum* cells towards the pumping sites in the inner membranes of the most basally located cells. Therefore, the intraepithelial potential jumps, according to these authors, could be due to the passage of sodium from the outer cells via the desmosome membranes (which would be highly sodium-selective) to the pumping sites in the innermost cell membrane (Ussing, 1965a). Since the epithelial cells are separated by large intercellular spaces which may communicate rather freely with the internal solution, it is conceivable that cells in the *stratum spinosum* may also

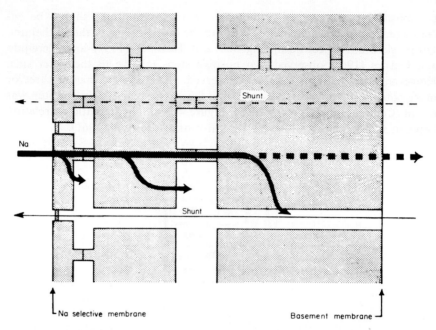

Figure 5. Schematic diagram of the skin showing the series arrangement of the epithelial cells, the active transport path, and the shunt paths through interspaces and through the cells. Modified after Ussing and Windhager (1964).

participate in the active transport of sodium by pumping ions into the interspace. The path of sodium across the epithelium is shown in figure 5.

Under short-circuiting conditions, a marked intracellular negativity has been observed in epithelia exposed to chloride-Ringer's solutions. In chloride-Ringer's solution, a potential well is to be expected from consideration of the behaviour of chloride ions. Chloride appears to be passively transported, and so, if cellular chloride concentration is lower than that of the bathing solution, chloride would be expected to enter the cells. The result would be swelling of these cells. However, appreciable swelling does not occur under these conditions; therefore, a negative well of sufficient magnitude to prevent chloride entry should be present.

Microelectrode studies of the effect of changing the internal potassium and external sodium concentration have shown that the simple hypothesis proposed by Koefoed-Johnsen and Ussing for the transepithelial potential difference must be modified. Cereijido and Curran (1965) found that changing the internal potassium concentration causes a change in transepithelial potential difference in the order of 50 to 60 mV per ten-fold change in potassium concentration. However, only 50 per cent of the change occurred at the inner barrier between the electrode and the inside solution; the remainder of the change occurred at the outer barrier of the epithelium. Several explanations may be offered to account for this finding.

i) It may be due to a rise in cell sodium caused by an increase in the sodium permeability of the external membrane or inhibition of the active transport step induced by a high inner potassium concentration (Hoshiko, 1961; Essig and Leaf, 1963).

ii) It may result from an increased shunting of the potential across the outer barrier caused by the high inner potassium concentration which would increase transepithelial conductance (Ussing and Windhager, 1964).

iii) Finally, if the epithelium were composed of several layers of cells functioning in a manner analogous to that proposed by Koefoed-Johnsen and Ussing (1958), the high internal potassium concentration would cause a depolarization of all inward facing membranes in all cell layers; the potassium-induced depolarization of the inward facing membrane of the cells between the microelectrode and the outer epithelial surface would be measured as a fall in potential difference between the electrode and the outer solution and would be erroneously ascribed to a change at a presumably sodium-selective barrier, which in fact would be composed of outward-facing sodium-selective membranes of the cells, plus the inward-facing potassium selective membranes between the electrode and the outer solution. However, the slope of the line relating the potential difference between the electrode and the inner solution to the inner potassium concentration is too low. Therefore, the inner membrane must be permeable to other ions in the system.

Although the modified Koefoed-Johnsen and Ussing model for the frog skin may serve to account for many of the experimental observations, several investigators have obtained evidence indicating that not all of the intra-epithelial sodium is involved in the process of sodium transport across the epithelium and in the generation of the potential difference. Thus Cereijido and Rotunno (1967) believe that only 37 per cent of the intraepithelial sodium is free in the cell water, thus accounting for the finding that under short-circuit conditions, sodium may enter the epithelium from external solutions containing 5–10 mM sodium, the 18–20 mV intracellular potential well being sufficient as a driving force. Moreover, MacRobbie and Ussing (1961) found that only 21 microns of the 58 micron thick epidermis is osmotically active.

Thus, although the original Koefoed-Johnsen and Ussing model offered a useful experimental hypothesis, more recent work has required modification of this simple explanation of the generation of the potential difference in relation to the sodium and potassium gradients. Moreover, when the ionic pump is considered below, it will be apparent that the one to one exchange of sodium for potassium at the basally located pump must be reevaluated.

The results obtained by Chowdhury and Snell (1965, 1966) are at variance with those presented above. In microelectrode studies on frog skin and toad bladder, these investigators found, using extremely fine tipped electrodes and a piezoelectric oscillator to vibrate the electrode in an axial direction so as to penetrate the tissue, that on entering from the outer or mucosal surface of the epithelium, the electric potential increases rather smoothly with distance once the presumed functioning cellular layer has been reached. Therefore, the smooth potential profile is taken to indicate a resistance distributed over the whole thickness of the epithelium and not confined to just two discrete boundaries. Moreover, the sodium and potassium sensitive functions of the epithelium do not appear to be located at discrete regions but are continuously distributed through the several cell layer in the frog skin or within one cell in the case of the toad bladder.

B. The Toad Bladder

The electrical potential profile across the toad bladder epithelium was investigated by Frazier (1962) who introduced glass micropipettes into the mucosal cells from the urinary surface. In the open-circuited bladder, the cell interior was electrically positive to the urine. Two positive potential steps, corresponding to the apical and basal epithelial surfaces were recorded when the micropipette tip was moved across the epithelium from the urinary side to the serosal side. The sum of the two steps seemed equivalent to the spontaneous transepithelial potential. A schematic diagram of the bladder wall and of the electrical potential profile across it are shown in figure 6.

The experiments of Leb and coworkers (1965) on the toad and bullfrog urinary bladder were similar to those of Koefoed-Johnsen and Ussing (1958) on the frog skin. They concluded that the mucosal surface of the bladder epithelium was sodium-selective and the serosal surface was potassium-selective, in agreement with the Koefoed-Johnsen and Ussing model for the frog skin. However, these experiments are subject to the same criterion as those performed on the frog skin, namely that changes in transmembrane potential difference, when the internal potassium concentration or the external sodium concentration is changed, fail to provide unequivocal evidence as to the location of the changes in potential difference at the inner or outer barriers respectively. This was shown to be so in the frog skin by Cereijido and Curran (1965). Moreover, experiments by Snell and Chowdhury (1965) have indicated that the contralateral effects of sodium and potassium on the transmembrane potential difference may be observed in toad bladder epithelia exposed to sulphate-containing bathing media.

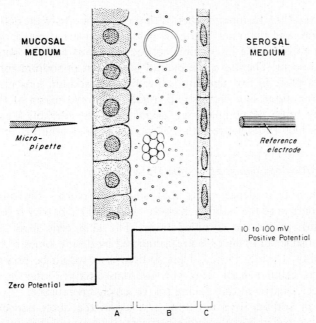

Figure 6. Schematic representation of the cross section of the toad bladder with its potential profile. Section a of the upper part of the Figure represents the epithelium; b represents the subepithelial connective tissue and smooth muscle layer; c is the serosa. The lower part of the figure shows the potential profile in the open-circuited bladder, with positive steps at the apical and basal surface of the epithelium. (From Leaf, 1965).

According to the Koefoed-Johnsen and Ussing model, these contralateral effects could be explained by changes in intracellular sodium and potassium concentrations. However, MacRobbie and Ussing (1961) have pointed out that such changes would not be expected if sulphate is used as the main anion. Chowdhury and Snell believe that the contralateral effects of sodium and potassium on the potential are mediated by their primary effect on the sodium transport system. Thus at concentrations in the range between 0 and 2·5 mM, potassium in the inner bathing solution activates the ion transport system but, at higher concentrations, potassium inhibits the system in a manner reminiscent of non-competitive inhibition of enzymes.

VII.   THE TRANSPORT OF SODIUM ACROSS FROG SKIN AND TOAD BLADDER

A. The Frog Skin

The asymmetric movement of sodium across the frog skin is related to the unequal characteristics of the outer and inner barrier of the epithelium. Thus it is believed that the inner barrier of the cells is the site of the active transport mechanism, and the outer barrier is passively traversed by the sodium ions, although considerable interaction between the ions and the outer barrier is known to occur. Therefore, the overall movement of sodium across the frog skin takes place in two separate steps: first the sodium ions must passively traverse the outer skin barrier and, once in the cytoplasm of the epithelial cells, they must be pumped out against the considerable electrochemical gradient which exists between the cells and the inner solution.

1. *Penetration of Sodium across the Outer Barrier*

Considerable evidence exists that sodium traverses the outer sodium-selective barrier of the skin in a passive fashion. This barrier is believed to be located just under the cornified layer. The skin cells form a functional syncytium since all living cells are connected by desmosomes of high sodium permeability (Ussing, 1965a). Thus sodium ions would be presented with a continuous cellular path from the sodium-selective outer barrier to the stratum germinativum cells facing the basement membrane.

However, sodium does not appear to cross the outer barrier by simple diffusion; considerable interaction between the ions and the outer membrane has been found (Cereijido and coworkers, 1964). This barrier appears to be rate-limiting for sodium transport and to be the site of action of numerous agents which modify the active transport of sodium.

The passive nature of sodium movement across the outer barrier was first postulated by Koefoed-Johnsen and Ussing (1958) when they first proposed their model for the frog skin. These authors based their assumption on the fact that the sodium content of the epithelial cells was low in comparison to that of the bathing solution and that the outer barrier was selectively permeable to sodium. Thus the passive movement of sodium into the cells was not hindered by a high intracellular concentration and sodium movement across the outer barrier took place down its concentration gradient. Removal of sodium across the inner barrier by an active transport process ensured the maintenance of a low intracellular sodium content. However, recent work in which the outer surface of the skin was exposed to low sodium concentrations seemed to indicate that the eletrochemical potential for sodium inside the cell was higher than that of the outer solution thus requiring an active step at this barrier. Thus Biber, and coworkers (1966) have found that net transport of sodium occurs when the outside sodium concentration is 1 mM and the inner is 115 mM. Estimates of epithelial sodium concentration suggested that if all of the epithelial sodium were in a single compartment, an active transport step across the outer barrier would still be required. However, these authors explained their results by an arrangement of sodium compartments in series. The outermost layer may have a sufficiently low sodium concentration to allow passive entry of sodium from an outer solution containing 1 mM sodium. The overall transport of sodium might then involve transport of this ion through successive cell layers with the deeper cells in the epithelium having progressively higher sodium concentrations. This model differs from that of Ussing and Windhager (1964) in that it suggests that sodium is transported from one cell to the next and that the epithelial sodium concentration is not uniform but increases towards the serosa. This is compatible with the finding of more than two potential steps across the skin.

Thus successive modifications of the original Koefoed-Johnsen and Ussing model for the frog skin have been required to account for passive sodium movement across the outer barrier and the multiple potential jumps found in about 20 to 30 per cent of microelectrode penetrations of the skin.

2. *Interaction of Sodium with the Outer Barrier*

The results obtained by many investigators suggested that the movement of sodium across the outer barrier of the frog skin may not be entirely due to simple passive diffusion of free sodium ions. Thus Kirschner (1955a) found that the short-circuit current across frog skin was not linearly related to the external sodium concentration but that it rose with increasing sodium concentration, passed through a maximum at about 35 mM and then decreased as the concentration was raised further. Cereijido and coworkers (1964) found that the sodium permeability of the outer barrier of the skin decreases markedly when the sodium concentration of the outer solution is increased from 7 to 115 mM. Unidirectional chloride fluxes were not affected, suggesting that the changes in membrane properties are relatively specific for sodium. This behaviour of the outer membrane towards sodium could be explained on the basis of a carrier-mediated facilitated diffusion (which is not an energy-requiring process), by an interaction of sodium with a limited number of sites in the membrane (Sjodin, 1959) or by a surface absorption layer of ions (Harris and Sjodin, 1961).

Several neurotropic compounds such as curare (Kirschner, 1955b), local anaesthetics (Skou and Zerahn, 1959), pilocarpine, atropine, pyridine-2-aldoxime methiodide (2-PAM) (Schoffeniels, 1960; Schoffeniels and Baillien, 1960; Tercafs and Schoffeniels, 1961b, 1961c, 1962; Schoffeniels and Tercafs, 1962) reversibly increase sodium transport in the frog skin by increasing the permeability of the outer barrier to sodium; the resulting increase in intracellular sodium concentration in turn induces an increase in the extrusion of sodium by the pump. All these agents contain a cationic nitrogen. Schoffeniels

B

has concluded that quaternary nitrogen derivatives enhance sodium permeability of the outer barrier, and if they are able to reach the active transport site, they then also inhibit this mechanism. This explains the secondary decrease in potential difference and sodium transport which generally follows the initial rise in sodium carriage.

The antidiuretic hormones (ADH) belong to a group of physiological agents whose action on the water balance of amphibia has been extensively studied (see Ewer, 1950, 1952a, 1952b). The isolation and synthesis of the neurohypophyseal hormones has led to considerable clarification of their physiological actions on the basis of their chemical structure (Du Vigneaud, 1956). It is beyond the scope of this chapter to discuss this aspect of the subject, so only the effect of ADH on ionic movement across toad bladder and frog skin will be touched upon here. Several excellent reviews on the subject have been published (Sawyer, 1961b; Leaf, 1965; Bentley, 1966).

Arginine vasotocin, an analogue of the mammalian neurohypophysial principles (Katsoyannis and Du Vigneaud, 1958), appears to be the natural antidiuretic hormone of non-mammalian vertebrates (Sawyer and coworkers, 1959, 1960; Jard and coworkers, 1960; Sawyer, 1961a). This hormone stimulates sodium transport in the isolated frog skin (Jard and coworkers, 1960; Sawyer, 1960). The many physiological investigations performed with Pitressin, a commercial preparation made from pooled beef and pig pituitaries, and therefore containing an uncertain mixture of arginine and lysine vasopressin (Sawyer, 1961b), indicate that the hormones are much less effective in promoting water reabsorption and sodium uptake in amphibian skin and bladder.

The effect of neurohypophyseal principles on frog skin was first studied by Fuhrman and Ussing (1951), who found that the unfractionated posterior pituitary of the blue whale, Insipidin (containing pressor and antidiuretic activity) and Pitupartin (containing oxytocic activity), increased the potential difference across the skin. The lower the initial spontaneous potential difference, the greater the effect of the hormone, but the maximum potential difference tended to be constant. Although these workers were unable to demonstrate a clear-cut effect of the hormone on sodium transport, their results (i.e. the increase of potential difference caused by the hormone) implied that the hormone must have had some effect on sodium transport. This is suggested by the observation that sodium influx was always higher than sodium outflux and that the difference between the two fluxes was greater when the potential difference was higher. They also observed that the hormone caused an increase in water movement from the outside to the inside in the presence of a ten-fold osmotic gradient, the outer solution being 1/10th frog Ringer's solution. It should be noted that at this time the relationship between the electric properties and the functioning of the sodium transport system were just beginning to be understood. Fuhrman and Ussing thought that the effect of these principles could be explained in terms of the lowering of the resistance to the passive diffusion of sodium through some layer in the skin. This hypothesis has been substantiated by more recent work (Curran and coworkers, 1963; Herrera and Curran, 1963).

Herrera and Curran (1963) found that calcium decreases and ADH (Pitressin) increases sodium transport across frog skin. They suggested that, although both agents presumably alter the sodium-permeability of the outer surface of the skin, they act independently on two different sites for sodium movement. Thus calcium causes a fall in sodium movement across the skin (Curran and Gill, 1962), accompanied by a decrease in chloride movement. On the other hand, ADH increases sodium movement without influencing chloride fluxes, suggesting that the changes brought about by ADH occur in a part of the barrier which is not involved in chloride movement. Both sites however may be involved in water movement since calcium decreases net water flow across the skin under an osmotic gradient (Gill and Neidergaard, cited in Herrera and Curran, 1963) while ADH increases it (Koefoed-Johnsen and Ussing, 1953). Therefore, the sites at which calcium and ADH act seem to be located on parallel paths of sodium movement and are probably in the outward-facing barrier of the skin.

Curran and coworkers (1963) extended these observations and found that ADH increases and calcium decreases the permeability of the outward-facing membrane of the cells to sodium. Neither of them appear to modify the active transport system itself which is assumed to be located at the basal surface of the cells. The rate of sodium transport is modified as a result of changes in the size of the cellular sodium pool caused by changes in the rate of sodium entry through the outer membrane. This movement of sodium across the outer barrier is not due to simple diffusion; moreover, considerable interaction between the ion and the barrier takes place (Cereijido and coworkers, 1964). Saturation of net sodium flux is observed when the sodium concentration of the outside solution is raised. This can be accounted for by the observation that as the sodium concentration is increased the sodium permeability of the outer barrier decreases and the rates of sodium entry into the cells does not increase proportionately to the outer sodium concentration, neither does the net active sodium flux, since this flux is dependent on the rate of sodium entry. The fact that the saturation effect is associated with the process of sodium entry into the cells is further substantiated by the fact that the intracellular sodium pool becomes virtually independent of external sodium at concentrations above 60 mM. If the saturation were determined by the active extrusion of sodium by the cell, then the sodium pool would be expected to increase following elevation of the sodium concentration above 60 mM.

An interesting observation of Ussing's (1965) is that swelling of the epithelium caused by exposure of frog skin to a hypotonic internal solution leads to an increase in active sodium transport. Swelling of the outermost living cell layer would decrease its resistance towards sodium ions, possibly by means of opening channels or stretching the membrane. This effect was further substantiated by the work of Ussing and coworkers (1965), who found that epithelial swelling caused by exposure of the inner side of the skin to physiological solutions in which sodium chloride was replaced by potassium chloride abolished the response to vasopressin. They concluded that epithelial swelling increased the permeability of the sodium selective barrier at the outward facing surface of the cells by stretching the cell membrane and altering the dimensions of the pathways through which sodium and water move, thereby mimicking the effects of vasopressin. It is interesting to note that when these authors added vasopressin to skin immersed in sodium sulphate Ringer's solution on the outside and potassium sulphate Ringer's solution on the inner side, swelling of the epithelium occurred simultaneously with an increase in the short-circuit current. When the potassium sulphate Ringer's solution was replaced by sodium sulphate Ringer's solution, the thickness of the epithelium decreased to approximately control values but the short-circuit current failed to return to initial values. The fact that in the presence of potassium sulphate Ringer's solution, vasopressin caused an increased current indicated that high internal potassium concentration *per se* does not abolish the responsiveness to vasopressin and that the lack of response observed with potassium chloride Ringer's solution may be related to the swelling of the epithelial cells which took place before the addition of vasopressin.

More recently, Aceves and coworkers (1968), studying the urodele *Ambystoma mexicanum* found that the short-circuit current, the potential difference and the inward sodium transport across the skin are markedly reduced when chloride in the external solution is replaced by sulphate. It would seem then that the entrance of sodium into the transporting compartment or pool requires the presence of chloride in the outer bathing medium.

3. *Penetration of Chloride*

The observation that sodium is actively transported by the skin served as an impetus to the study of the movement of chloride across the skin. When Koefoed-Johnsen and Ussing (1958) advanced their model, they suggested that both the outer and inner barriers of the epithelium were permeable to chloride, and hence this chloride permeability would serve as a partial shunt to the transepithelial potential difference. The replacement of chloride with a less permeant anion such as sulphate, leads to a rise in the transepithelial potential difference (Koefoed-Johnsen and Ussing, 1958). On the other hand, the presence of copper, silver or calcium in the external medium decreases chloride permeability of the skin (Koefoed-Johnsen and Ussing, 1949, 1958; Ussing, 1949; Curran and Gill, 1962).

Linderholm (1954) reported that chloride conductance decreased when sodium was removed from the external medium; a similar result was obtained by Macey and Meyer (1963) who found that passive chloride flux and conductance decreased when sodium in the external bathing solution was replaced by magnesium or potassium, the decrease being greater when the initial conductance was greater. They suggested that the chloride permeability of the outer barrier of the skin was sodium-dependent, decreasing in the face of a sodium lack.

There is evidence that net inward transepithelial transport of chloride takes place in the isolated frog skin at low internal concentrations of chloride in both *Rana pipiens* and *Rana esculenta* (Martin, 1964; Martin and Curran, 1966). Net inward movement of chloride occurs in the absence of an electro-chemical potential difference, thus fulfilling Rosenberg's criterion for active transport. It is inhibited by 2,4-dinitrophenol and iodoacetic acid, indicating that it is metabolically dependent. Active transport of chloride has been also described in the South American frog *Leptodactylus ocellatus*, by Zadunaisky and coworkers (1963) and Zadunaisky and De Fisch (1964). Chloride move-ment in *L. ocellatus* is inhibited by ouabain and by cupric ion but is insensitive to antidiuretic hormone. Further studies on this point are needed.

4. *The Epithelial Ionic Pool*

Two pathways within the epithelium must be considered when the move-ments of sodium and chloride through the epithelial layer are examined. The ions may enter the living cell layer situated under the cornified epithelium by way of the outer barrier which is permeable to sodium and chloride, and thence proceed to the interspaces. Alternatively, the ions may pass through the cell bridges into the deeper cell layers, where the interspaces are more or less in communication with the inner bathing solution. The role of the inter-cellular bridges as a diffusion path for sodium has been the object of study. Indirect evidence obtained by Zerahn (1965) suggests that they are open to sodium: thus lithium was found to accumulate in the skin, thereby displacing potassium which is pushed out into the inner bathing solution. The amount of potassium displaced is much greater than that contained in the outermost cell layers; this implies that the lithium must have passed through the bridges from one cell layer to the next. This observation suggests that these bridges are probably also open to sodium. Furthermore, Ussing and Wind-hager (1964) have shown that the cells are connected to each other through low resistance pathways represented by the desmosomes. The interspaces also represent ionic pathways through the epithelial layer serving as a shunt pathway for sodium, chloride and other ions. These intercellular channels show an increased shunt conductance towards sodium and sulphate ions when the osmolarity of the external sodium sulphate Ringer's solution is increased by the addition of urea. More recently, Cereijido and Rotunno (1967) have concluded that sodium in the epithelium occurs in two different compartments distributed uniformly throughout the epithelium. One compartment is involved directly in active transport and the other is not involved. The evidence for the uniformity of distribution of the transporting and non-transporting compartments rests on two main points. First, studies of the specific activity in the epithelium indicate that even the outermost cell layers have a compartment which is comparable to that of the rest of the cells. Secondly, the exchange of epithelial sodium with radio-sodium in the bathing solution is essentially complete within a matter of 10 minutes regardless of whether the tracer comes from the inner or the outer bathing solutions. Cereijido and Rotunno proposed that only 37 per cent of the sodium in the epithelium is in the free form, in agreement with the results of MacRobbie and Ussing (1961), who found that only 36 per cent of the epidermis is osmotically active. However, this agreement may be fortuitous, and the arguments seem to run counter to those proposed by Biber and co-workers (1966) who, on the basis of the behaviour of the electrical potential profile across the skin, suggest that the sodium compartments are arranged in accordance with an increasing concentration from the outer to the inner-most layer of the epidermis.

5. *Modification of the Intraepithelial Ionic Content*

Tercafs and Schoffeniels (1961a) found that immediately after interrupting the short-circuit current, the potential difference is much greater than it is before short-circuiting the skin. These workers interpret this finding as being due to the fact that under conditions of short-circuiting, sodium tends to enter the cells from the outer solution and potassium to leave across the inner epithelial barrier in the direction of the internal solution. These ionic movements cause an increase in intracellular sodium and a fall in potassium, which in turn cause stimulation of the pump, resulting in a fall in cell sodium to a level below that which prevailed before short-circuiting, and in a rise in cell potassium to a level above that before short-circuiting. This leads to an increase in potential difference. However, another explanation, which is not based on 'stimulation' of the pump following the fall in cell potassium and the rise in cell sodium, can also be offered. When the transepithelial potential difference is abolished, for example under short-circuit conditions, or when the inner surface of the skin is made negative relative to the outer surface, the electrical gradient opposing the movement of ions across the inner membrane is probably abolished, or rendered more favourable to cationic movements. Therefore, the work expended by the pump against the electrical gradient may now be employed in building up higher concentration gradients for sodium and potassium by increasing the intracellular potassium concentration and lowering the intracellular sodium concentration to levels beyond those which existed before short-circuiting. Tercafs and Schoffeniels did not determine the epithelial ionic content and this is important before this matter can be cleared up. These authors also showed that when the preparation was hyperpolarized, the potential difference was lower than previously. This result can be explained by the opposite argument, namely that increasing the electrical gradient prevents the pump from maintaining the concentration gradients at their normal levels. Thus the intracellular ionic concentration tends to approach those of the bathing solution, leading to a fall in the potential difference.

Many investigations have provided evidence that changes in intraepithelial sodium content influence the active transepithelial transport of sodium. Thus Curran and coworkers (1963) postulated that the increase in transepithelial sodium transport caused by antidiuretic hormone and the decrease caused by calcium are due to changes in the epithelial sodium transport pool. In the case of antidiuretic hormone, there would be an increase in the permeability of the outer epithelial barrier to sodium resulting in a rise of the intracellular sodium concentration and in turn to more sodium reaching the pump. Calcium would be expected to make the outer barrier more impermeable to sodium, and since the pump would continue to expel sodium, the intracellular sodium pool would decrease, ultimately leading to a decrease in the amount of sodium reaching the pump. This is why sodium transport falls. Cereijido and coworkers (1964) also showed that the rate of sodium transport across the skin is related to the magnitude of the cellular sodium pool, which depends on the sodium permeability of the outer barrier of the skin, which in turn is a function of the sodium concentration in the outer bathing solution. This is indicated by the fact that the permeability of the outer barrier decreases as the sodium concentration of the medium increases.

6. *The Transport of Ions across the Inner Barrier*

In the Koefoed-Johnsen–Ussing model for frog skin, the internal barrier of the epithelium is believed to be the site of active sodium and potassium transport. Here the sodium ions would be extruded against a considerable electrochemical gradient which can be conveniently represented as the free-energy change per equivalent of sodium expelled from the cells. This free energy change may be expressed by the relation

$$dG/dn = RT \ln [Na]_i/[Na]_c + E_m F$$

where dG/dn is the free energy change per equivalent of sodium extruded, E_m the potential difference across the basal epithelial membrane, $[Na]_i$ and $[Na]_c$ represent the sodium concentrations in the inner solution and the cells, respectively, and R,T and F have their usual meanings (see Conway and coworkers, 1961; Kernan, 1961).

Koefoed-Johnsen and Ussing suggested that the pump performed a 1:1 exchange of sodium for potassium and that therefore the pump did not bring about a net movement of charge, i.e. the pump itself did not contribute to the potential. Their model further implied that, since the passive leakage of potassium from the cells towards the inner solution is balanced by the active uptake of potassium from the inner solution by the cell, the movement of sodium through the pump is equivalent to the leakage of cell potassium across the inner barrier. Since the pump does not bring about a net movement of charge under short-circuit conditions, the current is presumed to cross the internal boundary of the epithelium in the form of potassium ions diffusing down a concentration gradient (Bricker and coworkers, 1963). However, if the inner potassium concentration is raised so that the electrochemical potential gradient for this ion is reduced or reversed, the short circuit current is not abolished. Hence, the ionic nature of the current carried across the inner barrier becomes a problem of considerable interest. After considering many alternatives to explain the persistence of the short-circuit current, these authors were forced to conclude that the active transport of sodium ions may account for net charge transfer across the inner border of the epithelium and hence that active sodium transport contributed to the short-circuit current. This indicates that the active transport mechanism is electrogenic and a strict 1:1 sodium for potassium exchange may not exist under the conditions of their experiments. It should further be noted that in these experiments a typical case of 'downhill active transport' for sodium took place, since this cation was almost completely replaced by potassium in the inner bathing solution (sodium concentrations ranged from 0 to 6 mM) and the potential difference across the skin was abolished by short-circuiting. The fact that the downhill movement of sodium was active was suggested by several lines of evidence: the flux ratios for sodium were consistently greater than could be accounted for on the basis of passive transport; strophanthidin, a specific inhibitor of active ion transport (see below), caused a fall in sodium influx and the bidirectional fluxes closely approximated to the sodium concentration ratios. Moreover, the effects of KCN and lowering of the environmental temperature were consistent with the continued operation of the active transport mechanism in the presence of low internal sodium concentration. These findings have been confirmed by Klahr and Bricker (1964) who have also suggested that a tight 1:1 sodium:potassium coupling does not appear to occur in their preparation. Electrogenic pumps are thoroughly discussed by Kernan in this volume.

Zerahn (1955) has shown that notwithstanding the ability of lithium to cross the outer sodium-permeable barrier of the epithelium, it is poorly extruded by the cells into the internal bathing medium. Instead, it tends to accumulate in the cell eventually blocking transport. Thus lithium does not substitute perfectly for intracellular sodium. Huf and Wills (1951) showed that transport of sodium across the skin depends on the presence of potassium in the solution. Rubidium and, to a lesser extent, caesium were substitutes for potassium but in none of their experiments was the inner solution potassium-free, since this ion was apparently discharged into the solution. Absence of potassium from the inner solution failed to completely abolish salt and water transport but reduced it to about two-thirds of the control values. Similarly, Fukuda (1942) had shown that removal of potassium from the inner solution inhibited the 'asymmetry potential' in frog skin. These results, together with the observation that the extrusion of sodium is potassium-dependent in many cell types, for example, red cells, nerve and muscle (Steinbach, 1952; Hodgkin and Keynes, 1953; Harris, 1954) suggested that the pump performed a forced exchange of sodium for potassium. However, recent work (such as that quoted above) has raised doubts on this point, especially on the 1:1 exchange ratio. The role of potassium in the inner solution can be explained in ways other than that of an obligatory exchange in a 1:1 ratio for intracellular sodium. For example, Ussing (1965a) suggested that removal of potassium from the inner bathing solution causes the cells to lose potassium chloride to the inner solution, resulting in the decrease in cell volume which brings about a reduction in active sodium transport due to an increased resistance of the active transport path. Since dilution of the potassium-free inner solution restores the active transport of sodium to control levels *pari passu* with the swelling of the epithelium, it is probable that the reduction in active sodium transport seen when the skin is exposed to a potassium-free inner solution is not caused by a fall in the intracellular potassium concentration (Ussing, 1965a). Moreover, it is quite possible that appreciable amounts of potassium remain in the interspaces in the skin epithelium when potassium is removed from the inner solution despite the communication of the interspaces with the subepithelial skin layers. Thus some ambiguity is introduced into any interpretation of the effects of inner potassium concentration on transepithelial sodium transport, and consequently, there is no conclusive evidence for a 1:1 sodium for potassium exchange. The evidence favouring the alternative possibility, that of the existence of an electrogenic mechanism, is not inconsiderable.

The active transport mechanism is driven by metabolic energy. Metabolic inhibitors reduce the rate of active ion transport (Huf and coworkers, 1957) with or without affecting the electrolyte equilibrium of the tissue. Thus 10^{-2} M fluoroacetate, 10^{-3} M azide and 10^{-3} M diethyl malonate inhibit active ion transport but do not affect the tissue electrolyte composition. The last two, at higher concentrations, also cause a gain in sodium and a loss of potassium from the skin. Leaf and Renshaw (1957) and Zerahn (1958) showed that about 20 moles of sodium are transported per mole of oxygen consumed. Similar figures have been reported for the toad bladder (Leaf and Dempsey, 1960) and other transporting tissues (Ussing, 1960).

Farquhar and Palade (1964) found that ATPase activity is associated with cell membranes facing the intercellular spaces, which presumably are open towards the inner side of the skin. This activity has been linked to active sodium and potassium transport by the studies of Skou (1957) and Post and coworkers (1960). Bonting and coworkers (1962) demonstrated the presence of a sodium and potassium dependent ATPase in homogenates of frog skin.

On the basis of work done primarily on nerve, kidney, brain, red blood cells and muscle, it has been proposed that this ATPase activity is intimately related to the active transport process (Post, 1960; Dunham and Glynn, 1961; Hess, 1962; Järnefelt, 1961; Auditore, 1962; Auditore and Murray, 1962; Post and Beaver, 1963; Kinsolving, 1965).

Skou (1965) has indicated that this transport enzyme should fulfil the following requirements. It should be located in the cell membrane, have a higher affinity for sodium than for potassium at a site on the inside of the membrane, have a higher affinity for potassium at a site on the outside of the membrane, be able to catalyze the hydrolysis of ATP and thus use the energy of ATP to generate an electrochemical gradient, hydrolyse ATP at a rate depending on intracellular sodium and extracellular potassium concentration, and be found in all cells capable of active sodium and potassium transport. Such a system has been found in many tissues and two additional properties have also been established, linking it further to the active transport system. First, there is a correlation between the effect of cardiac glycosides on cation transport in intact cells, and their effect on the enzyme system, with the enzyme system having the same quantitative relation to sodium and potassium as the transport system of the cell. That is to say, the sodium and potassium concentrations that give half-maximal activation of transport in red blood cells correspond to the concentrations that give half-maximal activation of the sodium and potassium activated enzyme system from red blood cells (Post and coworkers, 1960). Secondly, the stoicheiometry between the active transport of sodium and the hydrolysis of ATP due to the sodium–potassium activated enzyme system has been found to be close to 3 sodium ions transported per mole of ATP split (Bonting and Caravaggio, 1963; Sen and Post, 1964). In the frog skin, 1 mole of O_2 is consumed for every 20 equivalents of sodium transported (Leaf and Renshaw, 1957; Zerahn 1958); for a P/O ratio of 3 this means that approximately 3 sodium ions are transported per energy-rich bond split (Karnovsky, 1967; Keynes, 1967), in agreement with the Na/P ratio obtained for the enzyme system by Bonting and Caravaggio (1963) and Sen and Post (1964). The number of potassium ions transported per energy-rich phosphate bond hydrolysed seems to be lower than 3 (Post and Jolly, 1957). This agrees with the lack of positive evidence for a strict 1:1 exchange between sodium and potassium.

Insulin, which increases sodium transport across the bladder (Herrera, 1965) has no effect on the enzyme system (Bonting and coworkers, 1961). The absence of an effect *in vitro* of pitressin and oxytocin on the enzyme system is in agreement with the finding that these hormones modify the permeability of the outer barrier of the transporting epithelia (Leaf, 1960; Curran and coworkers, 1963).

The effect of cardiac glycosides on frog skin has been studied by Koefoed-Johnsen (1947) and Widdas (1961). Neither of these authors put forward any explanation as to the mechanism of action of glycosides. However, on the basis of the Koefoed-Johnsen and Ussing model, Grundy (1966) suggested that these agents, along with morphine, act primarily on the sodium pump mechanism. This has been confirmed in the toad bladder (Herrera, 1966). Bonting and Canady (1964) observed a parallelism between the effect of ouabain on sodium transport across the bladder, and its effect on the sodium–potassium activated ATPase system isolated from the bladder.

B. The Toad Bladder

As in frog skin, the movement of sodium across the epithelium of the toad bladder takes place in two steps: first by penetration of the outer barrier of the epithelial cells and then the basal epithelial barrier towards the serosal side of the epithelium. However, the number of layers traversed by the sodium ions in this case is somewhat less than in the frog skin. In fact, the epithelium of the bladder may be considered to be made up of a single cell layer, consisting of several cell types (see above).

The sodium transport system in the bladder is capable of establishing large chemical gradients between the urine and the body fluid (Leaf and coworkers, 1958); the sodium concentration of the urine may be reduced to less than 1 mequiv./l (as compared with approximately 115 mequiv./l in plasma), indicating that the bladder may establish or maintain such gradients.

1. *Penetration of Sodium across the Outer Barrier of the Epithelium*

There is a considerable body of evidence indicating that the movement of sodium across the outer cell barrier, despite involving interaction with the membrane, is passive in nature. A consideration of both the chemical and electrical gradients, existing across this barrier has borne out this view. Frazier and coworkers (1962) estimated the sodium pool involved in the active transport of sodium across bladder wall and compared it to the corresponding sodium concentrations of the mucosal medium when the latter was varied between 0·40 and 114 mequiv./l. In all instances the 'concentrations' of the sodium pool were lower than that of the mucosal medium, suggesting that the chemical gradient favoured the entry of sodium from the urine into the cells. Nevertheless, the electrical gradient must be also taken into consideration. This has been done by Frazier (1962) who with a micropipette pierced the outer epithelial barrier and determined that the cell interior is regularly positive with respect to the urine. At low urinary concentrations of sodium this step persisted but was reduced in magnitude—an observation which excludes the possibility that electrical forces contribute to the entry of sodium from the urine to the cell in the open-circuited bladder. However, it is very difficult to achieve a clear picture of the electrochemical gradient for sodium across this outer barrier. The state of the sodium in the transport pool is unknown. Very possibly the assumption that sodium ions are in the same state and have the same activity coefficient as in bulk solution on the urinary side is an oversimplification. However, if some of the sodium in the pool is in some way 'bound' or has a lower activity coefficient, this then could favour the entry of sodium into the cells and the estimation made by Frazier and Leaf (1963) and Frazier (1962) would represent the minimum estimate for the electrochemical gradient favouring sodium entry. Alternatively, if the 'sodium transport pool' were relegated to a small fraction of the intracellular volume, the chemical gradients may very well not exist. That the intracellular sodium content cannot be considered as made up of one compartment is shown by the fact that the 'sodium transport pool' represents less than one half of the sodium content determined chemically on the same bladders. However the location of the sodium outside the 'pool' is not definitely established, since it may be in the submucosa, within the epithelial cells, or in the spaces lying between the epithelial cells and the basement membrane.

2. *Interaction of Sodium with the Outer Barrier*

Several lines of evidence indicate that sodium entry across the outer barrier of the epithelial cells, though passive, entails interaction of this ion with the membrane. When sodium flux across the bladder is measured as a function of the concentration of sodium in the mucosal solution a curvilinear relationship in which the rate of increase in the current decreases at higher sodium concentration is found (Frazier and Leaf, 1964). This suggests the existence of a process that can be saturated. The rate-limiting step appears to be represented by the entry of sodium at the mucosal surface, since the size of the cellular 'transport pool' shows saturation when the sodium concentration in the mucosal solution is raised. Frazier's finding of competitive interference of the entry of sodium at this surface by other alkali metal ions, especially lithium and some organic ions such as guanidinium, which depress the labelling of the tissue sodium pool by ^{24}Na in the mucosal solution, reinforces the hypothesis of an interaction of sodium with the outer barrier (Frazier, 1964). The effect of vasopressin on the 'sodium transport pool' and on the active sodium transport across the bladder, was investigated by Frazier and coworkers (1962). They found that the tissue sodium pool was always higher in bladders treated with vasopressin than in the controls. However, in spite of the facilitated entry of sodium, the outer barrier still behaved as a major obstacle to the transbladder movement of sodium. Further evidence indicating that the outer cell barrier presents a certain restriction to sodium movement across it and that this restriction may be influenced by the serosal potassium concentration has been obtained by Essig and Leaf (1963). These workers found that removal of potassium from sodium Ringer's solution bathing the serosal bladder surface caused a marked depression of the sodium permeability of the apical surface of the epithelium accompanied by a decrease in sodium transport. Tissue labelling with mucosal ^{24}Na decreased, indicating that when potassium is removed from the serosal solution, the depression of mucosal permeability is quantitatively of greater significance than any inhibition of the sodium pump, which would tend to increase tissue labelling. Recently, evidence has been obtained indicating that the apical cell barrier is not a simple membrane, but may be more closely described by two permeability barriers in series. Amphotericin B, is a polyene antibiotic which interacts with the sterol group of cell membranes, especially cholesterol (Gottlieb and coworkers, 1958; Fiertel and Klein, 1959; Lamper and coworkers, 1960; Kinsky and coworkers, 1962; Ghosh and Ghosh, 1963; Butler and coworkers, 1965; Van Zutphen and coworkers, 1966). Lichtenstein and Leaf (1966) found that the addition of amphotericin B to the mucosal solution caused a rise in short-circuit current which was equivalent, at least during the first 30 minutes of exposure, to the increase in net sodium transport. This increase in net ion transport was ascribed to an increase in permeability of the apical epithelial surface. In contrast to vasopressin, the rise in permeability caused by amphotericin is not specific for sodium but also includes other small solutes. Thus potassium, thiourea and chloride can more easily traverse the bladder after exposure of the mucosal surface to amphotericin B. It appears that amphotericin B destroys the specific mucosal permeability barrier which accounts for the low permeability of the bladder to hydrophilic solutes. However, Lichtenstein and Leaf (1965) found that amphotericin B did not increase the permeability of the bladder to net water movement, as does vasopressin; only after vasopressin was added did the increase in net water movement appear, but without a further increase in sodium transport. Both agents were capable of increasing the permeability of the bladder to urea and appeared to affect permeability to this substance at the same sites at the mucosal barrier. To explain these results, Lichtenstein and Leaf proposed the existence of a double series barrier at the apical cell surface: a fine diffusion barrier, permeable to water but relatively impermeable to small solutes, and a deeper porous barrier upon which vasopressin exerts its action. This porous barrier is the major obstacle to water movement and vasopressin increases its water permeability. Sodium and urea movements are hindered by the dense diffusion barrier overlying the porous barrier and vasopressin acts by increasing the permeability of this dense barrier to them. Most small solutes such as thiourea, potassium and chloride are held back by this dense barrier and vasopressin has no effect on their penetration. However, amphotericin B appears functionally to remove this barrier and to allow an increased movement of sodium, chloride, urea and potassium across the apical cell surface.

Vasopressin is thought to have a dual site of action: on the porous layer with respect to water and on the dense diffusion layer for urea and sodium. A schematic representation of these barriers is presented in figure 7. The results of Petersen and Edelman (1964) also indicate that the mucosal cell surface is not a simple membrane but that more than one selective mechanism is in operation. Thus, at low vasopressin concentrations they found that a rise in calcium concentration depressed the hormonally induced increase in water flow caused by an osmotic gradient and the hormonally induced rise in urea permeability coefficient. However, no effect of calcium on the vasopressin-induced rise in sodium transport was observed. From these results it appears that calcium can dissociate the effects of vasopressin on water and urea movement from its effect on sodium transport. These results also indicate that sodium and urea movement are controlled at different sites. This is in disagreement with the work of Lichtenstein and Leaf.

On the other hand, Mendoza and coworkers (1967) found that amphotericin B caused a marked increase in osmotic water flow across the bladder,

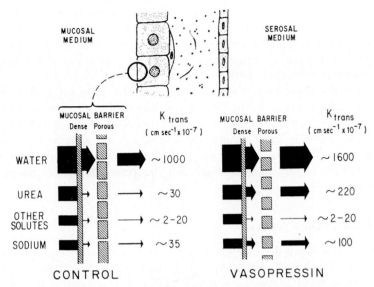

Figure 7. Diagram of the apical permeability barrier of the toad bladder epithelium. The urinary surface of the cells is represented as a dual barrier, a dense diffusion and a porous barrier in series. All substances, including water, are retarded at the diffusion barrier. Vasopressin increases the permeability of the bladder to urea and sodium by an effect on the diffusion barrier and to water by an effect on the porous barrier. (From Lichtenstein and Leaf, 1965).

thus suggesting that this agent under their experimental conditions could affect the porous barrier. Furthermore, they found that amphotericin B caused an increase in short-circuit current only after incubation overnight. These authors concluded that a simple site of action for amphotericin B could not be postulated, and that the effect of amphotericin B on bulk water flow may indicate alteration of the porous barrier. The results on short-circuit current do not permit a clear distinction between an effect of amphotericin B on passive penetration across the outer barrier or on the sodium pump.

The discrepancy in the results obtained by different investigators, using the same agents under similar experimental conditions, can probably be explained on the basis of the different geographical origin of the toads used. Thus, Davies and coworkers (1968) found that the bladders of Colombian but not Dominican toads exhibit rapid reversal of their membrane potential and short-circuit current, due presumably to active chloride transport, following exposure to potassium-Ringer's solution (Finn and coworkers, 1966a, 1967). Moreover, bladders of Colombian toads that have been depleted of endogenous substrate in the absence of aldosterone respond to the addition of pyruvate by an increase in sodium transport (Edelman and coworkers, 1963). Under identical conditions, bladders of Dominican toads do not respond to added substrate. However, in the presence of aldosterone, the addition of pyruvate causes an appreciable increase in sodium transport in bladders of toads from both sources.

There is evidence that the conductance of the mucosal surface of the epithelial cells is not uniform. This is not surprising since the epithelium is made up of at least three functional cell types. Gatzy and Clarkson (1965) found that there is small number of epithelial cells with a high mucosal permeability. This heterogeneity of the cells introduces uncertainties in the interpretation of any data dealing with the permeability of the mucosal barrier.

3. *The Intracellular Ion Pool*

Much of the work done on the mechanism of sodium transport across the toad bladder is based on the assumption that modifications in the rates of sodium movement across the mucosal and serosal barriers of the epithelial cells are reflected in changes in the sodium pool involved in sodium movement across the epithelium. However, several difficulties present themselves when the sodium content of the tissue is studied. Thus the sodium pool equilibrating with mucosal ^{24}Na is always lower than total tissue sodium (Frazier and coworkers, 1962), despite replacement of all the sodium in the serosal solution with choline. Under these conditions (i.e. in the steady state) the specific activity of the sodium being transported across the bladder should equal that in the mucosal solution (Leaf, 1965). The exact location of the radioactive sodium in the tissue is not yet precisely known; some of it may be located within the cells or maybe trapped in the collagen in the submucosal and serosal layers. This could be the case despite the fact that the composition of the medium in contact with the basal surface of the epithelium is probably identical with the serosal medium (Essig, 1965). Conceivably, an agent which increases the sodium permeability of the apical cell membrane causes an increase in the sodium content of the cells whereas an increase in the rate of movement of sodium across the basal barrier to the solution on the serosal side causes a fall in intracellular sodium concentration. The fact that vasopressin caused an increase in the intracellular sodium pool suggests therefore, that it acts by increasing the sodium permeability of the apical barrier (Frazier and coworkers, 1962). A similar mechanism has been proposed to explain the increase in water and urea permeability induced by vasopressin (Leaf, 1960; Hays and Leaf, 1962). Vasopressin regularly produces a decrease in the d.c. resistance of the bladder (Civan and coworkers, 1966). More recent work by Civan and Frazier (1968) has shown that 98 per cent of the fall in resistance in the bladder following vasopressin administration occurs at the apical permeability barrier. This result strengthens the view that vasopressin acts selectively at this barrier by increasing ionic mobility.

Tissue potassium content has also been extensively studied. Exposure of bladders to potassium-free serosal solutions not only causes a fall in short-circuit current but also a fall in tissue potassium and a rise in tissue sodium and a marked depression of mucosal sodium permeability (Hays and Leaf, 1961; Essig and Leaf, 1963). A potassium-free solution also reduces the tissue water content and the water permeability response to vasopressin (Finn and co-workers, 1966b). Cell potassium content rather than cell water content appeared to govern the permeability response of the mucosal side of the membrane. These results suggest that the vasopressin response requires the presence of a normal potassium content of the cells. This is in agreement with the theory put forward by Orloff and Handler (1962, 1964) to explain the action of vaso-pressin. They proposed that adenosine-3′,5′-monophosphate (cyclic-3′,5′-AMP) serves as an intracellular intermediate in the action of vasopressin. Like the hormone, cyclic-3′,5′-AMP causes an increase in water permeability and short-circuit current (Handler and Orloff, 1964) but unlike vasopressin, whose action is inhibited by reducing agents which react with sulphydryl groups to form covalent links, its effect is not modified by cysteine or thyo-glycollate (van Dyke and coworkers, 1942; Rasmussen and coworkers, 1960; Martin and Schild, 1962; Bentley, 1964; Handler and Orloff, 1964). However, both agents require that the potassium content of the cells be normal. The potassium dependent step probably occurs subsequent to the production of cyclic-3′,5′-AMP by the cell. Moreover, a metabolic dependence of the short-circuit current and the permeability response of toad bladder to vasopressin and cyclic AMP has been demonstrated by abolition of the response by iodoacetic acid, dinitrophenol, anaerobiosis, azide and fluoroacetate (Handler and coworkers, 1966).

In this context, it is interesting to note that Gachelin and Bastide (1968) have located two distinct Na^+-K^+-ATPases in the frog bladder with very different kinetic properties. Both are activated by potassium and inhibited by ouabain but whereas one is activated by high concentrations of sodium, the second is activated only by low concentrations and inhibited by high concentrations. This second enzyme, which is presumed to be located in the mucosal membrane of the bladder, seems to have a greater affinity for sodium in the presence of 3′,5′-AMP. Gachelin and Bastide believe that the first Na^+-K^+-ATPase is responsible for sodium pumping across the serosal side of the cell, whereas the second Na^+-K^+-ATPase is responsible for the medi-ated entry of sodium across the mucosal surface of the cell, a transfer that is accelerated by 3′,5′-AMP.

The role of cyclic-3′,5′-AMP in mediating the vasopressin effect on sodium transport across frog skin has not been as extensively studied as in toad bladder. None the less it is interesting to note that 5-hydroxytryptamine (5-HT) produces a significant increase in sodium transport across frog skin (Sayoc and Little, 1967). This action of 5-HT is probably mediated by cyclic-3′,5′-AMP, since 5-HT causes the formation of cyclic-3′,5′-AMP from ATP (Mansour, 1959). Furosemide, a potent diuretic which reduces sodium reabsorption in the renal tubule (Buchborn and Anastasakis, 1964), antagonizes the stimulatory action on sodium transport across toad bladder of arginine vasopressin, cyclic-3′,5′-AMP and caffeine. The fact that it acts as an antagonist of cyclic-3′,5′-AMP has been borne out by the work of Ferguson (1966). Furthermore, Finn and coworkers (1967) found that the transbladder potential difference and short-circuit current were reversed when the bladder was exposed to potassium-free solutions under circumstances known to decrease cell potassium concentration (Finn and coworkers, 1966b). This reversal of the potential difference and short-circuit current was attributed to a mucosal-to-serosal active movement of chloride which is absent or very small in magnitude when normal amphibian Ringer solution is used as the bathing medium. Reversal of the potential difference and short-circuit current could be abolished by dinitrophenol and sodium cyanide, both being potent inhibitors of oxidative metabolism.

Janáček and coworkers (1968), studying intracellular potassium and sodium activities by means of cation-sensitive glass microelectrodes, have rightly pointed out that the different methods used for sodium and potassium determination in amphibian epithelia are not free from error. The determination of sodium and potassium by direct chemical analysis of the tissue rests on the assumption that the extracellular space is freely accessible to inulin dissolved in the bathing solutions, so that the inulin space is considered a good estimate of the extracellular space. This is not necessarily so in all instances (Natochin and coworkers, 1965). Moreover, the determination of the intracellular ionic concentration is estimated from the difference of two figures which are similar in magnitude, namely the total tissue ionic content minus that estimated to exist in the extracellular space (as measured by inulin or other large molecular weight substances). Any small error in these determinations would cause a large error in the estimated cellular ionic concentration; this is especially evident in the case of sodium. This method has the further disadvantage that it does not distinguish between the ionic content of the epithelium and that of other cell types and it does not discriminate between ions associated with cytoplasmic constituents and those in free solution inside the cell. An alternative method based on the use of radioisotopes measures only the exchangeable pools (Schoffeniels, 1957; Hoshiko and Ussing, 1960; Andersen and Zerahn, 1963) and these measurements are affected by uncertainties introduced by diffusion delays in extra-cellular spaces (Hoshiko and coworkers, 1964). By using cation-sensitive glass microelectrodes (Eisenman and coworkers, 1957; Hinke, 1959, 1961; Eisenman, 1962), these difficulties may be obviated but others are introduced. Thus, although the intracellular potential recorded with a potassium-selective glass microelectrode is primarily a function of potassium activity, sodium ions also have an important effect which must be ascertained and subtracted; similarly, potassium ions modify the potential difference recorded with a sodium-sensitive glass micro-electrode. Furthermore, the change in potential occurring when the electrode passes from the extracellular medium into the cell results from the change in ionic activity plus the transmembrane potential. The latter must, therefore, be estimated and subtracted from the total potential difference. The uncertainties introduced by all the corrections required, are probably as great as those which affect the determinations performed by direct chemical analysis. However, it is interesting to note that the values obtained by means of the cation-sensitive microelectrode are compatible with those determined chemically (Finn and coworkers, 1966b; Herrera, 1968b). Intracellular potassium concentration was estimated as being 88·6 mequiv./l and intracellular sodium 28·4 mequiv./l. The intra-cellular sodium concentration is compatible with passive distribution of this ion. However, the intracellular potassium concentration is too high to be accounted for on the basis of the potential difference across the serosal epithelial surface. This is why Finn and coworkers suggest that potassium must be actively transported into the cells.

Further study of intracellular ionic pools is bound to be greatly simplified by the newly devised methods of isolating epithelial cells described by Hays and coworkers (1965), Perris (1966), and Gatzy and Berndt (1968). It has been found that epithelial cells dispersed by collagenase or hyaluronidase function normally. Their sodium and potassium content were 32 mequiv./kg cell water and 117 mequiv./kg cell water, respectively. They responded to ADH with a 50 per cent increase in Q_{O_2} and sodium free Ringer's solution depressed the Q_{O_2} by 40 per cent.

It has been proposed that the transbladder potential difference is equivalent to the sum of the two diffusion potentials, one generated by the movement of sodium across the sodium-permeable apical epithelial surface and the other by the movement of potassium across the potassium-permeable basal surface of the epithelium. The sodium pump is believed to be located at the basal epithelial surface. This model predicts that if the pump is inhibited by ouabain a rise in tissue sodium originating from the mucosal solution and a fall in tissue potassium concomitant with a decrease in potential difference should be observed. However, in bladders exposed to sodium-free serosal solutions, ouabain was found to cause a fall in active sodium transport, and the potential difference decreased in the absence of any significant change in tissue sodium and potassium. If, however, sodium was present in the serosal solution, inhibition of both active ion transport and the potential difference by ouabain was accompanied by a decrease in tissue potassium and a gain in sodium of serosal origin. It also was found that a large fraction of the changes in tissue sodium and potassium could be ascribed to changes in the subepithelial layers (Herrera, 1968a and b).

The model also predicts that if the serosa is made electrically negative with respect to the mucosa, sodium from the mucosal solution would enter the epithelium across the sodium-permeable apical surface and epithelial potassium would leave the cells across the potassium-permeable basal epithelial border. Ouabain could, indeed, cause a loss of potassium and a gain in sodium by the epithelium despite the lack of serosal sodium provided a 200 mV step (serosa negative with respect to mucosa) was applied across the bladder. Thus, although the mucosal and serosal surface of the bladder epithelium appear to be selectively permeable to sodium and potassium respectively, the transbladder potential difference does not appear to be entirely due to diffusion potentials but to be generated at least in part by the active transport mechanism. Therefore, it is possible that an electrogenic pump may coexist in parallel with the potassium channel at the basal border of the epithelium.

Aldosterone stimulates sodium transport across toad bladder (Crabbé, 1961a and b, 1963, 1964; Crabbé and Weer, 1964), but its mode of action is not yet clear. The present evidence points on the one hand to stimulation of sodium pumping independent of any effect on the sodium permeability of the mucosal surface of the epithelial cells (Edelman and coworkers, 1963; Fanestil and coworkers, 1967; Fimognari and coworkers, 1967), and on the other to a primary increase in sodium permeability of the apical cell surface, followed by an increase in the sodium pool which results in a secondary increase in the activity of the sodium pump. This demands an increase in substrate utilization of mainly pyruvate and acetoacetate (Sharp and Leaf, 1964a and b, 1965; Sharp and coworkers, 1965; Sharp and coworkers, 1966; Sharp and coworkers, 1966; Dalton and Snart, 1967).

Experiments by Sharp and coworkers (1966) indicate that increased sodium transfer across the bladder brought about by aldosterone and amphotericin B is associated with an increase in aerobic metabolism. This increase in metabolism depends on the presence of sodium in the mucosal solution, suggesting that the increased sodium entry caused by both agents results in an increased transport by the pump, which in turn results in increased aerobic metabolism; in the absence of sodium the Q_{O_2} diminishes drastically. Since amphotericin B and aldosterone increase the sodium pool and the sodium pool falls with decreasing mucosal sodium concentration, it is clear that changes in aerobic metabolism could be mediated by the level of the tissue sodium pool through changes in the rate of sodium transport by the sodium pump.

A hypothetical scheme linking the action of the neurohypophyseal hormone with sodium transport and energy metabolism has been proposed by Bentley (1966). This scheme can also explain the effect of aldosterone. The increased permeability of the cells to sodium induced by these agents leads to increased sodium entry into the cells. This activates the ATPase believed to be involved in sodium and potassium ion transport, thus resulting in the liberation of phosphate bond energy and the extrusion of sodium from the cell. ATP is then regenerated from ADP with accompanying utilization of oxygen. Whittam (1963) who has studied the possible relationship between sodium transport and energy metabolism suggests that ATP, which is one of the links between metabolism and ion transport, couples the two processes, especially with regard to respiration. The rate of respiration in mitochondria is regulated by the concentration of ADP which acts as the phosphate acceptor in oxydative phosphorylation (Chance and Williams, 1956). The formation of ADP is in turn partly governed by the rate of active ion transport. Increasing the sodium transport pool causes an increased rate in sodium transport which in turn induces increased formation of ADP from ATP. An increased ADP concentration would then cause an increased rate of respiration. If the sodium pool is lowered, the rate of transport will be decreased; the turnover of ATP will diminish and a fall in respiration will occur. Thus, Leaf and Dempsey (1960) found that vasopressin increased the Q_{O_2} of transporting toad bladder from $1\cdot06$ to $1\cdot76$ μl O_2/mg dry wt/hr. The respiratory response to vasopressin was abolished in sodium-free Ringer's solution. The increase in Q_{O_2} caused by vasopressin was secondary to its action on increased transport which in turn is a consequence of the increased sodium pool.

4. *Movement of Sodium across the Basal Surface of the Toad Bladder Epithelium*

Sodium is transported across the basal barrier of the epithelium from the cells towards the serosal solution against a considerable chemical and electrical potential gradient; therefore, sodium transport must be active in nature and must require the expenditure of metabolic energy (Frazier, 1962; Frazier and coworkers, 1962). The Koefoed-Johnsen–Ussing model (1958) assumed the existence of a mechanism located at the basal cell border of the frog skin epithelium which exchanged intracellular sodium for potassium from the inner bathing solution; this would represent an electroneutral exchange of sodium taking place without any separation of charge and therefore without any associated potential difference. The electrical potential difference step which exists across the basal membrane, was attributed to a diffusion potential created by the passive movement of intracellular potassium down its concentration gradient towards the inner solution. A similar scheme has been proposed for the toad bladder epithelium by Leb and coworkers, (1965). The hypothetical sodium for potassium exchange implies that the removal of serosal potassium would bring active sodium transport to a stop. The possible existence of a potassium diffusion potential across the basal epithelial surface could be tested by progressively increasing serosal potassium; this should bring about a reduction, abolition and finally, reversal of the potential difference, as the potassium concentration of the serosal medium was increased up to and beyond intracellular potassium concentration. However, when this was done, a significant potential persisted across the serosal surface of the epithelium, as measured with a microelectrode inside the epithelial cell, even when the potassium concentration was increased up to and beyond that estimated to exist inside the cell (114 mequiv./l). Moreover, it has been found that the potential difference across the serosal epithelial membrane is about 13 mV higher than can be accounted for by passive diffusion of potassium and chloride across the membrane (Essig and coworkers, 1963; Frazier and Leaf, 1963). Diffusion potentials arising from chloride and other ions as well as hydrogen ions could be ruled out as sources of the serosal potential step. By exclusion, it was concluded that an electrogenic pump was the source of the potential difference across the serosal barrier (Leaf, 1964). Thus sodium ions are believed to be transported from the cell interior to the serosal solution without a simultaneous exchange for hydrogen or potassium ions, and without any obligatory paired movement of chloride or other anions (Leaf, 1965). This means that the movement of sodium occurs against a considerable energy barrier.

The removal of serosal potassium results in a decrease in transbladder sodium movement. This was thought to indicate that an obligatory potassium movement from the serosal solution to the cells was produced by the pump simultaneously with the movement of sodium in the opposite direction. More recent work indicates that the removal of serosal potassium reduces the entry of sodium across the mucosal barrier (see above).

The basal location of the pump suggested by the electrochemical gradient existing across the bladder may be confirmed by treating the bladder with agents that selectively inhibit the active transport mechanism. For example ouabain, a selective inhibitor of active transport (Schatzmann, 1953; Glynn, 1967) in many cell types, including toad bladder cells (Koefoed Johnsen, 1957; Caldwell and Keynes, 1959; Bonting and Canady, 1964) acts at the serosal border of the epithelium (Herrera, 1966). Both the short-circuit current and potential difference were found to be depressed by ouabain. The movement of passively transported ions, for example chloride, were un-affected. The decrease in the short-circuit current and the mucosal-to-serosal sodium movement across the bladder appears to be due to inhibition of the sodium pump. Moreover, as shown in figure 8, the potential difference decreased at the same time as the current and, in the absence of serosal sodium, no change in tissue ion content could be observed. This suggests that ouabain inhibits an electrogenic sodium pump (Herrera, 1968b).

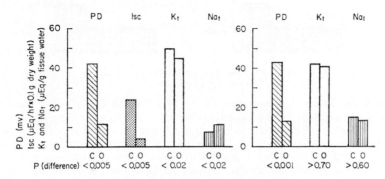

Figure 8. Effect of ouabain on toad bladder exposed to sodium Ringer's solution on the mucosal side and sodium-free Ringer's solution on the serosal side. The diagram on the left corresponds to short-circuited bladders and that on the right to open-circuited bladders. Values denoted by a C correspond to control bladder halves; those denoted by an O represent ouabain treated bladders. PD, potential difference; I_{sc}, short-circuit current; K_t, tissue potassium; Na_t tissue sodium. Ouabain decreases potential difference and short-circuit current but potassium and sodium content of the tissue remain practically unmodified. (From Herrera, 1968).

A schematic representation of the two hypotheses for sodium movement across the epithelium is shown in figure 9. Thier (1968), incidentally, from studies on amino acid transport in toad bladder, has concluded that two independent sodium pumping systems exist in the basal membrane of the epithelial cell.

Insulin has a stimulatory effect on sodium transport across the toad bladder, skin and intestine (Herrera, 1965; Crabbé, 1967). The hormone causes an appreciable increase in the short-circuit current and rate coefficient for sodium movement from the cells towards the serosal solution, with no significant change in the rate coefficient for sodium movement across the mucosal surface of the epithelial cells. From these observations it may be deduced that insulin stimulates the active transport step without modifying the permeability of the mucosal surface to sodium.

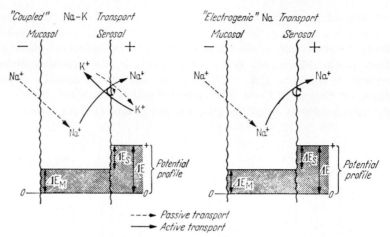

Figure 9. Schematic representation of the two hypotheses for sodium transport across polar epithelia. The diagram on the left represents the Koefoed-Johnsen–Ussing model in which intracellular sodium is ejected in exchange for potassium from the inner medium. On the right, the transport system is viewed as electrogenic; sodium is expelled towards the inner side without exchange for cations from the inner medium and without a paired movement of anions out of the cells. In both schemes, the potential step at the mucosal surface arises as a sodium diffusion potential. The potential step across the serosal surface is generated by the passive diffusion of potassium according to the diagram on the left, but due to direct separation of charge resulting from the transport of sodium according to the scheme on the right. (After Leaf, 1964).

Studies on the movement of sodium across the basal border of the epithelial cells of the toad bladder are sparse. With the exception of a few agents such as insulin and ouabain, most of the agents which are known to modify transport and whose effects have been studied on the bladder act at the apical cell surface.

The transporting mechanism located at the inner barrier resembles that of the frog skin and in general seems to fulfil the requirements for a sodium–potassium dependent ATPase system (see above). However, the role of potassium is not easy to assess at this time, since the main effect of potassium lack appears to be exerted at the mucosal surface and not at the pump site (Essig and Leaf, 1963).

The movement of sodium across the bladder is highly dependent on the presence of appropriate energy sources. Dramatic responses to the addition of exogenous substrates have been observed (Porter and Edelman, 1964; Sharp and Leaf, 1965); thus toad bladders depleted of exogenous substrate respond with a prompt and sustained stimulation of sodium transport following the addition of glucose or pyruvate to the bathing media. There is evidence from work carried out on the metabolic control reactions in toad bladder that bladders incubated in oxygenated Ringer's solution are in a semistarved state, since minimal reduction levels of pyridine nucleotide are observed (Davis and Canessa-Fischer, 1965; Canessa-Fischer and Davis, 1966).

Work by Zerahn on the frog skin (1966) and Essig and Leaf on toad bladder (unpublished data cited by Leaf, 1967) alters the simple picture of a serosal pump site. These authors found that the efflux of ^{24}Na across the inner or serosal surface of the epithelium is not modified by known metabolic inhibitors including potassium cyanide, dinitrophenol or iodoacetamide and they concluded that the sodium being eluted had already passed the site of transport. On the basis of these results and of indirect evidence indicating that the transport pool is small, and not increased by recirculation of serosal sodium, Leaf (1967) suggested that the cytoplasmic sodium pool involved in transport is very small. A low cytoplasmic sodium concentration may be explained if the large nuclei of the cells were rich in sodium and their sodium isolated from the cytoplasmic pool. This is not unlikely in view of the findings of Loewenstein and Kanno (1963) that the nuclear membranes of amphibian oocytes are highly resistant to the passage of ions, and of Allfrey and coworkers (1961) who put forward evidence that oocyte nuclei may be rich in sodium.

REFERENCES

Aceves, J., D. Erlij and C. Edwards (1968) *Biochim. Biophys. Acta.*, **150**, 744

Allfrey, V. G., R. Mendt, J. W. Hopkins and A. E. Mirsky (1961) *Proc. Natl. Acad. Sci. U.S.*, **47**, 907

Andersen, B. and H. H. Ussing (1957) *Acta Physiol. Scand.*, **39**, 228

Andersen, B. and K. Zerahn (1963) *Acta Physiol. Scand.*, **59**, 319

Auditore, J. V. (1962) *Proc. Soc. Exp. Biol. Med.*, **110**, 595

Auditore, J. V. and L. Murray (1962) *Arch. Biochem. Biophys.*, **99**, 372

Bentley, P. J. (1964) *J. Endocrinol.*, **30**, 103

Bentley, P. J. (1966) *Biol. Rev. Cambridge Phil. Soc.*, **41**, 275

Bentley, P. J. (1967) *J. Endocrinol.*, **37**, 349

Biber, T. U. L., R. A. Chez and P. F. Curran (1966) *J. Gen. Physiol.*, **49**, 1161

Bonting, S. L. and M. R. Canady (1964) *Amer. J. Physiol.*, **207**, 1005

Bonting, S. L. and L. L. Caravaggio (1963) *Arch. Biochem. Biophys.*, **101**, 37

Bonting, S. L., L. L. Caravaggio and N. M. Hawkins (1962) *Arch. Biochem. Biophys.*, **98**, 413

Bonting, S. L., K. A. Simon and N. M. Hawkins (1961) *Arch. Biochem. Biophys.*, **95**, 416

Bricker, N. S., T. Biber and H. H. Ussing (1963) *J. Clin. Invest.*, **42**, 88

Buchborn, E. and S. Anastasakis (1964) *Klin. Wochschr.*, **42**, 1127

Butler, W. T., O. W. Alling and E. Cotlove (1965) *Proc. Soc. Exp. Biol. Med.*, **118**, 297

Caldwell, P. C. and R. D. Keynes (1959) *J. Physiol. (London)*, **148**, 8P

Canessa-Fischer, M. and R. P. Davis (1966) *J. Cellular Physiol.*, **67**, 345

Capraro, V. (1967) *Protoplasma*, **63**, 21

Cereijido, M. and P. F. Curran (1965) *J. Gen. Physiol.*, **48**, 543

Cereijido, M., F. C. Herrera, W. J. Flanigan and P. F. Curran (1964) *J. Gen. Physiol.*, **47**, 879

Cereijido, M. and C. A. Rotunno (1967) *J. Physiol.* (*London*), **190**, 481
Chance, B. and G. R. Williams (1956) *Advances in Enzymology*, Vol. 17, Wiley–Interscience, New York. p. 65
Choi, J. K. (1965) *J. Cell Biol.*, **16**, 53
Chowdhury, T. K. and F. M. Snell (1965) *Biochim. Biophys. Acta.*, **94**, 461
Chowdhury, T. K. and F. M. Snell (1966) *Biochim. Biophys. Acta.*, **112**, 581
Civan, M. M. and H. S. Frazier (1968) *J. Gen. Physiol.*, **51**, 589
Civan, M. M., O. Kedem and A. Leaf (1966) *Amer. J. Physiol.*, **211**, 569
Conway, E. J. (1960) *Nature*, **187**, 394
Conway, E. J., R. P. Kernan and J. A. Zadunaisky (1961) *J. Physiol.* (*London*), **155**, 263
Crabbé, J. (1961a) *Endocrinology*, **69**, 673
Crabbé, J. (1961b) *J. Clin. Invest.*, **40**, 2103
Crabbé, J. (1963) *Nature*, **200**, 787
Crabbé, J. (1964) In J. de Graeff and B. Leijnse (Eds.), *Water and Electrolyte Metabolism*, Vol. 2, Elsevier, Amsterdam. p. 59
Crabbé, J. and P. de Weer (1964) *Nature*, **202**, 298
Crabbé, J. (1967) *Proc. Intern. Symp. Polypeptide and Protein Hormones*, Milan
Curran, P. F. and J. R. Gill, Jr. (1962) *J. Gen. Physiol.*, **45**, 625
Curran, P. F., F. C. Herrera and W. J. Flanigan (1963) *J. Gen. Physiol.*, **46**, 1011
Dalton, T. and R. S. Snart (1967) *Biochem. J.*, **105**, 24
Davies, H. B. F., D. G. Martin and G. W. G. Sharp (1968) *Biochim. Biophys. Acta.*, **150**, 318
Davis, R. P. and M. Canessa-Fischer (1965) *Anal. Biochem.*, **10**, 325
Du Bois-Reymond, E. (1848) *Untersuchungen über tierische Elektrizität*, Berlin
Dunham, E. T. and I. M. Glynn (1961) *J. Physiol.* (*London*), **156**, 274
Du Vigneaud, V. (1956) *Harvey Lectures Ser.*, **50**, 1
Edelman, I. S., R. Bogoroch and G. A. Porter (1963) *Proc. Natl. Acad. Sci. U.S.*, **50**, 1169
Eisenman, G. (1962) *Biophys. J.*, **2**, 259
Eisenman, G., D. O. Rudin and J. U. Casby (1957) *Science*, **126**, 831
Engbaek, L. and T. Hoshiko (1957) *Acta Physiol. Scand.*, **39**, 349
Essig, A. (1965) *Amer. J. Physiol.*, **208**, 401
Essig, A., H. S. Frazier and A. Leaf (1963) *Nature*, **197**, 701
Essig, A. and A. Leaf (1963) *J. Gen. Physiol.*, **46**, 505
Ewer, R. F. (1950) *J. Exptl. Biol.*, **27**, 40
Ewer, R. F. (1952a) *J. Exptl. Biol.*, **29**, 173
Ewer, R. F. (1952b) *J. Exptl. Biol.*, **29**, 429
Fanestil, D. D., G. A. Porter and I. S. Edelman (1967) *Biochim. Biophys. Acta.*, **135**, 74
Farquhar, M. G. and G. E. Palade (1964) *Proc. Natl. Acad. Sci. U.S.*, **51**, 569
Ferguson, D. R. (1966) *Brit. J. Pharmacol.*, **27**, 528
Fiertel, A. and H. P. Klein (1959) *J. Bacteriol.*, **78**, 738
Fimognari, G. M., G. A. Porter and I. S. Edelman (1967) *Biochim. Biophys. Acta.*, **135**, 89
Finn, A. L., J. S. Handler and J. Orloff (1966a) *Federation Proc.*, **25**, 567
Finn, A. L., J. S. Handler and J. Orloff (1966b) *Amer. J. Physiol.*, **210**, 1279
Finn, A. L., J. S. Handler and J. Orloff (1967) *Amer. J. Physiol.*, **213**, 179
Frazier, H. S. (1962) *J. Gen. Physiol.*, **45**, 515
Frazier, H. S. (1964) *J. Clin. Invest.*, **43**, 1265
Frazier, H. S., E. F. Dempsey and A. Leaf (1962) *J. Gen. Physiol.*, **45**, 529
Frazier, H. S. and A. Leaf (1963) *J. Gen. Physiol.*, **46**, 491
Frazier, H. S. and A. Leaf (1964) *Medicine*, **43**, 281
Fukuda, T. R. (1942) *Japan J. Med. Sci. Biol.*, **8**, 123
Fuhrman, F. A. and H. H. Ussing (1951) *J. Cellular Comp. Physiol.*, **38**, 109
Gachelin, G. and F. Bastide (1968) *Compt. Rend.*, **267**, 906
Galeotti, G. (1904) *Z. Physik. Chem.*, **49**, 542
Galeotti, G. (1907) *Z. Allg. Physiol.*, **6**, 99
Gatzy, J. T. and W. O. Berndt (1968) *J. Gen. Physiol.*, **51**, 770
Gatzy, J. T. and T. W. Clarkson (1965) *J. Gen. Physiol.*, **48**, 647
Ghosh, A. and J. J. Ghosh (1963) *Ann. Biochem.*, **23**, 113

Glynn, I. M. (1957) *J. Physiol.* (*London*), **136**, 148
Goldman, D. E. (1943) *J. Gen. Physiol.*, **27**, 37
Gonzalez, C., J. Sanchez and J. Concha (1966) *Biochim. Biophys. Acta.*, **120**, 186
Gonzalez, C., J. Sanchez and J. Concha (1967) *Biochim. Biophys. Acta.*, **135**, 167
Gottlieb, D., H. E. Carter, J. H. Sloneker and A. Ammann (1958) *Science*, **128**, 361
Grundy, H. F. (1966) *J. Pharm. Pharmacol.*, **18**, 694
Handler, J. S., and J. Orloff (1964) *Amer. J. Physiol.*, **206**, 505
Handler, J. S., M. Petersen and J. Orloff (1966) *Amer. J. Physiol.*, **211**, 1175
Harris, E. J. and R. A. Sjodin (1961) *J. Physiol.* (*London*), **155**, 221
Harris, M. (1954) *Symp. Soc. Exptl. Biol.*, **8**, 228
Hays, R. M. and A. Leaf (1961) *Ann. Internal Med.*, **54**, 700
Hays, R. M., B. Singer and S. Malamed (1965) *J. Cell Biol.*, **25**, 195
Herrera, F. C. (1965) *Amer. J. Physiol.*, **209**, 819
Herrera, F. C. (1966) *Amer. J. Physiol.*, **210**, 980
Herrera, F. C. (1968a) *J. Gen. Physiol.*, **51**, 261s
Herrera, F. C. (1968b) *Amer. J. Physiol.*, **215**, 183
Herrera, F. C. and P. F. Curran (1963) *J. Gen. Physiol.*, **46**, 999
Hess, H. H. (1962) *J. Neurochem.*, **9**, 613
Hinke, J. A. M. (1959) *Nature*, **184**, 1257
Hinke, J. A. M. (1961) *J. Physiol.* (*London*), **156**, 314
Hodgkin, A. L. and B. Katz (1949) *J. Physiol.* (*London*), **108**, 37
Hodgkin. A. L. and R. D. Keynes (1953) *J. Physiol.* (*London*), **120**, 46P
Hoshiko, T. (1961) In A. M. Shanes (Ed.), *Biophysics of Physiological and Pharmacological Actions.* Amer. Assn. Adv. Science, Washington. p. 31
Hoshiko, T., B. D. Lindley and C. Edwards (1964) *Nature*, **201**, 932
Hoshiko, T. and H. H. Ussing (1960) *Acta Physiol. Scand.*, **49**, 74
Huf, E. G. (1935) *Pfluegers Arch. Ges. Physiol.*, **235**, 655
Huf, E. G., N. S. Doss and J. P. Wills (1957) *J. Gen. Physiol.*, **41**, 397
Huf, E. G. and J. Wills (1951) *Amer. J. Physiol.*, **167**, 255
Janáček, K., F. Morel and J. Bourget (1968) *J. Physiol.* (*Paris*), **60**, 51
Jard, S., J. Maetz and F. Morel (1960) *Compt. Rend.*, **251**, 788
Järnefelt, J. (1961) *Biochim. Biophys. Acta.*, **48**, 104
Karnovsky, M. L. (1967) *Protoplasma*, **63**, 76
Katsoyannis, P. G. and V. du Vigneaud (1958) *J. Biol. Chem.*, **233**, 1352
Keller, A. R. (1963) *Anat. Record*, **147**, 367
Kernan, R. P. (1961) *Nature*, **190**, 347
Kernan, R. P. (1962a) *Nature*, **193**, 986
Kernan, R. P. (1962b) *J. Physiol.* (*London*), **162**, 129
Kernan, R. P. (1965) *Cell K*, Butterworths, London
Keynes, R. D. (1967) *Protoplasma*, **63**, 13
Kinsky, S. C., J. Avruch, M. Permutt, H. B. Rogers and A. A. Schonder (1962) *Biochem. Biophys. Res. Commun.*, **9**, 503
Kinsolving, C. R., R. L. Post and D. L. Beaver (1963) *J. Cellular Comp. Physiol.*, **62**, 85
Kirschner, L. B. (1955a) *J. Cellular Comp. Physiol.*, **45**, 61
Kirschner, L. B. (1955b) *J. Cellular Comp. Physiol.*, **45**, 89
Klahr, S. and N. S. Bricker (1964) *J. Clin. Invest.*, **43**, 922
Koefoed-Johnsen, V. (1957) *Acta Physiol. Scand.*, **42**, Suppl. 145, 87
Koefoed-Johnsen, V. and H. H. Ussing (1955) *Acta Physiol. Scand.*, **28**, 60
Koefoed-Johnsen, V. and H. H. Ussing (1958) *Acta Physiol. Scand.*, **42**, 298
Koefoed-Johnsen, V. and H. H. Ussing (1960) In C. L. Commar and F. Bronner (Eds.), *Mineral Metabolism*, Vol. 1, Part A, Academic Press, New York. Chap. 6, pp. 169, 203
Koefoed-Johnsen, V., H. H. Ussing and K. Zerahn (1953) *Acta Physiol. Scand.*, **27**, 38
Lampen, J. O., P. M. Arnow and R. S. Saferman (1960) *J. Bacteriol.*, **80**, 200
Leaf, A. (1960) *J. Gen. Physiol.*, **43**, Suppl., 175
Leaf, A. (1964) In J. de Graeff and B. Leijnse (Eds.), *Water and Electrolyte Metabolism*, Vol. 2, Elsevier, Amsterdam. p. 20

Leaf, A. (1965) *Ergeb. Physiol. Biol. Chem. Exptl. Pharmakol.*, **56**, 216
Leaf, A. (1967) *Proc. 3rd Intern. Congr. Nephrol. Washington*, Vol. **1**, Karger, Basel. p. 18
Leaf, A., J. Anderson and L. B. Page (1958) *J. Gen. Physiol.*, **41**, 657
Leaf, A. and E. F. Dempsey (1960) *J. Biol. Chem.*, **235**, 2160
Leaf, A. and A. Renshaw (1957) *Biochem. J.*, **65**, 82
Leb, D. E., T. Hoshiko, B. D. Lindley and J. A. Dugan (1965) *J. Gen. Physiol.*, **48**, 527
Lichtenstein, N. S. and A. Leaf (1965) *J. Clin. Invest.* **44**, 1328
Lichtenstein, N. S. and A. Leaf (1966) *Ann. N.Y. Acad. Sci.*, **137**, 556
Linderholm, H. (1954) *Acta Physiol. Scand.*, **31**, 36
Loewenstein, W. R. and Y. Kanno (1963) *J. Gen. Physiol*, **46**, 1123
Macey, R. I. and S. Myers (1963) *Amer. J. Physiol.*, **204**, 1095
MacRobbie, E. A. C. and H. H. Ussing (1961) *Acta Physiol. Scand.*, **53**, 348
Mansour, T. E. (1959) *J. Pharmacol. Exptl. Therap.*, **126**, 212
Martin, D. W. (1964) *J. Cellular Comp. Physiol.*, **63**, 245
Martin, D. W. and P. F Curran (1966) *J. Cellular Physiol.*, **67**, 367
Martin, P. J. and H. O. Schild (1962) *J. Physiol. (London)*, **163**, 51P
Martin, P. J. and H. O. Schild (1965) *Brit. J. Pharmacol.*, **25**, 418
Mendoza, J. A., J. S. Handler and J. Orloff (1967) *Amer. J. Physiol.*, **213**, 1263
Natochin, J. V., K. Janáĉek and R. Rybová (1965) *J. Endocrinol.*, **33**, 171
Orloff, J. and J. S. Handler (1962) *J. Clin. Invest.*, **41**, 702
Orloff, J. and J. S. Handler (1964) *Amer. J. Med.*, **36**, 686
Perris, A. D. (1966) *Can. J. Biochem. Physiol.*, **44**, 687
Petersen, M. J. and I. S. Edelman (1964) *J. Clin. Invest.*, **43**, 583
Porter, G. A. and I. S. Edelman (1964) *J. Clin. Invest.*, **43**, 611
Post, R. L. and P. C. Jolly (1957) *Biochim. Biophys. Acta.*, **25**, 119
Post, R. L., C. R. Merritt, C. R. Kinsolving and D. C. Allbright (1960) *J. Biol. Chem.*, **235**, 1796
Rasmussen, H., I. L. Schwartz, M. A. Schoessler and G. Hochster (1960) *Proc. Natl. Acad. Sci. U.S.*, **46**, 1278
Rosenberg, T. (1954) *Symp. Soc. Exptl. Biol.*, **8**, 27
Salée, M. L. (1967) *Ann. Endocrinol. (Paris)*, **28**, 706
Salée, M. L. and M. Vidrequin-Deliège (1966) *Ann. Endocrinol. (Paris)*, **27**, 526
Salée, M. L. and M. Vidrequin-Deliège (1967a) *Comp. Biochem. Physiol.*, **23**, 583
Salée, M. L. and M. Vidrequin-Deliège (1967b) *Arch. Intern. Physiol. Biochim.*, **75**, 501
Sánchez, J., C. González and C. Concha (1966) *Arch. Biol. Med. Exp.*, **3**, 79
Sawyer, W. H. (1960) *Nature*, **187**, 1030
Sawyer, W. H. (1961a) *Recent. Progr. Hormone Res.*, **17**, 437
Sawyer, W. H. (1961b) *Pharmocol. Rev.*, **13**, 225
Sawyer, W. H., R. A. Musnick and H. B. Van Dyke (1959) *Nature*, **184**, 1464
Sawyer, W. H., R. A. Musnick and H. B. Van Dyke (1960) *Circulation*, **21**, 1027
Sawyer, W. H., R. A. Musnick and H. B. Van Dyke (1961) *J. Gen. Comp. Endocrinol.*, **1**, 30
Sayoc, E. M. and J. M. Little (1967) *J. Pharmacol. Exp. Therap.*, **155**, 352
Schatzmann, H. J. (1953) *Helv. Physiol. Pharmacol. Acta.*, **11**, 346
Schoffeniels, E. (1957) *Biochim. Biophys. Acta.*, **26**, 585
Schoffeniels, E. (1960) *Arch. Intern. Physiol. Biochim.*, **68**, 231
Schoffeniels, E. and M. Baillien (1960) *Arch. Intern. Physiol. Biochim.*, **68**, 376
Schoffeniels, E., and M. L. Salée (1965) *Comp. Biochem. Physiol.*, **14**, 587
Schoffeniels, E. and R. R. Tercafs (1962) *Biochem. Pharmacol.*, **11**, 769
Sen, A. K. and R. L. Post (1964) *J. Biol. Chem.*, **239**, 345
Sharp, G. W. G., C. H. Coggins, N. S. Lichtenstein and A. Leaf (1966) *J. Clin. Invest.*, **45**, 1240
Sharp, G. W. G., C. L. Komack and A. Leaf (1966) *J. Clin. Invest.*, **45**, 450
Sharp, G. W. G. and A. Leaf (1964a) *Proc. Natl. Acad. Sci. U.S.*, **52**, 1114
Sharp, G. W. G. and A. Leaf (1964b) *Nature*, **202**, 1185
Sharp, G. W. G. and A. Leaf (1965) *J. Biol. Chem.*, **240**, 4816
Sharp, G. W. G., N. S. Lichtenstein and A. Leaf (1965) *Biochem. Biophys. Acta.*, **111**, 329

Sjodin, R. A. (1959) *J. Gen. Physiol.*, **42**, 983
Skou, J. C. (1957) *Biochim. Biophys. Acta.*, **23**, 394
Skou, J. C. (1965) *Physiol. Rev.*, **45**, 596
Skou, J. C. and K. Zerahn (1959) *Biochim. Biophys. Acta.*, **35**, 324
Snell, F. M. and T. K. Chowdhury (1965) *Nature*, **207**, 45
Steen, W. B. (1929) *Anat. Record*, **43**, 215
Steinbach, H. B. (1952) *Proc. Natl. Acad. Sci. U.S.*, **38**, 451
Tercafs, R. R. and E. Schoffeniels (1961a) *Arch. Intern. Physiol. Biochim.*, **69**, 459
Tercafs, R. R. and E. Schoffeniels (1961b) *Arch. Intern. Physiol. Biochim.*, **69**, 604
Tercafs, R. R. and E. Schoffeniels (1961c) *Science*, **133**, 1706
Tercafs, R. R. and E. Schoffeniels (1962) *Arch. Intern. Physiol. Biochim.*, **70**, 129
Thier, S. O. (1968) *Biochim. Biophys. Acta.*, **150**, 253
Ussing, H. H. (1949) *Acta Physiol. Scand.*, **19**, 43
Ussing, H. H. (1960) In O. Eichler and A. Farah (Eds.), *The Alkali Metals in Biology*, *Handbuch Der Experimentellen Pharmakologie*, Vol. **13**, Springer-Verlag, Berlin. p. 1
Ussing, H. H. (1965a) *Harvey Lectures Ser.*, **59**, (1963-1964), 1
Ussing, H. H. (1965b) *Acta Physiol Scand.*, **63**, 141
Ussing, H. H., T. U. L. Biber and N. S. Bricker (1965) *J. Gen. Physiol.*, **48**, 425
Ussing, H. H. and E. E. Windhager (1964) *Acta Physiol. Scand.*, **61**, 484
Ussing, H. H. and K. Zerahn (1951) *Acta Physiol. Scand.*, **23**, 110
Van Dyke, H. B., B. F. Chow, R. O. Greep and A. Rothen (1942) *J. Pharmacol.*, **74**, 190
Van Zutphen, H., L. L. M. Van Deenen and S. C. Kinsky (1966) *Biochim. Biophys. Res. Commun.*, **22**, 393
Voute, C. L. (1963) *J. Ultrastruct. Res.*, **9**, 497
Whittam, R. (1963) In J. F. Hoffman (Ed.), *The Cellular Functions of Membrane Transport*, Prentice Hall, Englewood Cliffs, N.J. p. 139
Whittembury, G., N. Sugino and A. K. Solomon (1961) *J. Gen. Physiol.*, **44**, 689
Widdas, W. F. (1961) In B. Uvnas (Ed.), *Proceedings of the First International Pharmacological Meeting, Stockholm*, Vol. 3. Pergamon Press Limited, Oxford. p. 239
Zadunaisky, J. A., O. A. Candia and D. J. Chiarandini (1963) *J. Gen. Physiol.*, **47**, 393
Zadunaisky, J. A. and F. W. de Fisch (1964) *Amer. J. Physiol.*, **207**, 1010
Zerahn, K. (1955) *Acta Physiol. Scand.*, **33**, 347
Zerahn, K. (1958) Ph.D. Dissertation, Copenhagen
Zerahn, K. (1966) *Proceedings of the Symposium on Membrane Biophysics, Smolenice, Czechoslovakia*

CHAPTER 2

Ion movements in renal tubules

Maurice B. Burg and Jared J. Grantham

Laboratory of Kidney and Electrolyte Metabolism,
National Institutes of Health,
Bethesda, Maryland 20014, U.S.A.

I. INTRODUCTION

The fundamental unit of the mammalian kidney is the nephron. Urine formation begins in the glomerulus where ultrafiltration of blood plasma results in the delivery of relatively large volumes of fluid, similar in composition to extracellular fluid, into the lumen of the renal tubule. As the ultrafiltrate flows along the tubule, a succession of anatomically and functionally distinct tubular segments transform it into urine. Normally more than 99 per cent of the original volume of ultrafiltrate is reabsorbed as the fluid composition is adjusted to maintain the fluid and electrolyte composition of the body in a state of physiologic balance. That both absorption and secretion of solutes occur in the kidney tubules was originally inferred from clearance studies and microscopical studies of dye transport. The anatomical sites of specific tubular transport processes were established using the micropuncture technique to sample urine from different nephron segments. Micropuncture together with clearance studies of renal function have continued to be the standard methods of renal physiology. Recently, another technique has been devised to study tubular transport. Single fragments of nephrons are dissected from the kidneys of rabbits and perfused *in vitro*. (Burg, Grantham, Abramow and Orloff, 1966). With the technique it is possible to study specific transport processes in various nephron segments in an artificial environment, thereby permitting control of many of the variables which complicate studies in the intact kidney. In this chapter the mechanisms of electrolyte transport in the different portions of the mammalian tubule will be considered with special emphasis on the results of recent studies using isolated perfused tubules.

II. PROXIMAL TUBULE

Approximately 70 per cent of the ultrafiltrate formed by the glomerulus is reabsorbed in this first segment of the nephron. The rate of fluid absorption in the mammalian proximal convoluted tubule was first measured directly by Walker, Bott, Oliver, and MacDowell (1941) in rats. They blocked the proximal tubule by inserting a drop of oil by micropuncture and then collected all the tubule fluid reaching the oil block. In order to determine how much of the fluid filtered at the glomerulus had been reabsorbed up to the point of collection they measured the increase in concentration of a marker substance (creatinine) which is filtered at the glomerulus but not reabsorbed. Approximately 60 per cent of the glomerular filtrate was absorbed to a point half way along the proximal tubule (only the first half of the proximal tubule reaches the surface and can be sampled by micropuncture). The glomerular filtration rate per nephron was calculated to be $1 \cdot 96 \ \text{mm}^3/\text{hr}$ or 33 nl/min. The original method has been improved by the use of inulin (Gottschalk, 1961) instead of creatinine and by the use of a dye, lissamine green, (Steinhausen, 1963) to identify the last surface loop of the proximal tubule before micropuncture.* The original results have been repeatedly confirmed. Although this method is probably the most reliable for measuring nephron filtration rate and fluid absorption, it is dependent on the unproved assumption that the presence of the oil drop and withdrawal of fluid by micropuncture do not alter the flow of fluid in the lumen. Erroneous results may be obtained if fluid leaks around the oil block as occurs under some conditions (Brenner, Bennett and Berliner, 1968).

Fluid absorption has also been studied *in vitro* using proximal tubules dissected from rabbit kidneys. The tubules were perfused as shown in figure 1. Albumin ^{131}I was used as a volume marker. The rate of absorption in the first half of the proximal tubule, the convoluted portion, was $1 \cdot 18$ nl/ mm tubule length per min (Burg and Orloff, 1969). The last half of the proximal tubule, the pars recta or proximal straight tubule which is inaccessible to micropuncture, absorbed fluid at a slower rate ($0 \cdot 47$ nl/mm/

---

*Lissamine green is filtered at the glomerulus but is poorly reabsorbed by the tubules. The beginning and end of the proximal tubule can be identified by observing the sequence of filling of its surface loops with the dye after intravenous injection.

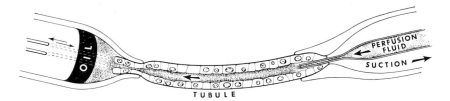

Figure 1. Arrangement for perfusing isolated kidney tubules. Two concentric pipets (on the right) are used to perfuse the tubule. The outer pipet holds the tubule over an inner pipet through which fluid is pumped. Fluid is collected by positioning the tubule within a single pipet (on the left). Mineral oil is layered over the fluid to prevent evaporation. A calibrated glass capillary is used to remove the fluid periodically and measure its volume (From Burg and Orloff, 1969).

min) (Burg and Orloff, 1969). This difference in net fluid absorption between the convoluted portion and pars recta had been inferred previously from micropuncture studies in rats (Walker, Bott, Oliver and MacDowell, 1941; Gottschalk, 1963). The pars recta also differs from the proximal convoluted tubule with regard to organic acid and glucose transport. In the pars recta *p*-aminohippurate transport is much more rapid than in the convoluted portion (Tune and Burg, 1969) whereas glucose absorption is less rapid (Tune and Burg, 1970), as is the rate of fluid absorption.

Walker, Bott, Oliver and MacDowell (1941) were also the first to note that the osmolality of luminal fluid does not change as fluid is absorbed. This has been repeatedly confirmed in recent years (Gottschalk, 1961). Since the total concentration of Na salts which constitute most of the osmotically active solute in the glomerular filtrate is also unchanged by the reabsorption of fluid in the proximal tubule (Gottschalk, 1961; Windhager and Giebisch, 1961, 1965), the fluid which is absorbed must be isosmotic with respect to Na salts. It is generally accepted that the sodium salts are 'actively' absorbed from the tubule fluid utilising energy from cellular metabolism, and that movement of water is 'passive' along the osmotic gradients created by salt transport (Gottschalk, 1961; Windhager and Giebisch, 1965). This view of the mechanism of sodium transport has been inferred from measurements of the transtubular electrochemical gradient for Na^+ under special experimental conditions.

A precise definition of the electrochemical gradient against which Na^+ is transported requires knowledge of the transtubular potential difference (PD). Conflicting results have been reported. Initially, a mean PD of -20 mV (lumen negative) had been found in rat proximal tubules (Windhager and Giebisch, 1965). However, Frompter and Hegel (1966) using a modification

c

of earlier techniques concluded that there is no measurable PD in this segment. They consider that in the earlier studies in which a PD was found, the tip of the electrode was actually in the tubule cells, rather than in the lumen as had been assumed. Further studies will be required to resolve this disagreement. In the isolated perfused rabbit proximal convoluted tubule, an electrical PD is present and can be measured reliably provided precautions are taken to prevent electrical leaks around the ends of the tubules. In early studies the perfusion pipet, serving as a bridge, was introduced up to 300 μ into the tubule through its broken end and no significant PD was found (Burg, Isaacson, Grantham and Orloff, 1968). Subsequently, the technique was improved by insulating the lumen of the tubule from the bath with a liquid dielectric compound (figures 2 and 3). Under these circum-

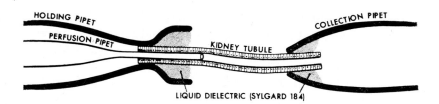

Figure 2. Arrangement for measuring electrical P.D. in isolated perfused kidney tubules. Pipets are similar to figure 1, except that Sylgard 184 is used to insulate the broken ends of the tubule from the bath. A chlorided silver wire or calomel half cell is used to make electrical contact with the fluid within the perfusion pipet and is connected to the input of an electrometer. The bath is grounded through a saline bridge with an identical wire or half cell (From Burg and coworkers, 1968).

stances a mean PD of $-3 \cdot 8 \pm 0 \cdot 5$ SEM mV stable for as long as 1 hour was found in 16 experiments (figure 4). When transport was inhibited with ouabain, the PD disappeared (figure 5), a finding which is consistent with the view that the PD is the result of electrolyte transport.

As indicated above, the Na concentration of tubule fluid is normally equal to that of plasma, and Na transport occurs without a detectable concentration gradient. However, under certain conditions the Na concentration in the lumen may be experimentally lowered. This was first demonstrated in rats infused intravenously with mannitol, a substance which enters the glomerular filtrate but is poorly reabsorbed (Windhager and Giebisch, 1961). Initially absorption of Na salts continues in the presence of mannitol. However, water is retained in the tubule lumen in consequence of an osmotic

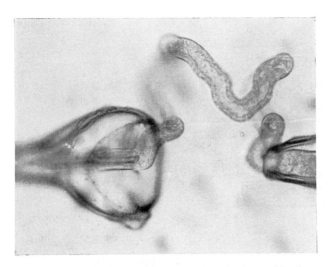

Figure 3. Proximal convoluted tubule during perfusion with the arrangement shown in figure 2.

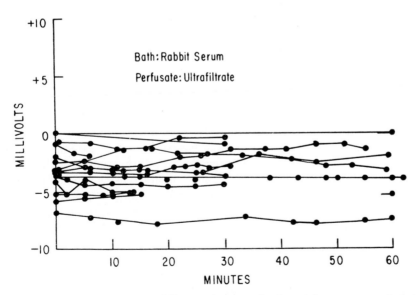

Figure 4. Electrical potential difference between bath and lumen across isolated perfused rabbit proximal convoluted tubules. The bath was grounded and served as reference. The bath contained rabbit serum; the perfusate contained ultrafiltrate of the same serum.

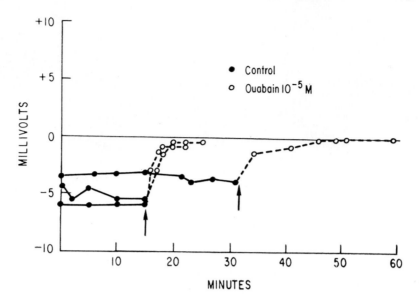

Figure 5. Effect of ouabain on the electrical potential difference across isolated perfused rabbit proximal convoluted tubules. Ouabain, added to the bath at the times indicated by the arrows, caused rapid depolarisation.

pressure created by the mannitol, and the Na concentration falls to a level at which net Na absorption ceases. In order to estimate the maximum Na concentration difference which can be maintained by the tubule, a column of fluid containing mannitol or some other poorly absorbed solute is placed in the proximal tubule between oil drops. After a few seconds the volume of fluid remains virtually constant and the Na concentration assumes a steady value some 35 mequiv./l less than in the plasma (Kashgarian, Stockle, Gottschalk and Ullrich, 1963). A similar result is observed using isolated perfused proximal tubules. When the tubules are bathed in rabbit serum and perfused with an ultrafiltrate of the same serum, fluid is absorbed without a change in the Na concentration (Kokko, Burg and Orloff, 1970). In the experiments shown in figure 6 isosmotic raffinose solution was added to the ultrafiltrate in different proportions so that the Na concentration of the perfusion solution was 14–62 mequiv./l less than in the serum. The fluid was perfused at a slow rate (approximately 3 nl/min) to facilitate detection of changes in Na concentration. The mean Na concentration in the collected fluid was 32·5 mequiv./l less than in the serum. This difference represents a steady-state value since the Na concentration in the collected tubular fluid usually decreased when

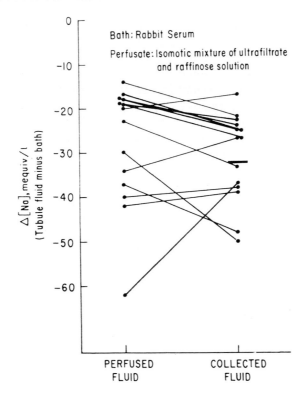

Figure 6. Sodium concentration difference [ΔNa] across isolated perfused rabbit proximal convoluted tubules (From Kokko, Burg and Orloff, 1970).

the initial difference between the perfusate and serum was less than 32·5 mequiv./l and increased when the initial difference was greater than 32·5 mequiv./l. Since under these circumstances both a PD (lumen negative) and a chemical concentration difference are present, net absorption occurs against an electrochemical gradient and is by definition 'active'.

The mechanism of chloride transport is less certain. The chloride concentration is generally higher in tubule fluid than in plasma because chloride is reabsorbed less rapidly than bicarbonate (Walker, Bott, Oliver and Mac-Dowell, 1941; Windhager and Giebisch, 1965). Ullrich and collaborators (1966) have measured the permeability to chloride of the intact rat tubule and found it to be so great that the Cl concentration difference alone accounted for Cl reabsorption. Since there is also a favorable electrical PD (lumen negative), it seems most likely that Cl transport is 'passive.' However, more detailed studies of chloride permeability and conductance are required to determine whether simple passive diffusion is the only mechanism of Cl transport in the proximal tubule.

The transtubular Na concentration difference which can be maintained by the proximal tubule is small despite a high rate of Na transport. This is due to the high permeability of the proximal tubule to Na and consequent back flux. The high permeability was first established in micropuncture studies in rats (Ullrich, 1966), and has been confirmed in isolated perfused rabbit proximal convoluted tubules (Kokko, Burg and Orloff, 1970). The Na permeability in the rabbit tubules is approximately 10^{-4} cm/sec (Kokko and coworkers). For such a high permeability, a concentration difference of 33 mequiv./l should result in a Na flux approximately equal to the normal Na transport rate. Thus, when the Na concentration in the lumen is 33 mequiv./l less than the bath net Na flux is reduced to zero.

Measurements of the electrical resistance of the proximal tubule wall are also consistent with a high permeability to electrolytes. Because of the tubular geometry, it is difficult to establish uniform current densities along the proximal tubule using micropipets. The tubule behaves as a cable conductor so that the current passed through the tubule wall from a micropipet in the lumen is maximum near the pipet and decays in more distant parts of the tubule in accordance with the space constant λ. (λ is the length of tubule over which the current density or the transtubular PD produced by the current decreases to $1/e$ or $0 \cdot 37$.) Using different methods, λ for the rat proximal tubule was calculated to be 55 μ (Windhager and Giebisch, 1961) and 86 μ (Hegel, Fromter and Wick, 1967) and the electrical resistance approximately 5 Ω cm^2 tubule area (Hegel and coworkers) or 800 ohm cm tubule length (for 20 μ diameter). Preliminary measurements of λ and electrical resistance in isolated rabbit proximal convoluted tubules are in good agreement with the results in the rat (Burg and Orloff, 1970).

Windhager, Boulpaep and Giebisch (1966) found that in *Necturus* the electrical resistance of the proximal tubule wall is less than the resistances of the individual tubule cell membranes. They interpreted this as indicative of a leak or shunt of electrolytes between the cells, presumably through the lateral intercellular spaces. It is conceivable that similar intracellular shunting accounts for the relatively high electrolyte permeability of mammalian proximal tubules.

Since the proximal tubules are highly permeable to Na salts, the effective osmotic pressure of these salts is less than ideal. This is expressed quantitatively in terms of the reflection coefficient (σ_{NaCl}) which is the ratio of the effective osmotic pressure of a test solute to that of an ideal solute. It can be expressed mathematically as

$$\sigma_{NaCl} = \frac{J_v}{L_p \, RT \, \Delta C}$$

where J_v is the flow of fluid induced by a difference in concentration of the Na salts (ΔC_{Na}) and L_p is the hydraulic conductivity of the tubule. If Na salts did not penetrate the tubule and excreted their full osmotic effect, σ_{NaCl} would be 1; actually it is less. σ_{NaCl} has been estimated for the renal tubule by use of a reference compound (raffinose) of sufficient molecular size so that its reflection coefficient (σ_{raff}) may be assumed to be 1. σ_{NaCl} determined by micropuncture studies of rat proximal tubules was 0·68 (Ullrich, 1966). The results for isolated perfused rabbit proximal convoluted tubule are shown in table 1 (Kokko, Burg and Orloff, 1970). Two different

Table 1. Calculation of reflection coefficient for sodium salts (σ_{NaCl}) across isolated perfused proximal tubule

Lumen	Bath	J_V (nl/mm/min)	ΔC (mOsm)	σ_{NaCl}
1. Ultrafiltrate	Serum	+1·59	0	
	Serum+NaCl	+3·72	+120	0·71±0·06[a]
	Serum+Raff.	+4·61	+118	
2. Raffinose	Serum	−2·89	0	0·68±0·06[b]
	Serum+NaCl	+0·90	+178	

$$^a\sigma_{NaCl} = \frac{(\Delta J_V/\Delta C)_{NaCl}}{(\Delta J_V/\Delta C)_{Raff}} \qquad ^b\sigma_{NaCl} = \frac{C_{Raff}}{C_{NaCl}} \text{ for } J_V = 0$$

Two methods (numbered 1 and 2) were used. In method 1 the increase in fluid absorption ($\Delta J_V = 3·72 - 1·59 = 2·13$) caused by NaCl added to the bath is divided by the increase caused by an equal concentration of raffinose ($\Delta J_V = 4·61 - 1·59 = 3·02$) in order to determine the reflection coefficient, $\sigma_{NaCl} = 2·13/3·02 = 0·71$. In method 2 raffinose solution of known concentration ($C_{Raff} = 298$) is perfused and J_V measured with and without NaCl added to the bath. The concentration of the bath ($C_{NaCl} = 435$) at which $J_V = 0$ is determined by extrapolation (figure 7) and $\sigma_{NaCl} = 298/435 = 0·68$.

methods were used. In the first, the perfusion fluid consisted of an ultrafiltrate of rabbit serum. In the control periods serum was present in the external bath and the control 'normal' rate of fluid absorption (J_v) due to active NaCl transport estimated. The bath was then made hyperosmotic by addition of NaCl or raffinose. The rate of absorption increased with both solutes, but the increase was greater with raffinose than with NaCl. From these data the reflection coefficient of NaCl was calculated using the formula shown. The mean reflection coefficient in 5 experiments was 0·71. This method of measuring reflection coefficient has the advantage that the luminal perfusion fluid is normal but an uncertainty is introduced by the unproved

assumption that the baseline rate of active fluid absorption is unaltered. In
the second method, the lumen was perfused with isosmotic raffinose solution,
eliminating Na from the perfusate and thus minimising any effect of active
Na transport on net fluid absorption. Net movement of fluid into the lumen
resulted as shown at the bottom of table 1. In the second portion of this
experiment NaCl was added to the bath making it hyperosmotic. The
reflection coefficient was then calculated from the concentration of NaCl
(obtained by extrapolation) required to reduce net water flow to zero, as
shown in figure 7. On the average NaCl had to be added to the bath until
the osmolality was 435 mOsm. From these results the reflection coefficient
was calculated to be 0·68 (table 1). There is good agreement between the
two methods for measuring reflection coefficient and the value arrived at is
similar to that reported by Ullrich (1966) for the rat proximal tubule *in vivo*.

It is generally accepted that water transport in the proximal tubule is
osmotically coupled to the active transport of sodium salts. However, no
measurable osmotic pressure difference across the tubule wall has been
observed, whereas from table 1 it is seen that the osmotic difference required
to absorb fluid at the normal rate is approximately 60 mOsm of NaCl which

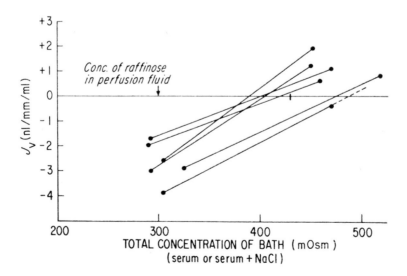

Figure 7. Measurement of the reflection coefficient for sodium salts (σ_{NaCl}) in
isolated perfused rabbit proximal convoluted tubules. The tubule lumen was perfused with raffinose solution. The bath contained rabbit serum or rabbit serum +
NaCl. The concentration of the bath at which $J_V = 0$ was determined by extrapolation. Calculation of σ_{NaCl} is shown in table 1.

should be easily measured.* Similar results in gallbladder have been explained by postulating that there is a region of high osmolality within the tissue, in the lateral intracellular spaces where it cannot be measured directly. Intercellular channels are present in the proximal convoluted tubule and it is conceivable that as suggested for gallbladder the fluid within them is ordinarily hyperosmotic. The most direct evidence implicating these channels was found in the gallbladder where coupled transport of salt and water was associated with distension of the lateral intercellular spaces (Kaye, Wheeler, Whitlock and Lane, 1966; Tormey and Diamond, 1967). In isolated perfused rabbit proximal convoluted tubules, no distension of intercellular spaces is apparent during normal fluid absorption (figure 8). The results of electron microscopic studies on rat kidneys fixed *in vivo* are conflicting; the intercellular spaces have been observed to be either distended or collapsed, depending on the technique used for fixation (Maunsbach, Madden and Latta, 1962).

A colloid osmotic pressure difference exists between the ultrafiltrate contained in the proximal tubule lumen and the plasma. Alterations in tubular fluid reabsorption have been noted in association with changes in the colloid content of the renal blood. It has been suggested that the colloid osmotic pressure difference may provide a driving force for the fluid absorption from this segment. However, Whittembury and collaborators (1959) calculated that in *Necturus* proximal tubule the colloid osmotic pressure difference could theoretically account for only 1 per cent of the fluid absorption. In order to verify this experimentally they erased the colloid osmotic pressure difference by placing albumin in the perfusion fluid and showed that fluid absorption remained. A similar calculation can be made for the rabbit proximal tubule. The water permeability of the rabbit proximal tubule is calculated from the experiments shown in table 1. It varies from $2 \cdot 9$ to $6 \cdot 3 \times 10^{-5}$ cm/sec/atm as an inverse function of the absolute osmolality of the experimental fluids. This variation with osmolality is similar to the effect of absolute osmolality on water permeability previously noted in the gallbladder (Diamond, 1966). Considering this measured water permeability, the colloid osmotic pressure of plasma can account for only 3 per cent of the normal net fluid absorption in the rabbit proximal tubule.

Experiments in which an effect of colloid osmotic pressure on fluid absorption has been demonstrated have involved changes in the colloid content of the plasma (Earley, Martino and Friedler, 1966; Brenner, Falchuk, Keimowitz and Berliner, 1969). When, as indicated above, colloid is added to the tubule fluid there is little effect. Lewy and Windhager (1968) provided a

*From table 1, 120 mOsm of NaCl added to the bath increased fluid absorption by 2 nl/mm/min, so that for the normal absorption rate of 1 nl/mm/min a difference of 60 mOsm of NaCl would be required.

plausible explanation for this discrepancy. The fluid which is absorbed by the renal tubule *in situ* is in turn absorbed from the intercellular space into the capillary circulation. Absorption into the capillaries depends on a balance of hydrostatic and colloid osmotic pressure between capillary lumen and interstitium. If absorption into the capillaries were decreased by a

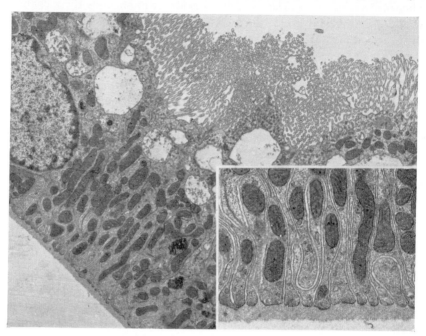

Figure 8. Electron micrograph of isolated perfused proximal tubule. Immediately prior to fixation, this specimen was actively pumping fluid from its lumen at the rate of 0·83 nl/mm/min. The lower magnification picture is a cross-section through portions of several tubular epithelial cells. Their appearance is in all respects identical with the best specimens obtained from intact kidneys. Towards the upper right hand corner, a typical well-developed brush border projects into the open tubular lumen. Numerous vesicles and vacuoles of various types and sizes are found in apical cytoplasm. The outer surface of the tubule, lower left hand corner, is bounded by a clean basement membrane. The cytoplasm of the basal half of the cells is characterized by numerous elongate mitochondria and by a complex labyrinth of plasma membrane folds. The higher magnification inset shows the mitochondria and membrane folds more clearly. The plasma membranes are paired, each pair separated by a narrow, electron-lucent space on the order of 200Å wide. These spaces are essentially lateral intercellular spaces; their complex folded pattern is the result of elaborate cytoplasmic interdigitations between adjacent cells. It is noteworthy that these intercellular spaces show no dilation or deformation attendant on the transcellular fluid transport. The spaces appear no wider here than in non-transporting controls. Magnification × 5,700. Inset magnification × 14,000. (Courtesy of Dr. John McD. Tormey.)

change in colloid osmotic pressure, fluid would accumulate in the intercellular spaces. It is possible that changes in volume and/or pressure of the intercellular spaces might in turn alter the rate of tubular fluid transport. The mechanism by which these physical changes might theoretically affect the active net fluid absorption is unclear at present.

Lewy and Windhager have also proposed that some such mechanism may be involved in glomerulotubular balance. The term glomerulotubular balance is used to designate a constant proportionate relationship between glomerular and tubular activity with respect to net fluid transport over a wide range of both filtration and reabsorption. In the earliest micropuncture studies it was noted that the *fraction* of the glomerular filtrate absorbed in the proximal tubules is approximately constant despite wide variations in filtration rate (Walker, Bott, Oliver and MacDowell, 1941). Thus, the absolute fluid absorption rate *in vivo* is directly proportional to the load of glomerular filtrate delivered to the tubule. Associated with variations of glomerular filtration rate *in vivo* there are changes in the concentration of protein in the peritubular capillaries due to ultrafiltration of larger or smaller fractions of the renal blood flow. These changes in colloid osmotic pressure in the peritubular capillaries account quantitatively for the accompanying changes in fluid absorption rate when filtration rate is varied *in vivo* (Falchuk, Brenner and Berliner, 1970). Perfusion–absorption balance is not evident in isolated perfused proximal tubules. In these the absolute rate of absorption is approximately constant independent of the perfusion rate (table 2). However, in contrast to the *in vivo* preparation the proximal tubule studied *in vitro* is not surrounded by capillaries and the colloid osmotic pressure of the serum is constant independent of perfusion rate. Similarly, when C.O.P.

Table 2. Lack of effect of flow rate on fluid absorption by isolated perfused rabbit proximal convoluted tubules

	Control	Experiment
Perfusion rate (nl/min)	8·0	19·0
Fraction absorbed	0·23	0·11
Absorption (nl/mm/min)	1·24	1·37

Each of eight tubules was perfused at two different speeds, alternating slow, fast, slow. There was no significant difference in absorption, comparing the mean of the control and experimental periods (pairing the results for each tubule the change in absorption was $+0·13 \pm 0·10$, $P > 0·2$). When there is glomerulotubular balance *in vivo*, absorption changes in proportion to filtration rate so that fractional absorption is constant. In the isolated perfused tubules, shown here, absorption did not change with perfusion rate, and fractional absorption decreased.

is maintained relatively constant in the rat kidney *in situ* and the rate of flow of tubule fluid is altered by pump perfusion, the absolute rate of fluid absorption does not vary with perfusion rate (Morgan and Berliner, 1969).

III.   LOOP OF HENLE

Each loop of Henle is divided into two limbs, a descending and an ascending limb. The epithelium lining the entire descending segment and papillary portion of the ascending segment is relatively flat, hence these portions of the loop are called the thin descending and ascending limbs. The epithelium of the ascending limb is thicker in the medulla, and this portion is called the thick ascending limb.

The descending limb is highly permeable to both water and solutes and, as a consequence, the urine in this segment becomes concentrated by equilibration of both salt and water with the hypertonic interstitial fluid (Morgan and Berliner, 1968). A short distance before the bend of the loop, the passive permeability to both water and electrolytes decreases and is maintained lower than in the thin descending limb from this point through the thin ascending segment. Urine flowing in the thin ascending portion of the loop becomes hypo-osmotic to the surrounding interstitium by virtue of the fact that sodium salts are transported out of the lumen more rapidly than is water (Jamison, Bennett and Berliner, 1967). A relatively low permeability to water of the ascending limb facilitates the establishment of the transtubular osmotic difference. Although the transport activity of the thick ascending limb has not been studied directly, it has been inferred that urinary dilution proceeds in this segment by means of a selective reabsorption of sodium chloride. The filtered potassium remaining at the bend of the loop is also reabsorbed in the ascending segment. As a consequence, the urine delivered into the distal convoluted tubules is virtually free of potassium and contains approximately 50–75 mM/l of sodium chloride (Malnic and coworkers 1964, 1966a; Bennett and coworkers, 1967, 1968).

IV.   DISTAL NEPHRON

The terminal segments of the nephron are the distal convoluted tubule, cortical collecting tubule and papillary collecting ducts. These segments establish large concentration differences for Na and K between urine and plasma and provide fine control of water excretion in response to antidiuretic hormone.

It was originally inferred from 'clearance' studies of urinary K excretion (Berliner, 1961) that most of the K appearing in the urine is secreted by the distal nephron. It was also concluded that net secretion of K is dependent on obligatory exchange of K for reabsorbed Na.

A. Distal Convoluted Tubule

Malnic, Klose and Giebisch (1964, 1966a, 1966b) have examined electrolyte transport mechanisms in the distal convoluted tubule of rats by means of micropuncture. They obtained direct evidence that sodium and water are removed from the fluid in distal convoluted tubules and that potassium is added in the same segment. In contrast to the proximal tubule discussed earlier, mean transtubular potential difference of approximately −50 mV was generally observed in the distal convoluted tubule. Since the concentration of Na in the luminal fluid decreases in this segment, Na absorption must occur against both an electrical and chemical gradient, i.e. by 'active' transport. In order to determine whether K secretion was also active, the quantitative relation between K secretion and PD was measured. The Nernst equation can be used for this comparison provided there is no net transfer of solute or water, a condition not readily obtained during the free flow of tubule fluid. A balanced state with no net transport was approximated by injecting a droplet of pure raffinose solution between oil columns. Raffinose, by virtue of its large molecular size penetrates the tubule poorly, and its osmotic effect in the tubule lumen minimises water absorption. Malnic, Klose and Giebisch (1966b) measured the concentration of Na and K in the tubule fluid and the electrical PD in this balanced state.

The PD was approximately −50 mV as it had been during free flow. Sodium concentrations were 36·7 and 31·4 mequiv./l, respectively, in samples from the beginning and end of the distal convoluted tubule. Both values are well below the theoretical value for electrochemical equilibrium of Na (approximately 1060 mequiv./l) confirming that the absorption of sodium is 'active'. The potassium concentrations in the same samples were 3·5 and 9·3 mequiv./l; also well below those for electrochemical equilibrium of K. They concluded that potassium enters the urine in the distal convoluted tubule passively along its electrical gradient but that the K concentration does not increase to the point of electrochemical equilibrium because of active reabsorption of potassium from the tubule lumen.

The precise origin of the electrical potential responsible for K secretion in the distal convoluted tubule is unknown. Giebisch, Malnic, Klose and Windhager (1966) consider it to be a diffusion potential approximately equivalent to the sum of the diffusion potentials at the luminal and peritubular border of the cells, a model similar to that previously proposed by Koefoed-Johnsen and Ussing (1958) for frog skin. Since the concentration of potassium in distal tubule cells is high compared to that in plasma (Burg and

Abramow, 1966), it was considered that the PD across the peritubular border of the cells may be determined by a predominant permeability to K of the membrane at this border. In support of this view it was found that increasing the K concentration in the plasma specifically depolarised the distal tubule (Giebisch and coworkers, 1966). The transtubular PD is also affected by alterations of the composition of the luminal fluid, but no specificity for either Na or K was noted. Instead the PD was directly related to the concentration of either NaCl or KCl in the tubule lumen and was increased by high concentrations of either salt. They concluded that the permeability of the luminal cell membrane to Na, K and Cl were similar and that the PD across this border was determined by the concentrations of all of these ions in the cells and tubule fluid (Giebisch and coworkers, 1966). Although the intra-luminal concentration of sodium decreases and that of potassium increases along the distal convoluted tubule in free flow, the sum of these ions remains essentially constant, and this, together with the equal permeabilities to sodium and potassium of the luminal membrane accounts for the observed constancy of the transtubular potential along the entire distal convoluted tubule.

Malnic, Klose and Giebisch (1966a) also examined the relationship between Na in the distal convoluted tubule fluid and the rate of K secretion. They were unable to find evidence for the previously postulated linked exchange of Na and K. The rate of Na delivery to the distal tubule was always much greater than the K secretion rate. Also, although K secretion was reduced by decreasing the concentration of Na in the perfusate to low levels (replacement of Na by choline), the PD also fell to -3 mV suggesting that reduction of the PD was responsible for the decreased K secretion rather than lack of Na as a counter ion (Malnic, Klose and Giebisch, 1966b).

B. Cortical Collecting Tubule

Cortical collecting tubules form from the junction of several distal convoluted tubules near the surface of the kidney. Detailed studies of electrolyte and water transport in the cortical collecting tubule have been carried out using isolated perfused rabbit renal tubules (Burg and coworkers, 1968; Grantham and coworkers, 1966, 1968, 1970). Although Na reabsorption and K secretion occur, as in the rat distal convoluted tubules, the mechanisms of transport differ in several important respects.

In order to measure PD and Na and K concentrations simultaneously in isolated collecting tubules, the arrangement shown in figure 9 was used. In general Na was absorbed from and K secreted into the lumen. As the contact time between the intratubular fluid and epithelium was prolonged by slowing the perfusion rate, the concentration differences for sodium and potassium

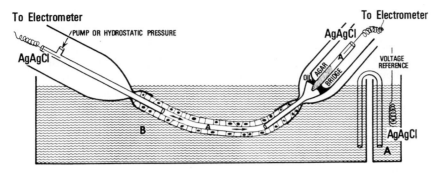

Figure 9. Arrangement for perfusing collecting tubules at slow rates and measuring transtubular PD at both ends. When the distal end of the tubule is cannulated as shown above, flow rate may be as low as ·01 nl/min using gravity perfusion. To deliver higher rates of flow (> 1 nl/min) the arrangement in figure 1 is used. To remove fluid from the collecting pipet for analysis of Na and K the voltage measuring bridge is withdrawn and the fluid under the oil layer is evacuated with a calibrated capillary pipet. The rate of perfusion is varied by changing the hydrostatic pressure at the perfusion end.

increased (figure 10). In other experiments in which the perfusate contained 40 mequiv. of K and 115 mequiv. of Na, the K concentration of the collected fluid was as high as 133 mequiv./l and the Na concentration as low as 6·5 mequiv./l. This K concentration is much higher and the Na concentration is much lower than that found in rat distal convoluted tubules under comparable

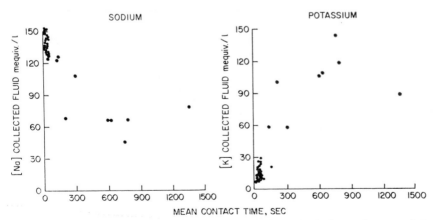

Figure 10. The relationship between mean contact time of tubular perfusate and the electrolyte concentration of the collected tubular fluid in isolated perfused rabbit cortical collecting tubules. The perfusate and bath contained 150 mM Na and 5 mM K.

conditions. Thus, tubule fluid in the cortical collecting tubule probably has a lower sodium concentration than in the distal tubule and the 'distal sodium dip' in stopflow patterns (Malvin, Wilde and Sullivan, 1958) may well correspond to the cortical collecting tubule rather than to the distal convoluted tubule, as had been supposed.

The mean lumen PD in the cortical collecting tubule determined with the arrangement shown in figure 9 was -47 mV at contact times greater than 100 sec. It was constant in any individual tubule and equal at the two ends. Since the luminal Na concentration decreased in all tubules despite the negative intraluminal PD, Na transport must be 'active'. In order to determine the mechanism of K secretion, the electrochemical potential for potassium $[E_K = -59 \log ([K]_{lumen}/[K]_{bath})]$ was compared with the measured transtubular potential, E_M. Figure 11 shows the difference between E_M and E_K as a function of mean contact time. At contact times exceeding 50 seconds, E_K consistently exceeded E_M, indicating uphill or 'active' transport of K into the tubule lumen. This mechanism of K secretion in the rabbit cortical collecting tubule is basically different from that found in the rat distal convoluted tubule. In the distal convoluted tubule E_K was always less

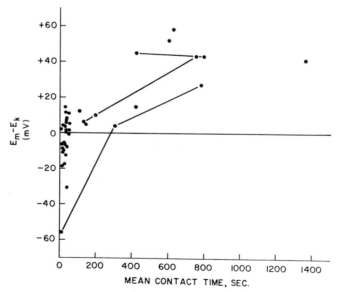

Figure 11. Relationship between mean contact time of tubular perfusate and the difference between transtubular PD (E_m) and the calculated potassium equilibrium potential (E_K) in isolated perfused rabbit cortical collecting tubules. See text for calculation of E_K. Perfusion fluid and bath contained 5 mM/L K. Positive values of E_M–E_K indicate that potassium accumulated in the lumen against an electrochemical gradient.

then E_M even in a balanced state, consistent with passive K secretion into the lumen.

The rate of secretion of K in the cortical collecting tubule depends on the concentration of intraluminal Na, although it is by no means clear that an obligatory Na–K exchange mechanism is involved. When solutions with Na concentration of 0–8 mequiv./l (containing sufficient raffinose to make them isosmotic) were perfused into rabbit cortical collecting tubules, potassium concentration rose only to 21·5 mequiv./l, a much lower concentration than with higher intraluminal Na concentrations (figure 10). At all luminal Na concentrations K transport was found to be active. Therefore, the inhibition of K secretion caused by lowering luminal Na concentration is most likely due to inhibition of active K transport rather than to decrease in PD, as in the rat distal convoluted tubule.

Normally, most of the K secretion in the distal nephron occurs in the distal convoluted tubule (Bennett and coworkers, 1967, 1968; Malnic and coworkers, 1964, 1966a). However, the increase in K secretion in the urine observed when Na excretion is increased probably occurs principally in the cortical collecting tubule and papillary collecting ducts. Ordinarily, the concentration of sodium in the urine leaving the distal convoluted tubule in the rat is approximately 30 mequiv./l, whereas higher concentrations are required for maximal active potassium secretion in the rabbit cortical collecting tubule. Thus a small amount of K probably is secreted by the cortical collecting tubule when Na and K balance is normal. When the concentration of sodium entering the cortical collecting tubule is increased, for example, by infusing sodium salts (Bennett and coworkers, 1967, 1968; Malnic and coworkers, 1964, 1966a; Landwehr and coworkers, 1967) or administration of diuretics which inhibit sodium reabsorption in the more proximal segments of the nephron, net secretion of potasisum is probably greatly accelerated in this segment and in the papillary collecting duct.

The precise origin of the electrical PD in the rabbit cortical collecting tubule is unknown. The nature of the PD appears to differ in at least one respect from that in the rat distal convoluted tubule. The transtubular PD in the cortical collecting tubule is specifically dependent on intraluminal Na concentration, whereas in the distal convoluted tubule the PD is the same whether equimolar NaCl or KCl is present in the lumen (Giebisch and coworkers, 1966). The result for cortical collecting tubules is illustrated in figure 12. The contact time was prolonged by reducing the perfusion pressure. The potential remained high as long as the intratubular sodium was above 20 mequiv./l. When the sodium concentration decreased below 20 mequiv.l, however, the potential fell independently of whether a high concentration of K was present in the lumen or not. This effect was fully reversible as was demonstrated by increasing the perfusion rate again. The PD always

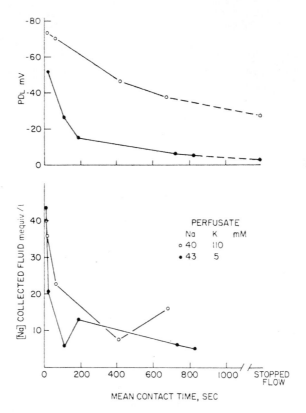

Figure 12. Relation between luminal Na concentration and electrical PD in rabbit cortical collecting tubules. At long contact times a larger fraction of the Na was absorbed, lowering its concentration in the lumen (lower graph) and causing a decrease in the electrical PD at the distal end of the tubule (upper graph).

decreased when flow was slowed despite replacement of Na by K. As shown in figure 12, substitution of raffinose for NaCl in the perfusate led to a similar result.

The PD in the cortical collecting tubule is also dependent on the concentration of K in the bath at the peritubular surface of the cells. As in the distal convoluted tubule, an increase in K concentration in the bath leads to depolarisation of the cortical collecting tubule. The relationship between the external potassium concentration and transtubular PD when potassium was substituted isotonically for sodium in concentrations as high as 95 mequiv./l is illustrated in figure 13. The dashed line is theoretical, based on the assumption that the tissue acts as a K electrode with the PD determined by the ratio of a fixed high K concentration [in the tubule cells (Burg and Abramow,

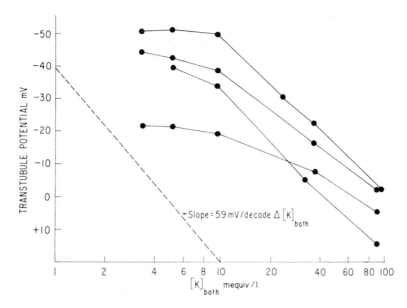

Figure 13. The relationship between transtubular PD and the K concentration of the external bath surrounding isolated perfused rabbit cortical collecting tubules. The perfusate contained Na 150 and K 5 (mM/L). Perfusion rate was greater than 1 nl/min and constant. KCl was substituted isotonically for NaCl in the bath. The PDs represent the steady state measured 2 minutes after changing the K concentration of the external bath. The mean slope of the linear portions of the curves (when K_{bath} was greater than 10 mequiv./l) was 45 mV per decade change in K_{bath}.

1966)] and the lower concentration present in the bath. At concentrations of K in the bath above 10 mequiv./l, the PD was directly related to external K concentration, the mean slope being 45 mV per decade change in external K. This result is consistent with a potential step at the peripheral border of the tubule cells largely dependent on the transmembrane K concentration ratio there [as also proposed for frog skin (Koefoed-Johnsen and Ussing, 1958) and distal convoluted tubule (Giebisch and coworkers, 1966)]. The lack of effect of external K between 3 and 10 mequiv./l on the PD may be due, as in muscle and nerve (Hodgkin, 1951), to the influence of other, less permeable ions, e.g. Na, when their concentrations become high relative to the K concentration.

It was originally proposed that the PD across frog skin is due to the sum of the Na and K diffusion PD's at the outer and inner cell surfaces, respectively, and that the active transport step consisted of electrically neutral Na for K exchange at the inside border of the cells (Koefoed-Johnsen and Ussing,

1958). This model for frog skin is difficult to fit in its simple form to the cortical collecting tubule when the results of removal of K from the bath are considered. Lowering the external K concentration to less than 1 mequiv./l caused the PD of collecting tubules to decrease more than 50 per cent within one minute, as illustrated by a representative study in figure 14. The half-time for isotopic exchange of ^{42}K in the cortical collecting tubule in the steady state is 12·6 minutes (Burg and Abramow, 1966). Thus, the initial phase of depolarisation in low potassium medium is much too fast to be explained by a simple loss of cellular potassium. More likely the initial rapid depolarisation is due to inhibition of an electrogenic sodium pump. If removal of

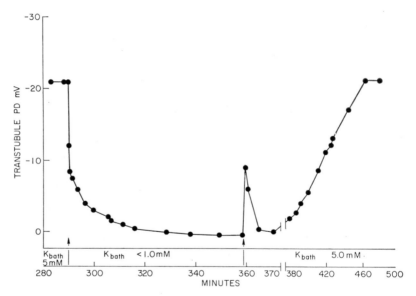

Figure 14. Effect of removing external K on transtubular PD of an isolated perfused rabbit cortical collecting tubule. The perfusate contained 5 mM of K. The PD fell rapidly when the K concentration of the bath was lowered. Replacement of K in the bath reversed the effect.

external K suddenly caused interruption of the outward current created by an electrogenic Na pump the passive sodium current from bath to cells would persist and cause immediate depolarisation of the tissue. Subsequent loss of cellular potassium would cause additional depolarisation but at a slower rate. According to this explanation external K might be required for Na transport, but not necessarily as a transported counter-ion.

Additional evidence against electroneutral exchange transport of Na and K at the peripheral border of tubule cells comes from comparison of the steady-state active flux of Na to that of K in mixed tubule suspensions (Burg, Grollman and Orloff, 1964; Burg and Orloff, 1966) and in isolated, non-perfused collecting tubules (Burg and Abramow, 1966). In these preparations the tubule lumens are collapsed and the observed transport is largely limited to the peritubular cell membranes. It was found that the calculated active efflux of Na greatly exceeded the influx of K making electroneutral (one for one) exchange unlikely. Although Na–K exchange transport at the peritubular border of the cells is considered unlikely, such an exchange might possibly be present at the luminal border of the cells and the theory of forced Na for K exchange in the distal nephron (Berliner, 1961) cannot be discarded.

As discussed earlier, there is active K reabsorption in the distal convoluted tubule (Malnic, Klose and Giebisch, 1966b). From comparison of distal tubule micropuncture samples with final urine, it was concluded that K reabsorption also occurs in the collecting ducts (Malnic and coworkers, 1964, 1966a; Bennett and coworkers, 1967, 1968). This conclusion lacks certainty, however, since the final urine reflects the activity of all the nephrons, whereas the micropuncture samples are by necessity from superficial nephrons which may differ in their function from deeper nephrons. K reabsorption was not found in the present studies. However, these studies were not specifically designed to stimulate K reabsorption, and the mechanism and even the existence of K reabsorption in the collecting tubules remains to be established.

The PD in the cortical collecting tubule is transiently increased by anti-diuretic hormone (ADH) (Burg, Isaacson, Grantham and Orloff, 1968), an effect analogous to the increase in PD caused by this hormone in frog skin (Fuhrman and Ussing, 1951) and toad bladder (Leaf and Dempsey, 1960). As in frog skin and toad bladder, this increase in PD in the collecting tubule is probably due to increased ion transport. Measurements of the electrical resistance of isolated perfused rabbit cortical collecting tubules support this conclusion. In order to measure electrical resistance the arrangement shown in figure 15 was used. The resistance of the cortical collecting tubule ranged between 1·7 and $26·5 \times 10^4$ ohm.cm tubule length (Helman and Grantham, 1969) which is high compared to the value of 800 ohm.cm tubule length previously found in proximal tubule. The higher resistance in the cortical collecting tubule is presumably due to the lower ionic permeability of the distal segments of the nephron. When antidiuretic hormone was added to the bath surrounding cortical collecting tubules, there was no change in their electrical resistance despite the transient increase in electrical PD (figure 16). Since the resistance of the collecting tubule is independent of current, this result indicates that the hormone transiently increases the transepithelial 'short-circuit current'. Thus, as in the frog skin and toad bladder, the short-circuit current (PD/resistance) of the collecting tubule is

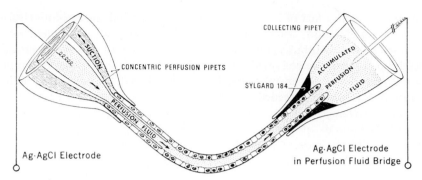

COLLECTING PIPET

CONCENTRIC PERFUSION PIPETS

SUCTION

ACCUMULATED

PERFUSION FLUID

SYLGARD 184

PERFUSION FLUID

Ag·AgCl Electrode

Ag·AgCl Electrode
in Perfusion Fluid Bridge

Figure 15. Arrangement for measuring membrane electrical resistance in renal tubules. On the left the outermost of two concentric pipets was used to hold the tubule in place. The inner large bore pipet was advanced 500μ into the tubule lumen electrically insulating the lumen from the bath. The Ag–AgCl wire in the perfusion pipette was used to inject current into the tubule lumen as well as to measure the resultant potential deflection at the proximal end of the tubule. The other end of the tubule was sealed into a single micropipet with Sylgard 184 to provide electrical insulation from the bath. Electrical contact with the lumen at this end was made using a fine capillary bridge and a second Ag–AgCl electrode. A third indifferent Ag–AgCl electrode served to ground the external bath. From measurement of the voltage deflections at both ends of the tubule during injection of current at the perfusion end the transepithelial resistance of the tubule was calculated using standard core conductor equations.

increased by vasopressin, presumably as the result of increased active ion transport.

The most important effects of ADH are on water permeability. It has been shown in direct studies that ADH increases the water permeability of the rat distal convoluted tubule (Ullrich, Rumrich and Fuchs, 1964), rabbit cortical collecting tubule (Grantham and Burg, 1966) and rat papillary collecting ducts (Morgan, Sakai and Berliner, 1968). The hormone is effective when present at the peripheral border of tubule cells but not in the lumen (Grantham and Burg, 1966). As in toad bladder (Orloff and Handler, 1967), both cyclic-3′,5′-AMP, the intracellular mediator of the action of vasopressin, and theophylline, an inhibitor of the degradation of the nucleotide, increase osmotic water flow in the collecting tubule in a manner indistinguishable from that due to vasopressin (Grantham and coworkers, 1966, 1968). In addition, prostaglandin inhibits the action of ADH on water flow in the collecting tubule (Grantham and Orloff, 1968), presumably on a competitive basis.

The effect of ADH on water permeability of cortical collecting tubules is manifest as an increased permeability to isotopic (THO) water, and, when

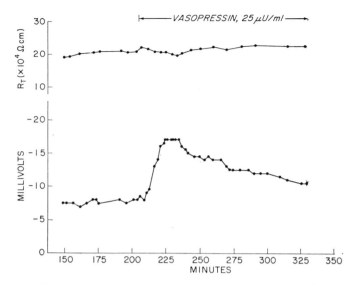

Figure 16. Effect of antidiuretic hormone on the spontaneous electrical PD and membrane resistance in an isolated perfused rabbit cortical collecting tubule. Addition of ADH caused the PD to increase and then to slowly decline towards baseline. The hormone had no effect on the transepithelial resistance.

the osmolality of the tubule fluid is less than that of the bath, as an increase in net fluid absorption (Grantham and Burg, 1966). Accompanying the increased net fluid absorption there is a visible change in the morphology of the tissue (Ganote, Grantham and coworkers, 1968). The cells swell and there is dilatation of the lateral intercellular spaces (figures 17 and 18). The

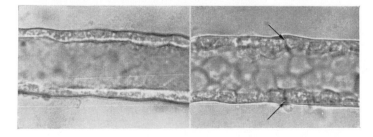

Figure 17. Photomicrographs of a functioning isolated perfused rabbit cortical collecting tubule. The focal plane is at the central axis of the tubule. Perfusate and bath are 130 and 290 mOsm/kg, respectively. On the left is its appearance in the absence of ADH. The same portion of the tubule is shown on the right after ADH had maximally increased the permeability to water. The cells are swollen, and the intercellular markings (arrows) are more prominent than in the control.

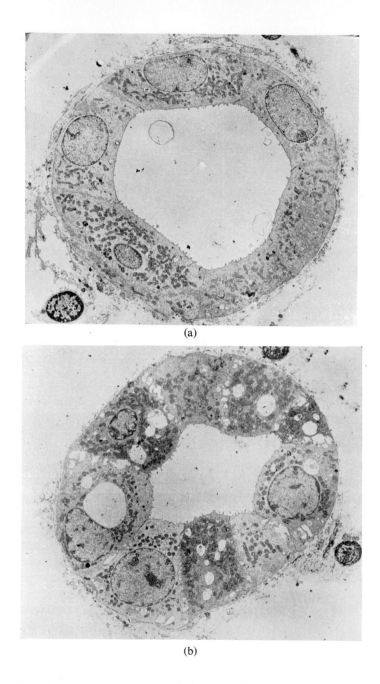

(a)

(b)

Figure 18. Electron micrographs of isolated perfused rabbit cortical collecting tubules before and after treatment with antidiuretic hormone. The osmolality of the perfusate and bath was 130 and 290 mOsm/kg, respectively. Before adding ADH (a) the cells are relatively flat and there is no distention of the intercellular spaces. After applying ADH (b) the cells are swollen and the intercellular spaces are widely dilated.

cells swell because ADH specifically increases the permeability of the luminal cell membranes to water allowing more rapid entry into the cells of dilute fluid from the tubule lumen. The water which enters the cells flows into the external bath both across the basal surface of the cells and through the lateral surface of the cells and the lateral intercellular spaces (figure 18) (Grantham, Ganote and coworkers, 1969). Movement of water through the

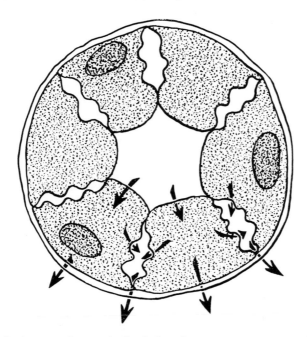

Figure 19. Pathways of osmotically induced net water movement in collecting tubules treated with ADH. The arrows indicate the paths of bulk water flow.

lateral intercellular spaces is accompanied by increased hydrostatic pressure which causes dilatation of these spaces.

Papillary collecting ducts have been studied directly by means of retrograde microcatheterisation (Hierholzer, 1961; Hilger, Klumper and Ullrich, 1958), micropuncture (Gottschalk and Mylle, 1959), and perfusion *in vitro* (Morgan and Berliner, 1968; Morgan, Sakai and Berliner, 1968). Na reabsorption and K secretion have been observed (Hierholzer, 1961), as well as a direct effect of ADH on water permeability (Morgan and Berliner, 1968). The electrical PD (Windhager and Giebisch, 1965; Marsh and Solomon, 1965) is less than in the distal convoluted tubule and cortical collecting tubule, but studies of the transport mechanisms are not sufficiently detailed for comparison with the other distal nephron segments.

V. CONCLUDING REMARKS

With the development of specialised microtechniques it has become possible to study directly the transport mechanisms in renal tubules. As a result of these studies, the location of the major sites of ion and water transport have been identified and considerable insight has been gained concerning the cellular transport mechanisms involved.

Acknowledgement

The authors thank Dr. Jack Orloff for his valuable suggestions and discussion in the preparation of this manuscript.

REFERENCES

Bennett, C. M., B. M. Brenner and R. W. Berliner (1968) *J. Clin. Invest.*, **47**, 203
Bennett, C. M., J. R. Clapp and R. W. Berliner (1967) *Am. J. Physiol.*, **213**, 1254
Berliner, R. (1961) *Harvey Lectures*, **55**, 141
Brenner, B., C. Bennett and R. Berliner (1968) *J. Clin. Invest.*, **47**, 1358
Brenner, B., K. Falchul, R. Keimowitz and R. Berliner (1969) *J. Clin. Invest.*, **48**, 1519
Burg, M. and M. Abramow (1966) *Am. J. Physiol.*, **211**, 1011
Burg, M., J. Grantham, M. Abramow and J. Orloff (1966) *Am. J. Physiol.*, **210**, 1293
Burg, M., E. F. Grollman and J. Orloff (1964) *Am. J. Physiol.*, **206**, 483
Burg, M., L. Isaacson, J. Grantham and J. Orloff (1968) *Am J. Physiol.*, **215**, 788
Burg, M. and J. Orloff (1966) *Am. J. Physiol.*, **211**, 1005
Burg, M. and J. Orloff (1969) *J. Clin. Invest.*, **47**, 2016
Burg, M. and J. Orloff (1970) Unpublished observation.
Diamond, J. (1966) *J. Physiol.*, **183**, 58
Earley, L., J. Martino and R. Friedler (1966) *J. Clin. Invest.*, **45**, 1668
Falchuk, K., B. Brenner and R. Berliner (1970) Personal communication
Fromter, E. and U. Hegel (1966) *Pflügers Arch.*, **291**, 107
Fuhrman, F. and H. Ussing (1951) *J. Cell. Comp. Physiol.*, **38**, 109
Ganote, C. E., J. J. Grantham, H. L. Moses, M. B. Burg and J. Orloff (1968) *J. Cell Biol.*, **36**, 355
Giebisch, G., G. Malnic, R. M. Klose and E. E. Windhager (1966) *Am. J. Physiol.*, **211**, 560
Gottschalk, C. W. (1961) *Physiologist*, **4**, 35
Gottschalk, C. (1963) *Harvey Lectures*, **58**, 99
Gottschalk, C. W. and M. Mylle (1959) *Am. J. Physiol.*, **196**, 927
Grantham, J. and M. B. Burg (1966) *Am. J. Physiol.*, **211**, 255
Grantham, J. J., M. B. Burg and J. Orloff (1970) *J. Clin. Invest.*, **49**, 1815
Grantham, J. J., C. E. Ganote, M. B. Burg and J. Orloff (1969) *J. Cell Biol.*, **41**, 562
Grantham, J. and J. Orloff (1968) *J. Clin. Invest.*, **47**, 1154
Hegel, U., E. Fromter and T. Wick (1967) *Pflügers Arch.*, **294**, 274
Helman, S. I. and J. J. Grantham (1969) *Fed. Proc.*, **28**, 524
Hierholzer, K. (1961) *Am. J. Physiol.*, **201**, 318
Hilger, H. H., J. D. Klumper and K. J. Ullrich (1958) *Pflügers Arch.*, **267**, 218
Hodgkin, A. L. (1951) *Biol. Rev.*, **26**, 339
Jamison, R. L., C. M. Bennett and R. W. Berliner (1967) *Am. J. Physiol.*, **212**, 357
Kashgarian M., H. Stockle, C. Gottschalk and K. Ullrich (1963) *Pflügers Arch.*, **277**, 89
Kaye, G., H. Wheeler, R. Whitlock and N. Lane (1966) *J. Cell Biol.*, **30**, 237
Koefoed-Johnsen, U. and H. H. Ussing (1958) *Acta Physiol. Scand.*, **42**, 298
Kokko, J., M. Burg and J. Orloff (1970) *J. Clin. Invest.*, (in press)

Landwehr, D. M., R. M. Klose and G. Giebisch (1967) *Am. J. Physiol.*, **212**, 1327
Leaf, A. and E. Dempsey (1960) *J. Biol. Chem.*, **235**, 2160
Lewy, J. and E. Windhager (1968) *Am. J. Physiol.*, **214**, 943
Malnic, G., R. M. Klose and G. Giebisch (1964) *Am. J. Physiol.*, **206**, 674
Malnic, G., R. M. Klose and G. Giebisch (1966a) *Am. J. Physiol.*, **211**, 529
Malnic, G., R. M. Klose and G. Giebisch (1966b) *Am. J. Physiol.*, **211**, 548
Malvin, R. L., W. S. Wilde and L. P. Sullivan (1958) *Am. J. Physiol.*, **194**, 135
Marsh, D. J. and S. Solomon (1965) *Am. J. Physiol.*, **208**, 1119
Maunsbach, A., S. Madden and H. Latta (1962) *J. Ultrastruct. Res.*, **6**, 511
Morgan, T. and R. W. Berliner (1968) *Am. J. Physiol.*, **215**, 108
Morgan, T. and R. Berliner (1969) *Am. J. Physiol.*, **217**, 992
Morgan, T., F. Sakai and R. W. Berliner (1968) *Am. J. Physiol.*, **214**, 574
Orloff J. and J. S. Handler (1967) *Am. J. Med.*, **42**, 757
Steinhausen, M. (1963) *Pflügers Arch.*, **277**, 23
Tormey, J. and J. Diamond (1967) *J. Gen. Physiol.*, **50**, 2031
Tune, B. and M. Burg (1969) *Am. J. Physiol.*, **217**, 1057
Tune, B. and M. Burg (1970) (In preparation)
Ullrich, K. (1966) *Proc. 3rd Int. Cong. Nephrol.*, **1**, 48, Karger, Basel
Ullrich, K. J., G. Rumrich and G. Fuchs (1964) *Pflügers Arch.*, **280**, 99
Walker, A., P. Bott, J. Oliver and M. MacDowell (1941) *Am. J. Physiol.*, **134**, 580
Whittembury, G., D. Oken, E. Windhager and A. Solomon (1959) *Am. J. Physiol.*, **197**, 1121
Windhager, E., E. Boulpaep and G. Giebisch (1966) *Proc. 3rd Int. Cong. Nephrol.*, **1**, 35 Karger, Basel
Windhager, E. and G. Giebisch (1961) *Nature*, **191**, 1205
Windhager, E. and G. Giebisch (1961) *Am. J. Physiol.*, **200**, 581
Windhager, E. and G. Giebisch (1965) *Physiol. Rev.*, **45**, 214

CHAPTER 3

Water and ionic transport by intestine and gall bladder

C. J. Edmonds

Medical Research Council Department of Clinical Research,
University College Hospital Medical School,
University Street, London W.C.1., England

I. INTRODUCTION AND TERMINOLOGY

A large quantity of fluid enters the alimentary tract of an animal every day
the greater part due to secretions from glands associated with the gut. In an

adult man, for example, the total quantity is about eight to nine litres, representing nearly 20 per cent of the total body water. The conservation of so much water and of the electrolytes associated with it is clearly of great importance in the organism's fluid economy. The fact that water and electrolyte loss in the faeces is so small indicates the efficiency of the mechanisms which prevent wastage. The introduction of new techniques and of new ideas about transport processes have considerably stimulated interest in the way in which the epithelium of the alimentary tract is able to perform its important functions in salt and water metabolism. A great deal of investigation has been undertaken during recent years and it has become evident that the phenomena involved have an interest beyond gastrointestinal physiology alone.

Since very considerable fluxes of water and electrolytes take place across the gut epithelium in both directions, it is particularly important that terminology should be clearly defined. In the past considerable confusion has arisen because of lack of an agreed terminology and it is only recently that some uniformity has begun to emerge although unfortunately there still tends to be some dichotomy between that used for *in vivo* and that for *in vitro* experiments. The terminology used in the present chapter and discussed below is commonly employed.

The flows are always considered with respect to the animal's internal environment. Thus flow from the lumen to the extracellular fluid, blood or lymph is called influx or insorption and conventionally given a positive sign while flow from the animal to the lumen is called efflux or exsorption and given a negative sign (Code, 1960). The symbols J_i and J_o are frequently used to represent influx and efflux rates but in the present chapter J_{LS} and J_{SL} will be used, L and S standing for lumen and serosa respectively, since this is less ambiguous and more consistent with *in vitro* terminology. The net flux, J_n, is given by $J_n = J_{LS} + J_{SL}$. J_n may be negative or positive; the former is associated with the term 'secretion' and the latter with the term 'absorption'. These terms do not in any way refer to the nature of the processes occurring. Whether absorption or secretion is taking place will depend on many factors, including the concentrations of ions in the lumen, electrical effects, the presence of active transport, etc. Another concept of some use concerns the luminal concentration of an ion at which J_n for the ion is zero. The concentration is sometimes called an equilibrium concentration but this is misleading since it cannot usually be directly observed but must be deduced indirectly from plots of changes of J_{LS} and J_{SL} with various luminal concentrations (for example, see Curran and Schwartz, 1960; Edmonds, 1967b; Fordtran and coworkers, 1968). A better term, the critical luminal concentration, will be used in this chapter.

For *in vitro* studies a rather different terminology has evolved; the luminal surface of the epithelium is generally called the mucosal surface and the unidirectional fluxes described as mucosa to serosa (J_{MS}) or serosa to mucosa (J_{SM}). These expressions will be adopted here although a wide variety of other expressions have been used (Barry and Smyth, 1960).

In some instances, for example where a substance is metabolized in the tissue, as may be the case with bicarbonate, it is difficult to assign precise meaning to statements about transepithelial fluxes. In such cases, the flux values across each face of cell layer must be considered separately.

A problem closely related to terminology is concerned with the choice of parameter to which flux values are to be related; this is particularly important when different parts of the gut or different species are to be compared. Unfortunately, although the choice is wide, there is no general agreement since no parameter is entirely satisfactory. Those used include the length of the gut, the surface area, the wet or dry weight, the protein or the potassium content. Any one chosen must be used in a carefully standardized way, in which case there is probably little to choose between them. The subject is discussed in some detail by Levin (1967).

II. METHODOLOGY

Many different methods have been employed in the study of water and electrolyte transport in the intestine and gall bladder. An extensive review of the historical development and variety of methods has been recently given (Parsons, 1968) and the present discussion will be only a brief summary necessary for understanding the significance of the observations.

The methods fall naturally into two groups: those carried out *in vivo* and those carried out *in vitro* on tissues removed from animals.

A. In Vivo Experiments

Balance studies involve an input–output analysis of some region of the gut as, for example, by comparing the input into the colon from the ileum with the output emerging in the faeces. Such studies are valuable because they can even be done on conscious animals, and indicate the major changes in the luminal constituents produced by an intestinal segment. However the information obtainable is limited (see, for example, its application to colonic absorption, Edmonds, 1967a).

Perfusion studies usually involve passing or recirculating solutions through segments of gut which have been cannulated in anaesthetized experimental animals. Absorption and secretion can be observed if an unabsorbed marker, for example, polyethylene glycol (PEG 4000) is included in the perfusion solution. If, in addition, appropriate radionucleides are present, the influx and efflux rates can be determined. In man similar experiments have been done during surgical operations and, by employing multilumen tubes which can be swallowed, it is possible to perfuse segments of gut in conscious subjects (see, for example, a recent review of human methodology by Duthie, 1967). The surgical construction of isolated loops of intestine for perfusion studies, usually called Thiry–Vella loops, has been a method employed for many years although technically studies are difficult to perform satisfactorily. Furthermore, the extent to which lengths of bowel deprived of their proper contents possess an epithelium whose function is like that of the normal tissue is uncertain. There is indeed evidence that the state of the epithelium is partially dependent on the gut contents (Jervis and Levin, 1966).

Results of perfusion studies often show considerable variation which is not surprising in view of the many unsatisfactory features of the *in vivo* method. Mixing within the gut lumen is one problem. It is certainly not very good at low flow rates where there must exist an unstirred layer of considerable thickness. Increasing the flow rate to create turbulence and improve mixing is generally not practical as the gut becomes too distended. In experiments in which flow rates have been varied, some alarming changes in flux rates have been recorded (for example, see Cramer and Haqq, 1963; Dawson and McMichael, 1968). These are probably attributable to several factors, alteration in mixing, gut distension producing changes in surface area, changes of concentration gradient along the perfused segment, etc. Many experiments have been done in which a segment of gut or the gall bladder has been cannulated, a solution placed in the lumen and then removed after some interval. Obviously in this type of experiment there is a possibility of serious error from lack of adequate mixing.

B. In Vitro Experiments

Detailed reviews of the techniques have been written by Wiseman (1961) and Smyth (1963). A number of different types of method are used.

i) Fisher and Parsons (1949–50) method involves circulating oxygenated fluid through an isolated segment of intestine suspended in a saline medium.

ii) A similar preparation has been used with the modification that the serosal fluid is eliminated and drops of fluid are collected from the serosal surface (for example, Smyth and Taylor, 1957, with intestine and Diamond, 1962a, with gall bladder). Alternatively, liquid paraffin may be substituted on the serosal side.

iii) The 'everted sac' technique was introduced by Wilson and Wiseman (1954). A segment of intestine is gently turned inside out using a glass rod so that the luminal surface is exterior. The sac may then be tied to form a small sausage, filled with fluid and suspended in a flask containing oxygenated fluid or alternatively it may be cannulated. Various other modifications of the technique have been used.

iv) The method introduced by Ussing and Zerahn (1951) for frog skin studies has been applied to intestinal epithelium. The tissue is mounted as a sheet in a chamber with oxygenated solutions bathing both sides. It is sometimes possible to strip off the muscle coat so that only epithelium with a thin layer of connective tissue is left. Then there is good access to both faces of the epithelium, mixing is good and it is possible to eliminate the trans-epithelial potential difference (p.d.) by short circuiting the preparation.

There are many obvious advantages to these *in vitro* techniques but there are some serious and often inadequately assessed difficulties. The active life of *in vitro* epithelium is limited, the transmucosal p.d. for example, usually starts to fall within about fifteen minutes of setting up and by 1–2 hours is low. Probably, therefore, the preparations should not be used beyond one hour after isolation and even less than that is desirable. A number of observations suggest that *in vitro* tissue does not function in exactly the same way as *in vivo*. In small intestine, for example, permeability to inulin and phenol red was increased (Smyth and Taylor, 1957) and in the gall bladder solute and water fluxes *in vitro* were only about 25 per cent of the *in vivo* values (Whitlock and Wheeler, 1964).

A variety of other methods have been employed. For example, preparations have been used with the vascular system perfused. Absorption studies have been done *in vivo* by analysis of blood levels or using whole-body counting techniques but these and other methods have played relatively little part so far in the development of ideas on epithelial function. There is, however, one further method which has exciting possibilities. It involves combining the physiological approach with the structural, which implies in the main electron microscopy to reveal ultrastructure. The union of the two methods has already given some interesting results concerning water transport (Kaye and coworkers, 1966; Tormey and Diamond, 1967).

A final point concerns the calculation of the flux values. To determine J_{LS} and J_{SL}, radionucleides or chemical tracers are used and it is obviously essential that the epithelial layer should be in a steady state with respect to the tracer when the measurements are made. If this is so, a two compartment model can be taken as the basis for the calculations (Visscher and coworkers, 1944a; Curran and Solomon, 1957). Where flux rates are considerable compared with the size of the intermediate pool in the cell layer, as with sodium for example, the time required to reach the steady state is very short so that during the period of observation the two compartment model is appropriate. But where the intermediate pool is large compared with the flux rates, as with potassium, a considerable time can elapse before the steady state is reached and a two compartment model may therefore be invalid for the period of observation (Edmonds, 1967c).

D

III. INTESTINE

A. Anatomy and Structure

The first part of the intestine, the duodenum, is short. It is less than 1 per cent of the total length, fixed to the posterior abdominal wall, has no mesentery and is relatively inaccessible. The rest of the small intestine, a long tube formed into many loops and attached to the posterior abdominal wall by the mesentery, is divided into an initial two-fifths called the jejunum and a remainder called the ileum. The jejunum and ileum differ in appearance. The former is thicker, has a larger diameter and is more vascular, but the dividing line between them is arbitrary and the change is gradual. The ileum leads to the large intestine, the first part of which is a structure of variable size called the caecum. This is followed by the ascending and descending colon and in some species there is a transverse segment. Each part has a nearly complete outer peritoneal investment although the ascending and descending segments are fairly closely attached to the posterior abdominal wall. There is finally a pelvic segment which leads to the rectum and anus.

All parts of the intestine have an outer smooth muscle coat and an inner mucosal layer. The latter consists of a very thin sheet of smooth muscle, the muscularis mucosa; a middle layer of connective tissue, the lamina propria, and the epithelial layer itself. The epithelium is formed of a sheet of columnar absorptive cells, 50–100 μ tall, and is penetrated by the orifice of numerous crypts lined by a mixture of cell types including undifferentiated and goblet cells. The crypts are not only important in secreting mucus but also as the site of cellular proliferation. In the small intestine the epithelium is also thrown into many finger-like processes, the intestinal villi, each with a connective tissue core and containing a variety of cellular elements in addition to a network of capillaries and lacteals, both of which lie in very close relationship to the basal parts of the epithelial cells.

The correlation of structure and function is becoming increasingly important. A recent review has surveyed in some detail the ultrastructure of the intestinal epithelium (Trier, 1968), hence only the main features will be considered here.

At the apex of the absorptive cell is the brush or striated border which consists of many fine hair-like projections, the microvilli (figure 1). Their average length is 1 μ and diameter about 0.1μ. Their presence considerably increases the absorptive area of the cell but in addition it has been shown that a number of enzymes are localized to the brush border (Miller and Crane, 1961), so possibly the microvilli have metabolic significance. The microvilli are covered by a coating of fine filaments, the so-called 'fuzz'. It is firmly attached to the plasma membrane and probably is a sulphated mucopoly-saccharide and formed by the cell to which it is fixed (Ito and Revel, 1964;

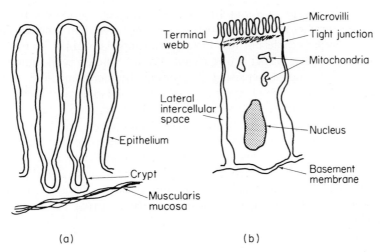

(a)   (b)

Figure 1. (a) Diagram of small intestinal mucosa at low magnification showing villi and crypts. (b) Diagram of an absorptive cell at high magnification (electron microscopy).

Ito, 1965). Immediately beneath the microvilli is a differentiated part of the cell consisting of interlacing filaments called the terminal web. The cytoplasm contains an endoplasmic reticulum, Golgi material, lysosomes and scattered mitochondria and a nucleus located towards the base of the cell. A variety of other structures of uncertain significance such as microtubules and vesicles have been observed. The cell base lies close to a continuous layer of homogeneous material referred to as the basement membrane and probably a protein–carbohydrate complex. It is about 300 Å wide and covers both the cells and intercellular spaces. In the case of many cells, the lateral plasma membranes are closely applied towards the apical aspect of the cell, but separated by spaces up to 3μ wide towards the basal region. At the extreme apical region, the outer leaflets of the lateral plasma membranes of adjacent cells fuse to form the 'tight junction' so that over $0\cdot1-2\mu$ no intracellular space exists (Farquar and Palade, 1963). Closely applied to the basement membrane is a layer of connective tissue and fibroblasts undergoing active replication and migration (Kaye and coworkers, 1968). Finally, it must be noted that the epithelium is not a static population but that cell division (in the crypts) and migration and shedding of cells are continuous processes and that the life of the intestinal cell is probably only one to two days (Leblond and Messier, 1958; Lipkin and coworkers, 1963).

The structural studies indicate that for absorption to occur a substance must either cross the tight junction and pass through the intercellular region or pass through the cells, crossing the apical membrane (a typical trilaminar unit type membrane of about 100 Å thick), the cell cytoplasm, the lateral plasma membranes and the basement membrane. The tight junction appears fused and impermeable but this histological impression may be misleading (Dewey and Barr, 1964; Lowenstein, 1966). Studies with lipid-soluble substances have shown increasing absorption with increasing lipid solubility, suggesting that they are crossing lipoidal barriers (Höber and Höber, 1937; Schanker and coworkers, 1958). Water and water-soluble substances are also absorbed, so presumably the barriers are penetrated by water-filled hydrophilic channels. Changes of electrical conductivity with temperature are consistent with the hypothesis that ions move across the tissue by diffusion through an aqueous solution (Schultz and Zalusky, 1964a; Edmonds and Marriott, 1968b). No discontinuities in the apical membrane are visible on electron microscopy, presumably because resolution is not yet adequate.

B. Water

It has been known for many years that intestine could transfer water from the mucosal to serosal side, even against an hydrostatic pressure gradient (Reid, 1892, 1901) and even when the colloidal osmotic pressure difference is eliminated, as for example, if the animal's own plasma is placed in the lumen (Visscher and coworkers, 1945). Experiments with deuterated water have shown that water movement is not limited to absorption but that considerable fluxes occur in both directions across intestinal epithelium (Visscher and coworkers, 1944b). A number of effects on water absorption have been studied in recent years and the results must be accommodated in any mechanism to explain water transport.

1. *Effects of Classical Osmosis*

Classical osmosis is the term generally used to describe the fluid flow produced by differences of osmotic pressure between solutions bathing either side of a tissue. An important finding has been that intestine can absorb water against an osmotic gradient of this sort. Thus solutions considerably hypertonic to plasma (>100 mosmoles difference) can be absorbed both by the small (Parsons and Wingate, 1958; Hakim and coworkers, 1963) and large intestine (Goldsmith and Dayton, 1919; Parsons and Paterson, 1965).

A number of experiments have been carried out to determine the osmolality of the absorbate in relationship to that of the luminal solution. *In vivo* experiments on the jejunum have shown that when the luminal fluid is isotonic with plasma, the absorbate is also isotonic (McGee and Hastings, 1945; Pearson, 1958). However with the ileum there has been less agreement.

Both isotonic and hypertonic absorbates have been described while in the colon the absorbate generally appears to be hypertonic (Powell and Malawar, 1968). The effect of varying the tonicity of the luminal fluid on the absorbate has been chiefly studied *in vitro* on small intestine and it has been found that the absorbate tends to be isotonic with the luminal solution whether osmolality is varied by using sodium chloride, sucrose or mannitol. In these experiments the serosal fluid was either absent (or replaced paraffin oil) or itself had the same tonicity as the luminal fluid (Clarkson and Rothstein, 1960; Lee, 1968). McHardy and Parsons (1957) found that *in vivo*, both in rat ileum and jejunum, the absorbate was approximately isotonic with the luminal solution.

It is interesting to note that when fluid was obtained from natural intestinal contents, for example from ileostomies or from faeces, it was found to be isotonic, or nearly so, with the plasma (McGee and Hastings, 1945; Kanaghinis and coworkers, 1963; Wrong and coworkers, 1967). This suggests that although the osmolality of the absorbate may be manipulated in experimental procedures, very little osmotic gradient can ever naturally develop across the mucosa.

2. *Hydrostatic Pressure*

Considerable pressure may develop in the gut lumen as a result of smooth muscle contractions and the possibility arises that this could be important in absorption. However, experiments in which the intraluminal pressure of intestine was varied up to 40–50 cm of water showed that absorption was little affected (Fisher and Parsons, 1949; Fisher, 1955), a finding particularly impressive as a pressure of only 4 cm of water applied to the serosal side reduced absorption significantly (Wilson, 1956b).

3. *Effects of Luminal Sugars*

For a number of years it has been known that fluid absorption depends on sugars (Fisher, 1955; Smyth and Taylor, 1957). This dependence is most marked in the jejunum and upper ileum and is much less developed in the lower ileum and colon (Barry and coworkers, 1961; Parsons and Wingate, 1961). Sugars could be related to fluid transport in several ways.

i) Diffusion of glucose along a concentration gradient might entrain a water flow, a process usually called codiffusion (Meschia and Setnikar, 1959).

ii) Metabolizable sugars could be substrates involved in energy production necessary for fluid transport.

iii) Some specific interaction between glucose and fluid transport could be taking place. It was early evident that simple sugar diffusion was not alone responsible, since glucose was found to stimulate fluid movement even when on the serosal side only (Lifson and Parsons, 1957). Experiments using *in vitro* preparations of rat jejunum and a number of different sugars helped considerably to elucidate the relationships of sugar and fluid transport. It was found that when placed in the lumen, fructose could stimulate fluid transport though less well than glucose; galactose was slightly effective while 3-methyl-D-glucose and mannose were ineffective. In contrast, on the serosal side, both fructose and mannose were almost as stimulating as glucose while again galactose and 3-methyl-D-glucose were almost ineffective (Barry and coworkers, 1964; Duerdoth and coworkers, 1965). The different handling of these sugars suggested an explanation. Glucose, galactose and 3-methyl-D-glucose were all actively absorbed from the lumen but only glucose was metabolized by intestinal cells. Fructose was metabolized but not actively transferred while mannose could only penetrate the cells when it was on the serosal side. It was suggested therefore that to influence fluid transport a sugar had to penetrate the cells and be metabolized.

Experiments done *in vivo* on jejunum and upper ileum have shown that glucose in the lumen will also stimulate fluid absorption although in this case adequate metabolic substrate should be coming from the blood (figure 2). Furthermore, although fructose is well metabolized and absorbed, it does not stimulate fluid transfer *in vivo* (Holdsworth and Dawson, 1964). It seems likely therefore that some specific process is also involved and the evidence, reviewed below, demonstrating the close interrelation of the absorption of certain sugars and of sodium provides the probable explanation. This is consistent with the fact that in the lower ileum, luminal glucose has little effect on sodium or fluid absorption and in the colon none at all (Powell and Malawar, 1968). It is also of interest in this connexion, although of uncertain significance at present, that the metabolic energy of the upper small intestine appears to derive principally from glycolysis while in the lower ileum and colon it comes predominantly from the citric acid cycle (Parsons and Paterson, 1960; Matthews and Smyth, 1961).

4. *Relationship of Water and Solute Transport*

An important characteristic of water transport established by a number of investigations is that fluid absorption only occurs when there is solute absorption (Curran and Schwartz, 1960; Clarkson and Rothstein, 1960; Duthie and Atwell, 1963). Changes of fluid absorption are found in association with changes of solute absorption in the absence of any gradient of water activity; a close relationship between fluid and solute absorption has been demonstrated in both the small and large intestine (figure 3).

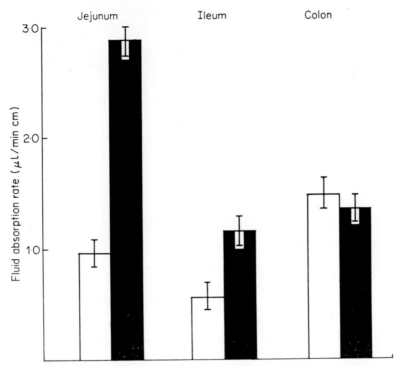

Figure 2. Effect of luminal glucose on fluid absorption by jejunum, ileum and colon of normal rats. Solid columns show absorption rate when luminal fluid contained glucose at a concentration of 56 mM. (Figure prepared from data of Powell and Malawar, 1968).

5. *Theories of Water Transport*

The finding that water was absorbed against its activity gradient led at first to the view that water was actively transported by intestinal epithelium. Other evidence tended to support this hypothesis; for example, the effect of the removal or supply of metabolic substrates and the action of metabolic inhibitors in preventing absorption. Peters and Visscher (1939) proposed an hypothesis that aimed to unify the phenomena of water and solute transport and developed as the so-called 'fluid circuit theory'. This has attracted attention intermittently since. Two streams of water were envisaged as crossing the tissue in opposite directions. Each stream carried solute but the flow from lumen to plasma carried more than that from plasma to lumen (Visscher and coworkers, 1944b). Recent developments in irreversible thermodynamics have enabled this to be described more satisfactorily. The relationship between solute flow, J_S, and volume flow, J_V, is given by $J_S = J_V \overline{C}_S (1 - \sigma_S)$ where σ_S is Staverman's reflexion coefficient for the solute

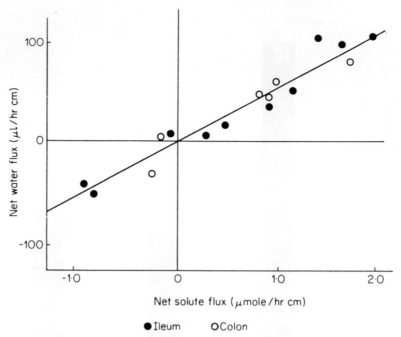

Figure 3. Relationship of water to solute absorption in ileum and colon of normal rats. Water activity was held constant but solute activity was varied by altering the luminal NaCl concentration. (Figure prepared from data of Curran and Solomon, 1957, and Curran and Schwartz, 1960).

and $\overline{C}_S$ is the mean concentration across the membrane (Kedem and Katchalsky, 1958). Thus, if the fluid flows in each direction were through different channels, having different values for σ_S, then net solute movement could occur even in the absence of net fluid movement. As it is impossible to account for absorption by classical osmosis and diffusion, some form of active water movement appears to be necessary. Recently, however, a similar hypothesis has been applied to jejunal absorption, without active water transport being postulated (Fordtran and coworkers, 1968). They proposed that water, with solute, is absorbed from the lumen by a local osmotic effect (figure 4a).

The alternative view, that water absorption is passive, has to meet the objection that there are no obvious forces to account for the net flow that occurs against an osmotic gradient. Curran (1960) suggested that the development of a local region of concentrated solute in the mucosa could provide an answer to this problem. The hypothesis, often referred to as the double-membrane hypothesis, has been developed by Curran and his associates over several years and will be briefly summarized here. It is proposed that solute movement is primary and that water follows from an osmotic effect due to a region of local higher concentration in the epithelial layer. A simple model has been described (Curran and McIntosh, 1962) to show how this could be effected (figure 4b). The essential features are two barriers A and B with differing reflexion coefficients, σ_A and σ_B such that $\sigma_A \gg \sigma_B$, a mechanism for transferring solute actively from compartment C1 to C2 and a compartment C2 with relatively rigid walls. As a result of transfer of solute into C2, the concentration of solute exceeds that in C1 and thus an osmotic pressure π_A is developed across A given by

$$\pi_A = \sigma_A RT\Delta C_A \qquad (1)$$

where ΔC_A is the steady state concentration difference between C1 and C2,

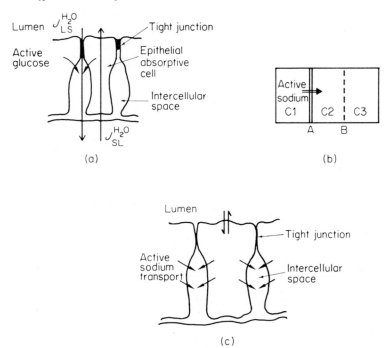

(a)

(b)

(c)

Figure 4. Models of water transport. a: a version of the fluid circuit theory as suggested by Fordtran and coworkers (1968). It was suggested that bulk flow of fluid took place through the tight junction. This was considered to have large pores and flow resulted from the presence of an hyperosmotic solution in the intercellular spaces, a consequence of glucose 'pumping'. The flow $J_{SL}^{H_2O}$ was thought to be largely through the cells where the pores were smaller, and so much less solute was entrained. b: diagram to illustrate the double membrane hypothesis. c: diagram showing how the double membrane hypothesis can be applied to intestine. The intercellular plasma membrane is regarded as 'A' and solute is actively moved into the intercellular spaces to create the hyperosmotic compartment C2.

and R and T have their customary significance. The flow of fluid resulting produces a rise in hydrostatic pressure in C2 and hence the volumes flowing across A and B in the steady state J_{VA} and J_{VB}, respectively, are given by

$$J_{VA} = L_{PA}(\Delta P_A - \sigma_A RT\Delta C_A) \tag{2}$$

$$J_{VB} = L_{PB}(\Delta P_B - \sigma_B RT\Delta C_B) \tag{3}$$

where ΔP_A and ΔP_B are the hydrostatic pressure differences and L_{PA} and L_{PB} are the hydraulic conductivity coefficients of the membranes. In the steady state $J_{VA} = J_{VB} = J_V$ and if C1 and C3 are similar in pressure and concentration, it can be easily shown that equations (2) and (3) reduce to

$$J_V = L_P(\sigma_A - \sigma_B)RT\Delta C_A \tag{4}$$

where L_P is a measure of hydraulic conductivity of the combination of A and B. A detailed theoretical analysis by Patlack and coworkers (1963) has shown that by using various membrane parameters and solutions in C1 and C2, isotonic and anisotonic flows can be produced and Curran and McIntosh (1962) have described a physical model which behaves as predicted.

The anatomical equivalents of the barriers has excited considerable interest. Curran (1960) suggested that the hyperosmotic solution might be in a subepithelial region. In view of the extensive capillary network, it seems unlikely that a high concentration could be maintained there but the discovery of the well-developed intercellular spaces has suggested an alternative site and during water absorption these spaces have been observed to be dilated (Williams, 1963). Solute is envisaged as passively entering the cells from the gut lumen and then being actively extruded into the intercellular spaces. The lateral plasma membranes constitute barrier A (figure 4c). Neither barrier need, of course, be a membrane but could simply be a region possessing differentiated properties and in fact barrier B may be formed by the connective tissue layer which closely invests the epithelium (Kaye and coworkers, 1968).

At present the concept of active water transport has been discarded for the intestine and probably also for all other animal tissues and the idea that water movement depends on solute transport dominates. The details of the mechanism are, however, still uncertain. The double-membrane hypothesis currently enjoys most popularity for the intestine but it has been challenged by the closely related 'standing-gradient local osmosis' theory. Since the latter was initially proposed for the gall bladder, it will be considered in that section but at present the choice between the two views remains open.

C. Pore Size

Various attempts have been made to apply the concept of pore size to intestinal transport. The principal methods of investigation have involved the use of water-soluble molecules of various sizes either to study the degree to which they can cross the epithelium or to observe their osmotic effect when in the lumen. Höber and Höber (1937) found that if the molecular radius of a water-soluble molecule exceeded 4Å little absorption occurred. Lindemann and Solomon (1962), using the method of measuring water movement under the osmotic gradient produced by several non-electrolytes of various molecular size, found a similar equivalent pore size for rat ileum. Recently the method has been applied to human intestine and an appreciable difference has been found between jejunum and ileum; the former has a pore radius of 8–9 Å and the latter only 3–4 Å (Fordtran and coworkers, 1965). This difference is probably the explanation of the apparent discrepancy found when water permeability and diffusion rates are compared in jejunum and ileum. Measurement of water permeability under osmotic gradients has shown that the hydraulic conductivity of the jejunum considerably exceeds that of the ileum. Fordtran and coworkers (1965) for example, found almost a ten-fold difference. Diffusion studies using tritiated or deuterated water have shown however, that there is little difference between jejunum and ileum. Water can cross the epithelium both by bulk flow and diffusion, although with pores of the above dimensions bulk flow will be most important in net movement. Hydraulic conductivity depends on bulk flow and is related to r^4 while diffusion is related to r^2, where r is the pore radius. Differences in pore radius are therefore particularly apparent in hydraulic conductivity measurements.

Some experiments have indicated that the pore wall may be charged. Curran and Schwartz (1960) observed that colonic Cl^- fluxes exceeded the Na^+ fluxes and suggested that the difference could be explained if the ions were diffusing through a positively charged barrier. Edmonds and Marriott (1968a) were, however, unable to produce any significant changes in p.d. by altering fluid movement by varying mannitol concentration in the luminal solution in rat colon. Similar experiments done *in vitro* on rat ileum demonstrated considerable potential changes, probably due to streaming potentials and consistent with fluid flow through negatively charged pores. This is especially interesting since fructose can stimulate fluid movement without producing a change in p.d., suggesting that more than one set of channels is concerned in fluid movement across the epithelium.

D. Sodium

Visscher and coworkers (1944b) using radionucleides demonstrated that considerable amounts of sodium crossed intestinal epithelium in both directions. Since a small p.d. is present across the mucosa, with the serosal side nearly always positive with respect to the lumen, electrical forces must always be considered in analysing the movement of charged particles.

In vivo experiments, in which the effects of sodium concentration and p.d. have been studied, have shown clearly that sodium is moved against the electrochemical gradient (Curran and Solomon, 1957; Curran and Schwartz, 1960; Edmonds, 1967a, b) and with *in vitro* techniques on ileum it was shown that ouabain reduced the net sodium flux by reducing J_{MS}^{Na} while leaving J_{SM}^{Na} unchanged (Schultz and Zalusky, 1964a). These results suggest an active absorption of sodium but cannot exclude the possibility that sodium might be carried in a water stream by solvent drag. Certainly some molecules are carried in this way, for example, urea (Hakim and Lifson, 1964). However, there is evidence against such an explanation. Thus, sodium absorption can occur even when fluid movement is zero (see, for example, Green and coworkers, 1962). Also *in vitro* experiments have shown that in the short-circuit condition a net sodium flux from the mucosal to serosal side takes place without an accompanying chloride flux, even though the tissue is very permeable to chloride, a result difficult to explain in terms of solvent drag (Schultz and Zalusky, 1964a; Clarkson and Toole, 1964). These experiments were done on ileum and colon and it is possible that in the case of the jejunum which appears to have larger pores, solvent drag does play a significant role in sodium absorption (Fordtran and coworkers, 1968). In general, however, it seems highly probable that an active transport mechanism for sodium is present in the absorptive cells over the whole length of intestine.

1. *Interrelationship of Sodium, Sugar and Amino Acid Absorption*

Experiments on a variety of animals have shown that absorption of glucose and some amino acids against their concentration gradient depends on the presence of sodium in the gut lumen, a specific action as the substitution of lithium, magnesium or choline was ineffective (Riklis and Quastel, 1958; Csáky, 1961; Schultz and Zalusky, 1964a). The corollary also appears true; sodium absorption is considerably stimulated by the presence of actively transported sugars or amino acids even when non-metabolizable compounds such as 3-methyl-D-glucose are used. Sugars which are metabolized but not actively transported, for example fructose, are ineffective (Schultz and Zalusky, 1964b). The interaction of sugar, amino acid and sodium absorption, which is only found in the jejunum and upper ileum, suggested that there might be a carrier in the plasma membrane adjacent to the lumen having an affinity for both sodium and glucose (or amino acid). In the absence of sodium or glucose, the carrier is supposed to be less efficient in moving its substrate so that when intracellular sodium concentration is less than that of the lumen, there is a net flow of substrate into the cell (Crane, 1965). An alternative view is that the sodium–sugar interaction takes place at the basal pole of the cell and is suggested by the observation that ouabain blocks sodium and glucose absorption only when present on the serosal side (Csáky, 1963; Csáky and Hara, 1965). However, the latter finding could be interpreted as a consequence of the rise of intracellular sodium concentration following ouabain treatment. At present, there is no unequivocal evidence for either hypothesis and recently a further complication has been introduced by the suggestion that glucose may affect NaCl absorption by influencing bulk flow of water (Fordtran and coworkers, 1968).

2. *Diffusion and Critical Luminal Concentration of Sodium*

Apart from active transport, considerable movement of sodium occurs in both directions across the epithelium, presumably due largely to passive diffusion. The flux J_{SL}^{Na} is not however completely independent of luminal sodium concentration, suggesting that there is some exchange diffusion (figure 5). The flux J_{LS}^{Na} shows much more dependence on luminal sodium

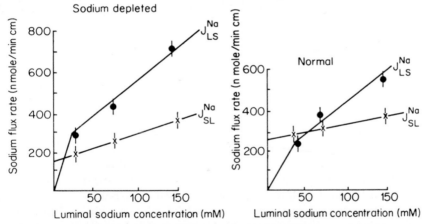

Figure 5. Influx (J_{LS}^{Na}) and efflux (J_{SL}^{Na}) of sodium measured with various concentrations of sodium chloride solution in the lumen of ascending colon. The data came from studies on six normal and six sodium-depleted rats and results are given as mean $\pm$ 1 S.E. Note that the critical luminal sodium concentration ($J_{LS}^{Na} = J_{SL}^{Na}$) is considerably reduced by sodium depletion. (From Edmonds, 1967b).

concentration and there is no clear evidence of saturation kinetics over the physiological concentration range. The critical luminal sodium concentration falls progressively along the intestine from a value of about 100 mequiv./l in the jejunum to less than 50 mequiv./l in the lower ileum and colon and it can be reduced to a very much lower level by sodium depletion of the animals (Curran and Schwartz, 1960; Edmonds, 1967b; Fordtran and coworkers, 1968).

3. *Sodium within the Epithelial Layer*

Little is known about the details of how sodium crosses the cellular layer. It is possible that an appreciable proportion bypasses the cells, perhaps passing through large shunts transitorily created by cell shedding (Clarkson, 1967). Much sodium presumably enters the absorptive cells in part to diffuse back to the lumen, but mainly to be pumped into the lateral intercellular spaces as the major contributor to the hyperosmotic solution. It would seem likely that this sodium pump would be similar to that postulated for many tissues. It has certainly some features in common, for example, ouabain-sensitivity. Several investigators have also detected Na^+–K^+ sensitive ATPases in the intestinal mucosa (for example Smith, 1967) and surprisingly, Berg and Chapman (1965) localized a high proportion to the striated membrane. It raises the question whether part of the flux J_{SL}^{Na} might be due to active sodium secretion and it is of interest in this connection that Ferreira and Smith (1968) have found that in some circumstances J_{SL}^{Na} may exceed J_{LS}^{Na} in toad colon in the absence of any electrochemical gradient to account for it.

E. Chloride and Bicarbonate

Early observations using *in vivo* recirculation perfusion showed that chloride concentration in the colon could fall to 60–70 mequiv./l (D'Agostino and coworkers, 1953) and ileum to less than 20 mequiv./l, though in the jejunum very little fall in chloride concentration occurred (Ingraham and Visscher, 1936; Lifson, 1939; Dennis and Visscher, 1940). Bicarbonate concentration rose to 60 mequiv./l in the colon and 40 mequiv./l in the ileum. The rise of bicarbonate concentration cannot definitely be attributed to bicarbonate secretion since secretion of hydroxyl ions or hydrion absorption would be indistinguishable. Equilibrium concentrations of bicarbonate determined in dog and man have shown very low values for jejunum, about 5 mequiv./l, higher values for ileum, about 40–70 mequiv./l, and for colon about 75 mequiv./l (Swallow and Code, 1967; Phillips and Summerskill, 1967). Since the p.d. across small intestine is rarely observed to be greater than 15 mV (serosa positive), it seems unlikely that this could be responsible. An estimate can be made from the Nernst equation, since if the epithelium behaved passively and there was no significant entrainment of anion in water flow, the critical luminal concentration, C_L, would then be given by $C_L = C_S \exp(\Delta V F / RT)$ where C_S is the serosal concentration, ΔV is the p.d. and R, T and F have their usual meaning. In the small intestine, taking the p.d. as 15 mV, the expected chloride concentration would be about 50 m equiv./l and bicarbonate about 14 mequiv./l, while in the colon (p.d. not greater than 20–40 mV) the expected values would be 25–50 mequiv./l and 6–14 mequiv./l, respectively. These results suggest that in the small intestine both chloride and bicarbonate transport cannot simply be passive; in the large intestine the same is probably true for bicarbonate but chloride transport may be passive.

Measurements of transmucosal fluxes *in vivo* have tended to support these findings. The flux ratio equation (Ussing, 1949) is useful in this situation (figure 6), as it gives the ratio expected if ionic movements were simply due

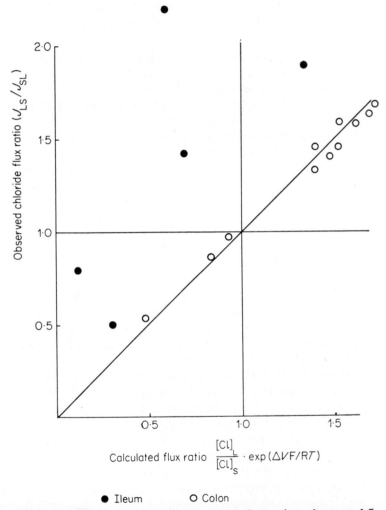

$$\text{Calculated flux ratio} \quad \frac{[Cl]_L}{[Cl]_S} \cdot \exp(\Delta VF/RT)$$

● Ileum   ○ Colon

Figure 6. Relationship between observed chloride flux ratio and expected flux ratio in ileum and colon of normal rats. The expected ratio was calculated using the flux ratio equation as described in the text. (Figure prepared from data of Curran and Solomon, 1957; Curran and Schwartz, 1960; Edmonds and Marriott, 1967).

to the electrochemical gradient. It can be stated as : $J_{LS}/J_{SL} = C_L/C_S$ exp $(\Delta VF/RT)$. In the ileum, the flux ratio was not consistent with passive transport alone, active chloride absorption was also present but in the colon chloride movements could be explained by the observed electrochemical forces (Curran and Solomon, 1957; Curran and Schwartz, 1960; Edmonds and Marriott 1967). *In vitro* experiments by contrast have given results which indicated passive handling of chloride and bicarbonate in both ileum and colon (Cooperstein and Hogben, 1959; Clarkson and Toole, 1964; Schultz and coworkers, 1964). The discrepancy between *in vivo* and *in vitro* studies is unexplained at present. A further curious observation is that chloride sometimes appears to be actively secreted by the jejunum (Tidball, 1961; Taylor and coworkers, 1968). The significance of this is uncertain but it may reflect the activity of specific glands.

The effect of carbonic anhydrase inhibitors has been studied in a number of experiments although the results have not been very helpful. Both chloride and bicarbonate absorption were depressed in the jejunum in rats but in dogs only chloride was affected; in the ileum and colon bicarbonate secretion and chloride absorption were both depressed (Parsons, 1956; Kinney and Code, 1964). There may well be some relationship between chloride absorption and bicarbonate secretion but present evidence is inadequate and the effect of carbonic anhydrase inhibitors may well not be specific.

F. Potassium

There is relatively little information on potassium transport by intestine. The potassium concentration of the fluid occurring naturally in the terminal ileum is about 20–25 mequiv./l; that is, it is considerably greater than expected if potassium were passively distributed across the epithelium. Probably, however, the high concentration reflects relatively rapid fluid absorption, since the transmucosal potassium flux ratio in ileum has been shown to be consistent with passive transport both *in vivo* and *in vitro* (Gilman and coworkers, 1963; Phillips and Code, 1966). In the colon, the critical luminal concentration in normal dogs and rats was 20–30 mequiv./l, a value considerably higher than expected from the electrochemical gradient. (Phillips and Code, 1966; Edmonds, 1967c). The results of *in vivo* [42]K studies indicated that potassium movements across the mucosa involved passive diffusion along the electrochemical gradient but in addition potassium appeared to be actively secreted into the lumen. The origin of the secreted potassium is uncertain; it may come from leakage from the mucosal cells as they contain potassium in high concentration. Recent investigations by Barnaby and Edmonds (1969) have indicated that most of the potassium crossing the mucosa mixes with only a small proportion of the mucosal potassium suggesting that only a few of the mucosal cells are involved in potassium transport.

G. Electrical Potential and Short-circuit Current

A p.d. is generally present across the intestinal epithelium even when the solutions bathing each side are similar; the serosa is nearly always positive with respect to the lumen. The p.d. varies in different parts of the intestine and in the jejunum and upper ileum and it is considerably influenced by certain sugars and amino acids in the lumen. Intestinal p.d.s are small, not usually exceeding 10 mV, while in the colon they tend to be higher, up to 20 mV in rat and up to 30–40 mV in dog (Cooperstein and Brockman, 1959; Edmonds, 1967a,c). In regions where sugars are effective, they cause a rapid rise of p.d. up to 15 mV, depending on the sugar concentration present (Barry and coworkers, 1964). In general, actively transported sugars like glucose, galactose and 3-methyl-D-glucose stimulate the p.d. It is not necessary that the sugars be metabolized and fructose which is well metabolized but not actively transported does not stimulate the p.d. in rat jejunum. Amino acids and disaccharides hydrolysed by the mucosa also stimulate p.d. (Baillien and Schoffeniels, 1961; Kohn and coworkers, 1966).

The rise of p.d. produced by sugars is associated with a rise of s.c.c. while tissue resistance is unchanged. Comparison of net sodium transfer and s.c.c. has shown fairly good agreement for ileum and colon (Schultz and Zalusky, 1964a; Clarkson and Toole, 1964; Cofré and Crabbé, 1967). On the other hand, considerable discrepancies have been observed with jejunum and it has been suggested that there may be a neutral sodium pump, moving sodium with an associated anion, involved in absorption (Barry and coworkers, 1965) and in secretion (Taylor and coworkers, 1968).

Although the p.d. is closely related to sodium transport, and can be conveniently described in terms of tissue resistance and active ionic movements with other ions following passively, this gives little insight into the detailed mechanism by which the p.d. originates. As the adjacent cells appear to be firmly joined at the terminal bar, the transmucosal p.d. would be expected to arise from the sum of two p.d.s, one at the luminal and the other at the serosal membrane of the epithelial cells. Microelectrode studies have in fact demonstrated such potentials and shown that the rise of p.d. induced by glucose seems to be due to a change at the serosal membrane (Baillien and Schoffeniels, 1965; Wright, 1966; Edmonds and Nielsen, 1968). Another method of studying the problem, namely by varying the composition of the bathing solutions (figure 7), showed that the transmucosal p.d. was practically independent of the serosal potassium concentration in rat ileum and colon and dominated by the luminal sodium concentration (Schultz and Zalusky, 1964a; Edmonds and Marriott, 1968b). On the basis of these and other results, it was suggested that the p.d. arose from an electrogenic sodium transport mechanism on the serosal side. It is not clear how these findings are explained in terms of the model having two p.d.s in

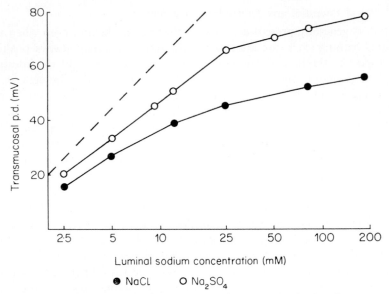

Figure 7. Dependence of colonic transmucosal p.d. on luminal sodium concentration. The p.d. was relatively high as the rat had been on a low sodium diet. The luminal solutions were all made isosmotic by addition of mannitol. The interrupted line is the theoretical relationship if mucosa behaved as a perfect sodium electrode. (From Edmonds and Marriott, 1968a).

series and plainly further investigation is needed to bring the variety of information about the p.d. into a consistent scheme.

H. Hormonal Effects

In lower mammals antidiuretic hormone at physiological levels appears to stimulate fluid absorption by the intestine but in man the reverse appears to be true; fluid and sodium absorption are depressed (Blickenstaff, 1954; Aulsebrooke, 1961; Soergel and coworkers, 1968). Adrenal steroids have tended to give more consistent results although with the small intestine, stimulation of sodium absorption and potassium secretion by adrenal steroids has only sometimes been observed (Berger and coworkers, 1960; Clarke and coworkers, 1967). With the colon, however, there has been general agreement: sodium depletion (Edmonds, 1967b), DOCA (Berger and coworkers, 1960) and aldosterone (Levitan and Ingelfinger, 1965; Edmonds and Marriott, 1967) all stimulate sodium absorption and potassium secretion. The critical luminal sodium concentration is much lowered by sodium depletion, and the descending colon can absorb sodium when the luminal sodium concentration is less than 15 mequiv./l (Edmonds, 1967b). Aldosterone also has a striking effect on the electrical activity of the colon,

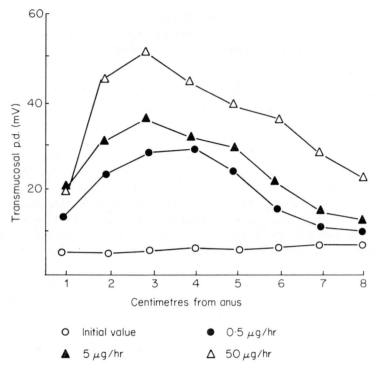

Figure 8. P.d. change at different points along rat descending colon due to five-hour intravenous infusion of aldosterone. The results are means taken on five rats. (Edmonds and Marriott, unpublished data).

an effect dependent on dose and seen principally in the descending colon (figure 8). It is obviously of importance to homeostasis that there should be regulation of electrolyte loss; sodium loss for example, can then be reduced when sodium intake is low (table 1). It is less certain, however, whether

Table 1. Sodium, potassium and chloride in the stools of normal and sodium-depleted rats

	Normal		Sodium-depleted	
	Total daily loss mequiv./rat/24 hr	Concentration mequiv./kg of stool water	Total daily loss mequiv./rat/24 hr	Concentration mequiv./kg of stool water
Na	0·16	48	0·03	12
K	0·28	64	0·28	94
Cl	0·08	28	0·07	24

Rats were sodium-depleted either by dietary restriction or by peritoneal dialysis. (From Edmonds, 1967a)

antidiuretic hormone plays any physiological role in gut function in normal animals, though obviously it is a potentially disturbing variable in investigations where states of hydration are uncontrolled.

I. Conclusion

Intestine varies strikingly in its transport functions along its length. A common feature throughout is the ability to absorb sodium actively but even here there is not uniformity. In the upper intestine, sodium absorption shows dependence on sugars and amino acids in the lumen but not at all in the lower ileum and colon. In contrast, adrenal steroids mainly act on the distal colon and their effect progressively diminishes proximally. The movements of potassium, chloride and bicarbonate also show differences in various parts of the intestine. These variations cannot be chance effects; it must be significant that glucose is normally only found in the lumen in the upper intestine. Also the fall of critical luminal sodium concentration along the intestine could ensure that sodium absorption continued even as the intraluminal sodium concentration fell. Revealing observations can therefore be made simply by treating the epithelium as a 'black box' but increasingly attention is being devoted to the functional and structural mechanisms in the cells which are responsible for the effects. Comparatively little is known of this aspect but a start has been made with water and sodium transport and there is little doubt that here is the fertile ground of future development.

IV. GALL BLADDER

The liver secretes bile continuously at a pressure of about 15–25 cm of water and in man, a volume of about one litre is produced daily. The bile passes into the gall bladder where it is reduced to about a tenth of its initial volume. Various changes in composition are produced. The principal ones are an increase in sodium and bile acid concentration and a reduction in chloride concentration (table 2). There is considerable micelle formation so

Table 2. Composition of bile

	Hepatic bile (mequiv./1)	Gall-bladder bile (mequiv./1)
Na	174	250
K	6·6	9·1
Cl	64	17
HCO_3	34	4·4
Bile acids	76	250
pH	7·4	6·4

Both hepatic and gall-bladder bile are isosmotic with plasma. The values given above are averages obtained from dog by Reinold and Wilson (1934) and Wheeler and Ramos (1960). Measurements on fish, canine, rabbit and human bile have given similar results.

that the fluid remains isosmolar with plasma despite the high concentrations of sodium and bile acids. Various other components of bile such as bilirubin, bile acids, phospholipids and cholesterol are concentrated as a consequence of fluid absorption and of the impermeability of the epithelium to these substances.

A. Structure

The gall bladder wall has three layers; an outer layer of peritoneum resting on loose connective tissue, a layer of smooth muscle and an inner mucous membrane formed of an epithelium, with connective tissue containing capillaries, lacteals, etc. Basically the gross structure is therefore similar to intestine, and ultrastructure studies of the epithelial cells have also shown that they closely resemble intestinal absorptive cells (Kaye and coworkers, 1966). The epithelium consists of a single layer of tall columnar cells which have many microvilli and numerous mitochondria, especially at the base of the cell, and other subcellular organelles are also present. There is a tight junction at the apical region while the lateral plasma membranes are only loosely attached and may be separated by wide intercellular spaces.

B. Water

Fluid absorption appears to be the most important function of the gall bladder. The characteristics of the process have many parallels with those already outlined for intestine. It can be shown by using tritiated water that water crosses the epithelium in either direction but normally there is a net transfer from luminal to serosal side, at a rate of about 15 μl/min cm^2 in fish, up to 45 μl/min cm^2 in rabbit. These rates are considerably faster than those found in intestine. Absorption of water takes place not only where no osmotic gradient is present across the mucosa, as when bile is naturally present in the lumen, but also against an osmotic gradient (Diamond, 1962a; Grim, 1963). Furthermore, hydrostatic pressure has relatively little effect unless it is applied to the serosal side when even pressures as low as 5 cm of water will rapidly stop transport (Dietschy, 1964). A number of metabolic inhibitors, for example cyanide, 2,4-dinitrophenol and iodoacetate were able to reduce or stop water absorption, indicating its dependence on metabolism. Ouabain also had this effect but only when it was on the serosal side (Diamond, 1962a). Finally, as with intestine, fluid absorption does not occur in the absence of solute absorption (Whitlock and Wheeler, 1964).

A number of hypotheses have been advanced to explain how water is absorbed. They have been reviewed by Diamond (1965) and for the most part will only be briefly discussed here. The facts about water transport militate against explanations based on classical osmosis or filtration by hydrostatic pressure. Pinocytosis, advanced as a possibility by Grim (1963), seems improbable since the relatively few vesicles seen in the cytoplasm of the epithelial cells appear inconsistent with the transport rate and further it leaves unexplained the selectivity of absorption (Wheeler, 1963). Diamond (1962c) also considered electroosmosis but this obviously failed to account for the volume of fluid transported and he concluded at that time that water transport was an example of codiffusion. Thus water transport was linked with the well-recognized fact that the flux of one substance along its concentration gradient can induce the flux of another, even in inanimate systems, probably as a consequence of frictional forces. However, between 1960 and 1963, the double-membrane hypothesis developed and offered a possible explanation of water absorption. Later Diamond (1964b) suggested a related scheme referred to as 'local osmosis' which he and collaborators subsequently developed (Tormey and Diamond, 1967; Diamond and Bossert, 1967). The important features are shown diagrammatically in figure 9. It was suggested that water absorption depends on solute absorption. Salt and water pass into the cells from the lumen and salt is then actively extruded into the lateral intercellular spaces. If salt pumping were located towards the

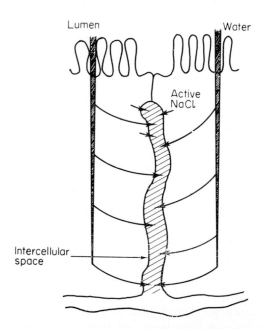

Figure 9. Diagram to illustrate the standing-gradient osmotic flow hypothesis. Solute is actively moved into the intercellular spaces and water enters as a result of the hypertonic solution formed. A gradient of concentration is developed along the length of the intercellular space.

apices of the cells, an hypertonic solution would be formed there, its osmolality decreasing as it passed towards the base due to equilibration with the cell fluid. Thus there would be a gradient of osmolality present along the intercellular space in contrast with the well-mixed intermediate compartment of the double-membrane hypothesis. The choice between these various views is not easy. That codiffusion cannot alone be responsible seems clear from studies in which the osmolality of the luminal solution was varied; it was found that the osmolality of the absorbate changed much more than predicted by the codiffusion hypothesis (Diamond, 1964b). However, when the concentration in the lumen of a non-permeant substance like sucrose was altered, the absorbate was found by Whitlock and Wheeler (1964) to be hypertonic but Diamond (1964b) found it to be isotonic with the luminal solution; the former result agrees with the predictions of the double-membrane hypothesis and the latter with those of local osmosis. Recent electron-microscopy studies showing the dilatation of intercellular spaces during water transport (Kaye and coworkers, 1966; Tormey and Diamond, 1967) have provided further support for the view that the hyperosmotic solution develops in these spaces but the exact mechanism of its formation and operation are problems that are still unresolved.

C. Transport of Ions

In intestine as well as in many other tissues, active ionic transport is generally associated with the presence of a p.d. across the tissue. In the gall bladder, however, the observed p.d. is very small, usually less than 1 mV. Nevertheless there is a good deal of evidence indicating that sodium and chloride are actively absorbed. With fish gall bladder, sodium chloride absorption was found even when the luminal solution had a sodium chloride concentration about one-tenth that of the serosal solution and also the transport mechanism appeared to be saturable as absorption rate fell off at higher luminal sodium concentrations (Diamond, 1962a). In contrast with water absorption, sodium absorption was relatively unaffected by opposing osmotic pressures (Whitlock and Wheeler, 1964). This essentially agrees with the trivial effect solvent drag appears to exert (Diamond, 1962c). Radionucleide studies showed that sodium moved in both directions across the epithelium so that diffusion was presumably also involved. These experiments, on rabbit gall bladder, also showed that the ratio J_{MS}/J_{SM} both for sodium and chloride considerably exceeded the flux ratio expected from the electrochemical gradient as calculated by Ussing's equation (Wheeler, 1963). These findings together with the absence of a transepithelial p.d. suggest that sodium chloride is absorbed by a neutral pump. An alternative possibility, that independent sodium and chloride pumps are present and operating equally so as to produce no net transfer of charge, is untenable since the substitution of a non-transported ion like sulphate for chloride practically arrested sodium absorption while producing no change in p.d. (Diamond, 1962a, b). Thus it appears that sodium and chloride are linked in some way during active transport. The linkage appears to be specific since no cation can replace Na^+ and although Br^- can substitute for Cl^-, the use of other anions stops transport (Diamond, 1962a). A variety of substances have been examined to see whether sodium chloride absorption can be influenced. The presence of glucose in the lumen is without effect but various metabolic inhibitors, 2,4-dinitrophenol, cyanide, iodoacetate, etc. and ouabain on the serosal side stop or reduce absorption (Diamond, 1964a; Dietschy, 1964). Oxytocin at relatively high concentration on the serosal side stopped transport but ADH had no effect (Diamond, 1962a).

The bicarbonate concentration of gall bladder bile is relatively low, suggesting that this anion may also be actively absorbed. Wheeler (1963) was able to show that net transport of bicarbonate from luminal to serosal side was in fact against its electrochemical potential. In the case of potassium however, although its concentration in gall bladder bile may be appreciably greater than that of plasma, measurement of potassium flux ratios showed that they were entirely consistent with passive transport (Wheeler, 1963). The relatively high potassium concentration is explained by the finding that the activity coefficient for potassium in gall bladder bile is much lower than in hepatic bile or plasma. This is presumably a consequence of the high concentration of bile acids with micelle formation (Dietschy and Moore, 1964).

D. Potential Difference

As already pointed out, in the absence of significant ionic gradients between mucosal and serosal solutions, the p.d. across gall bladder epithelium is very small. However, if the solutions differ in composition, as for example when an impermeant solute like sucrose is substituted for some electrolyte on one side, considerable diffusion potentials appear. This effect was employed by Diamond (1962b) to estimate relative passive ionic permeability. The gall bladder has rectifying properties in relation to these potentials. The observed p.d. was found for example by Whitlock and Wheeler (1964), to be dependent on the side on which sucrose substitution was made, indicating that p.d. arose from a serial membrane arrangement. No doubt diffusion potentials are also responsible for the charge observed across the gall bladder wall, the serosa here being positive, when normal concentrated bile is in the lumen as this differs considerably in composition from the blood.

Another possible source of a gall bladder p.d. comes from streaming potentials created by flow of fluid through pores having fixed charges.

Diamond and Harrison (1966) demonstrated the phenomenon in rabbit gall bladder by altering the osmolality of one of the bathing solutions by adding sucrose. The magnitude of the p.d. was proportional to osmolality, the relationship averaging about 0·04 mV/mosmole, and diffusion and streaming potentials appear to be simply additive. In fish and rabbit gall bladder, the surface towards which water flowed became more positive, indicating that the pores were negatively charged.

E. Conclusion

The principal function of the gall bladder, namely to concentrate the bile, is served by the presence in the epithelium of a mechanism for transporting sodium and chloride which has the characteristics of a 'neutral' pump generating no transport p.d. It is uncertain whether this is unique or whether a similar mechanism may be present in some parts of the intestine. With the possible exception of bicarbonate, the movement of water and the observed concentrations of other substances in the bile appear to result from sodium chloride absorption. Thus gall bladder epithelium is considerably more simple than intestine in its transport properties but here only too little is known of the cellular functions underlying the observed effects.

REFERENCES

Aulsebrook, K. A. (1961) *Endocrinology*, **68**, 1063
Baillien, M. and E. Schoffeniels (1961) *Biochim. Biophys. Acta.*, **53**, 537
Baillien, M. and E. Schoffeniels (1965) *Arch. Intern. Physiol. Biochim.*, **73**, 355
Barnaby, C. F. and Edmonds, C. J. (1969) *J. Physiol. (London)*, **205**, 647
Barry, R. J. C. and D. H. Smyth (1960) *J. Physiol. (London)*, **152**, 48
Barry, R. J. C., J. Matthews and D. H. Smyth (1961) *J. Physiol. (London)*, **157**, 279
Barry, R. J. C., S. Dikstein, J. Matthews, D. H. Smyth and E. M. Wright (1964) *J. Physiol. (London)*, **171**, 316
Barry, R. J. C., D. H. Smyth and E. M. Wright (1965) *J. Physiol. (London)*, **181**, 410
Berg, G. G. and B. Chapman (1965) *J. Cellular Comp. Physiol.*, **65**, 361
Berger, E. Y., G. Kanzaki and J. M. Steele (1960) *J. Physiol. (London)*, **151**, 352
Blickenstaff, D. D. (1954) *Amer. J. Physiol.*, **179**, 471
Clarke, A. M., M. Miller and R. Shields (1967) *Gastroenterology*, **52**, 846
Clarkson, T. W. and A. Rothstein (1960) *Amer. J. Physiol.*, **199**, 898
Clarkson, T. W. and S. R. Toole (1964) *Amer. J. Physiol.*, **206**, 658
Clarkson, T. W. (1967) *J. Gen. Physiol.*, **50**, 695
Code, C. F. (1960) *Perspectives Biol. Med.*, **3**, 560
Cofré, G. and J. Crabbé (1967) *J. Physiol. (London)*, **188**, 177
Cooperstein, I. L. and S. K. Brockman (1959) *J. Clin. Invest.*, **38**, 435
Cooperstein, I. L. and C. A. M. Hogben (1959) *J. Gen. Physiol.*, **42**, 461
Cramer, C. F. and T. A. Haqq (1963) *Can. J. Biochem. Physiol.*, **41**, 127
Crane, R. K. (1965) *Federation Proc.*, **24**, 1000
Crane, R. K. (1968) In *Handbook of Physiology*, Vol. 3, Williams and Wilkins, Baltimore, p. 1323
Csáky, T. Z. (1961) *Amer. J. Physiol.*, **201**, 999

Csáky, T. Z. (1963) *Federation Proc.*, **22**, 3
Csáky, T. Z. and Y. Hara (1965) *Amer. J. Physiol.*, **209**, 467
Curran, P. F. (1960) *J. Gen. Physiol.*, **43**, 1137
Curran, P. F. and A. K. Solomon (1957) *J. Gen. Physiol.*, **41**, 143
Curran, P. F. and G. F. Schwartz (1960) *J. Gen. Physiol.*, **43**, 555
Curran, P. F. and J. R. McIntosh (1962) *Nature*, **193**, 347
D'Agostino, A., W. F. Leadbetter and W. B. Schwartz (1953) *J. Clin. Invest.* **32**, 444
Dawson, A. M. and H. B. McMichael (1968) *J. Physiol. (London)*, **196**, 32P
Dennis, C. and M. B. Visscher (1940) *Amer. J. Physiol.*, **129**, 176
Dewey, M. M. and L. Barr (1964) *J. Cell. Biol.*, **23**, 553
Diamond, J. M. (1962a) *J. Physiol. (London)*, **161**, 442
Diamond, J. M. (1962b) *J. Physiol. (London)*, **161**, 474
Diamond, J. M. (1962c) *J. Physiol. (London)*, **161**, 503
Diamond, J. M. (1964a) *J. Gen. Physiol.*, **48**, 1
Diamond, J. M. (1964b) *J. Gen. Physiol.*, **48**, 15
Diamond, J. M. (1965) *Symp. Soc. Exp. Biol.*, **19**
Diamond, J. M. and S. C. Harrison (1966) *J. Physiol. (London)*, **183**, 37
Diamond, J. M. and W. H. Bossert (1967) *J. Gen. Physiol.*, **50**, 2061
Dietschy, J. M. (1964) *Gastroenterology*, **47**, 395
Dietschy, J. M. and E. W. Moore (1964) *J. Clin. Invest.*, **43**, 1551
Duerdoth, J. K., H. Newey, P. A. Sanford and D. H. Smyth (1965) *J. Physiol. (London)*, **176**, 23P
Duthie, H. L. (1967) *Brit. Med. Bull.*, **23**, 213
Duthie, H. L. and J. D. Atwell (1963) *Gut*, **4**, 373
Edmonds, C. J. (1967a) *J. Physiol. (London)*, **193**, 571
Edmonds, C. J. (1967b) *J. Physiol. (London)*, **193**, 589
Edmonds, C. J. (1967c) *J. Physiol. (London)*, **193**, 603
Edmonds, C. J. and J. Marriott (1967) *J. Endocrinol.*, **39**, 517
Edmonds, C. J. and J. Marriott (1968a) *J. Physiol. (London)*, **194**, 457
Edmonds, C. J. and J. Marriott (1968b) *J. Physiol. (London)*, **194**, 479
Edmonds, C. J. and O. Nielsen (1968) *Acta Physiol. Scand.*, **72**, 338
Farquhar, M. G. and G. E. Palade (1963) *J. Cell Biol.*, **17**, 375
Ferreira, H. G. and M. W. Smith (1968) *J. Physiol.*, **198**, 329
Fisher, R. B. (1955) *J. Physiol. (London)*, **130**, 655
Fisher, R. B. and D. S. Parsons (1949–50) *J. Physiol. (London)*, **110**, 36
Fordtran, J. S., F. C. Rector, M. F. Ewton, N. Soter and J. Kinney (1965) *J. Clin. Invest.*, **44**, 1935
Fordtran, J. S., F. C. Rector and N. W. Carter (1968) *J. Clin. Invest.*, **47**, 884
Gilman, A., E. S. Koelle and J. Ritchie (1963) *Nature*, **197**, 1210
Goldschmidt, S. and A. B. Dayton (1919) *Amer. J. Physiol.*, **48**, 440
Green, K., B. Seshadri and A. J. Matty (1962) *Nature*, **196**, 1322
Grim, E. (1963) *Amer. J. Physiol.*, **205**, 247
Hakim, A., R. G. Lester and N. Lifson (1963) *J. Appl. Physiol.*, **18**, 409
Hakim, A. A. and N. Lifson (1964) *Amer. J. Physiol.*, **206**, 1315
Hober, R. and J. Hober (1937) *J. Cellular Comp. Physiol.*, **10**, 401
Holdsworth, C. D. and A. M. Dawson (1964) *Clin. Sci.*, **27**, 371
Ingraham, R. C. and M. B. Visscher (1936) *Amer. J. Physiol.*, **114**, 676
Ito, S. (1965) *J. Cell Biol.*, **27**, 475
Ito, S. and J. P. Revel (1964) *J. Cell Biol.*, **23**, 44A
Jervis, E. L. and R. C. Levin (1966) *Nature*, **210**, 391
Kanaghinis, T., M. Lubran and N. F. Coghill (1963) *Gut*, **4**, 322
Kaye, G. I., H. O. Wheeler, R. T. Whitlock and N. Lane (1966) *J. Cell Biol.*, **30**, 237
Kaye, G. I., N. Lane and R. P. Pascal (1968) *Gastroenterology*, **54**, 852
Kedem, O. and A. Katchalsky (1958) *Biochim. Biophys. Acta.*, **27**, 229
Kinney, V. R. and C. F. Code (1964) *Amer. J. Physiol.*, **207**, 998
Kohn, P. G., D. H. Smyth and E. M. Wright (1966) *J. Physiol. (London)*, **185**, 47P

Leblond, C. P. and P. Messier (1958) *Anat. Record*, **132**, 247
Lee, J. S. (1968) *Gastroenterology*, **54**, 366
Levin, R. (1967) *Brit. Med. Bull.*, **23**, 209
Levitan, R. and F. J. Ingelfinger (1965) *J. Clin. Invest.*, **44**, 801
Lifson, N. (1939–40) *Amer. J. Physiol.*, **128**, 603
Lifson, N. and D. S. Parsons (1957) *Proc. Soc. Exp. Biol., Med.*, **95**, 532
Lindemann, B. and A. K. Solomon (1962) *J. Gen. Physiol.*, **45**, 801
Lipkin, M. P., P. Sherlock and B. Bell (1963) *Gastroenterology*, **45**, 721
Lowenstein, W. R. (1966) *Ann. N.Y. Acad. Sci.*, **137**, 441
Matthews, J. and D. H. Smyth (1961) *J. Physiol. (London)*, **158**, 13P
McGee, L. C. and A. B. Hastings (1945) *J. Biol. Chem.*, **142**, 893
McHardy, G. J. R. and D. S. Parsons (1957) *Quart. J. Exp. Physiol.*, **42**, 33
Meschia, G. and I. Setnikar (1959) *J. Gen. Physiol.*, **42**, 429
Miller, D. and R. K. Crane (1961) *Biochim. Biophys. Acta.*, **52**, 293
Parsons, D. S. (1956) *Quart. J. Exp. Physiol.*, **41**, 410
Parsons, D. S. (1968) In *Handbook of Physiology*, Vol. 3, Williams and Wilkins, Baltimore, p. 1177
Parsons, D. S. and D. L. Wingate (1958) *Biochim. Biophys. Acta.*, **30**, 666
Parsons, D. S. and C. R. Paterson (1960) *Biochim. Biophys. Acta.*, **41**, 173
Parsons, D. S. and D. L. Wingate (1961) *Biochim. Biophys. Acta.*, **46**, 170
Parsons, D. S. and C. R. Paterson (1965) *Quart. J. Exp. Physiol.*, **50**, 220
Patlack, C. S., D. A. Goldstein and J. F. Hoffman (1963) *J. Theoret. Biol.*, **5**, 426
Pearson, J. W. (1958) *J. Appl. Physiol.*, **13**, 313
Peters, H. C. and M. B. Visscher (1939) *J. Cellular Comp. Physiol.*, **13**, 51
Phillips, S. F. and C. F. Code (1966) *Amer. J. Physiol.*, **211**, 607
Phillips, S. F. and W. H. J. Summerskill (1967) *J. Lab. Clin. Med.*, **70**, 686
Powell, D. W. and S. J. Malawer (1968) *Amer. J. Physiol.*, **215**, 49
Reid, E. W. (1892) *Brit. Med. J.*, **1**, 1133
Reid, E. W. (1901) *J. Physiol. (London)*, **26**, 435
Reinold, J. G. and D. W. Wilson (1934) *Amer. J. Physiol.*, **107**, 378
Riklis, E. and J. H. Quastel (1958) *Can. J. Biochem. Physiol.*, **36**, 347
Schanker, L. S., D. J. Tocco, B. B. Brodie and C. A. M. Hogben (1958) *J. Pharmacol.*, **123**, 81
Schultz, S. G. and R. Zalusky (1964a) *J. Gen. Physiol.*, **47**, 567
Schultz, S. G. and R. Zalusky (1964b) *J. Gen. Physiol.*, **47**, 1043
Schultz, S. G., R. Zalusky and A. E. Gass (1964) *J. Gen. Physiol.*, **48**, 375
Smith, M. W. (1967) *Biochem. J.*, **105**, 65
Smyth, D. H. (1963) In R. Creese (Ed.) *Recent Advances in Physiology*, 8th ed, Churchill, London, p. 36
Smyth, D. H. and C. B. Taylor (1957) *J. Physiol.*, **136**, 632
Smyth, D. H. and E. M. Wright (1966) *J. Physiol. (London)*, **82**, 591
Soergel, K. H., G. E. Whalen, J. A. Harris and J. E. Greene (1968) *J. Clin. Invest.*, **47**, 1071
Swallow, J. H. and C. F. Code (1967) *Amer. J. Physiol.*, **212**, 717
Taylor, A. E., E. M. Wright, S. G. Schultz and P. F. Curran (1968) *Amer. J. Physiol.*, **214**, 836
Tidball, C. S. (1961) *Amer. J. Physiol.*, **200**, 309
Tormey, J. McD. and J. M. Diamond (1967) *J. Gen. Physiol.*, **50**, 2031
Trier, J. S. (1968) Morphology of the epithelium of the small intestine. In 'Handbook of Physiology', Sec. 6, Vol. III, on Intestinal absorption; C. F. Code, ed.; Amer. physiol. Soc., Washington, D.C.
Ussing, H. H. (1949) *Acta Physiol. Scand.*, **19**, 43
Ussing, H. H. and K. Zerahn (1951) *Acta Physiol. Scand.*, **23**, 110
Visscher, M. B., R. H. Varco, C. W. Carr, R. B. Dean and D. Erickson (1944a) *Amer. J. Physiol.*, **141**, 488
Visscher, M. B., E. S. Fletcher, C. W. Carr, H. P. Gregor, M. S. Bushey and D. E. Barker (1944b) *Amer. J. Physiol.*, **142**, 550

Visscher, M. B., R. R. Roepke and N. Lifson (1945) *Amer. J. Physiol.*, **144**, 468
Wheeler, H. O. (1963) *Amer. J. Physiol.*, **205**, 427
Wheeler, H. O. and O. L. Ramos (1960) *J. Clin. Invest.*, **39**, 161
Whitlock, R. T. and H. O. Wheeler (1964) *J. Clin. Invest.*, **43**, 2249
Williams, A. W. (1963) *Gut*, **4**, 1
Wilson, T. H. (1956a) *J. Appl. Physiol.*, **9**, 137
Wilson, T. H. (1956b) *Amer. J. Physiol.*, **187**, 244
Wilson, T. H. and G. Wiseman (1954) *J. Physiol.* (*London*), **123**, 116
Wiseman, G. (1961) In *Methods in Medical Research*, Vol. 9, Yearbook Medical Publishers,
    Chicago p. 287
Wright, E. M. (1966) *J. Physiol.* (*London*), **185**, 486
Wrong, O. M., A. Metcalfe-Gibson, R. B. I. Morrison, S. T. Ng and A. V. Howard (1967)
    Clin. Sci., **28**, 357

CHAPTER 4

Hydrochloric acid secretion by gastric mucosa

John G. Forte

Department of Physiology–Anatomy,
University of California,
Berkeley, California, U.S.A.

I. INTRODUCTION

The mechanism of hydrochloric acid secretion by the stomach falls within the purview of ion transport by virtue of the fact that gastric juice is produced in relatively large volumes, at the same time being highly concentrated with respect to H^+. The gastric secretory product may achieve acidity as high as

111

160 mM (pH$\simeq$0·8) as compared to the blood or nutrient fluid of pH 7·4, thus representing a concentration gradient of more than a million-fold. Although the concentration gradient for Cl^- is much less than that for H^+, energy is required in order to maintain Cl^- secretion.

The minimum free energy (ΔG) required for the transport of an ion may be estimated from the concentration and electrical gradients, and is described by the following equation (ignoring such factors as solvent drag and internal resistance of the transport machinery):

$$\Delta G = nRT \ln\frac{a_1}{a_2} + nzFE \qquad (1)$$

where n is the number of gram ions secreted, R is the gas constant, T is the absolute temperature, a_1 and a_2 are the activities of the ion on either side of the limiting membrane, z is the net charge of the ion, F, the faraday constant, and E is the electrical potential difference across the limiting membrane.

Knowing the concentrations of the major ionic constituents of the gastric

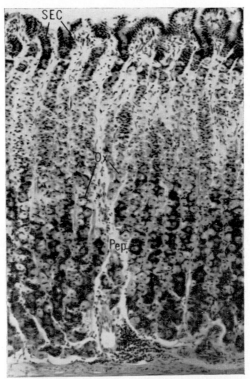

Figure 1. Light micrograph of rabbit gastric mucosa. Surface epithelial cells (SEC) extend over the mucosal surface and extend into the gastric crypts (GC). Oxyntic cells (Ox) and chief cells (Pep) which secrete pepsinogen, are found deeper within the gastric glands. Stained with hemotoxylin and eosin, magnification $\times$ 200.

secretory product and the transmural potential difference it can be estimated that about 8,000 to 10,000 calories/mole are required for HCl secretion. Since the output of gastric juice can also be high it is immediately obvious that HCl secretion by the stomach constitutes a significant calorific requirement for the tissue, and the systemic components responsible for the deposition of HCl are likely to be present in great abundance and have a relatively high turnover rate. The work performed in the secretion of HCl and the production of electric current by the stomach is derived from metabolism within gastric tissue. The cells of the gastric epithelium which secrete acid are called oxyntic cells, after the Greek word *oxyntos*, making acid. The fact that secretion may be sustained for long periods against large electrochemical gradients establishes the oxyntic cells amongst the most metabolically active cells of the body.

Unlike many other exocrine secretory organs, the stomach is accessible and a relatively easy preparation to handle, so that detailed studies of electrical properties and ion flux analysis are possible. The two preparations most frequently referred to in this chapter will be the isolated amphibian gastric mucosa and the *in situ* dog stomach flap, since both have been extensively used for studies of gastric ion-transport. The current status of the mechanisms of HCl secretion will be treated along three major experimental lines. These are first, a morphological approach, specifying the structural components related to the secretory process; secondly, a biophysical approach, including the analysis of ion flux and related electrical properties of gastric tissue; and thirdly, a biochemical approach, involving the study of the metabolic and enzymic events associated with secretion. Where applicable, working models and schemes will also be included.

II. STRUCTURAL BASIS OF GASTRIC H⁺ SECRETION

A. General Morphology of Gastric Mucosa and Fine Structure of Acid Secreting Cells

Interpretation of the electrical and secretory data requires occasional reference to morphological details. Therefore it is useful to consider both the general histology of the gastric mucosa and its ultrastructure as visualized by electron microscopy. The gastric mucosa is not a simple flat sheet of epithelial cells, but rather consists of numerous glandular inpocketings or depressions extending below the secretory surface. This morphological orientation, which is further differentiated into regional accumulations of specific cell types, provides several problematic considerations when one attempts to interpret electrical data, as will be seen later.

Micrographs of longitudinal sections from the main body of rabbit and frog stomachs are shown in figures 1 and 2, respectively. The surface

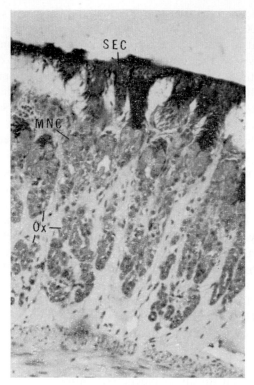

Figure 2. Phase-contrast micrograph of bullfrog gastric mucosa. The major epithelial cell-types of the mucosa are shown: surface epithelial cells (SEC) mucous neck cells (MNC); and oxyntic cells, Ox. Magnification × 295.

epithelial cells, which secrete mucus, cover the entire mucosal surface and extend into the gastric pits. Gastric glands lead or open into the gastric pits: usually there are about four to five glands per pit. Mucous neck cells are located at the juncture between the neck region of the gastric pit and gland. A major difference is evident in the cellular constituents of the glandular epithelium when one compares amphibians and mammals. Mammalian gastric glands contain chief cells which secrete pepsinogen and also contain oxyntic cells.* In the main body of amphibian gastric mucosa, chief cells are absent and only oxyntic cells are present. Langley (1881) was one of the first to present evidence for functional specification of various cell types in the gastric epithelium; oxyntic cells were specified as the source of HCl

---

*Oxyntic cells of mammalian gastric mucosa are often referred to as parietal cells, a name based on the unusual shape and histological position of the cell.

primarily because no other activity was associated with them. However, since then much additional evidence has served to confirm the original view.

Studies of oxyntic cells with the electron microscope have yielded some interesting results. Oxyntic cells contain numerous mitochondria, a fact which is consistent with biochemical data showing high rates of oxidative metabolism for gastric tissue (Davies, 1948). Even more interesting is the appearance of extensive smooth-surfaced endoplasmic reticulum (SER) within the apical region of the cells. These details are shown in figures 3 and 4 for the amphibian and mammalian oxyntic cells, respectively. An additional feature of cytological specialization for the mammalian oxyntic cell is the investment of the apical cell membrane into deep infoldings, or intracellular cannaliculi as they are called.

B. Structural Correlates with H^+ Secretion

Profound structural changes at the apical cell surface have most definitely been correlated with the HCl secretory activity of mammalian and amphibian oxyntic cells (Vial and Orrego, 1960; Ito, 1961; Sedar, 1961a, 1961b; Sedar and Forte, 1962). A possible role for the SER in these structural alterations has been suggested by several authors (Sedar, 1965; Forte, 1966; Forte and coworkers, 1969). It is pertinent to briefly discuss the evidence which implicates these membranous cytoplasmic elements in the secretory process.

The secretory surface of the resting (non-secreting) oxyntic cell contains relatively few, short, microvilli projecting into the lumen of the gastric gland. The apical region of the cytoplasm contains numerous elements of the SER which, under conditions of aldehyde fixation, appear as long tubular elements with a regular diameter of 500–600 Å. The uniformity of tubular diameter and the vast number of these membranous structures in the apical cytoplasm often result in a geometrical grouping suggestive of a hexagonal array; this may be seen in the insert to figure 3.

Activation of gastric mucosa to secrete acid, either by neural or humoral stimulation, typically results in a large increase in 'apical surface activity' of oxyntic cells. Oxyntic cells from secreting bullfrog gastric mucosa are shown in figures 5 and 6. Long, often anastomosing, extensions of plasma membrane project into the glandular lumen. Since the SER is seen in close proximity to the cell surface, and often in direct apposition to the plasma membrane (figure 6), it has been suggested that these tubular elements contribute to the increased surface area of the plasma membrane during HCl secretion (Sedar and Forte, 1962; Sedar, 1965; Forte, 1966). The concept of the inter-relationship between SER and the apical oxyntic cell membrane was originally formulated by examination of static electron micrographs from mucosae in varying states of secretory activity; however, additional support

E

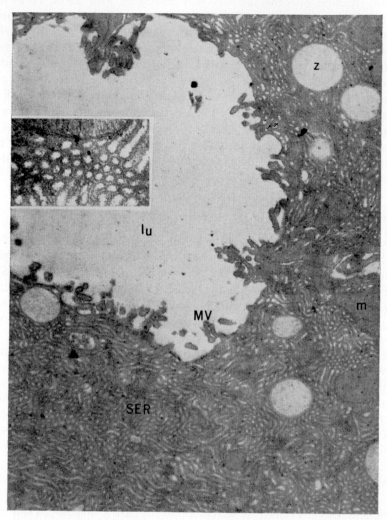

Figure 3. Electron micrograph of apical region of two adjacent oxyntic cells from a non-secreting bullfrog gastric mucosa. There is an abundance of closely packed tubular elements of the smooth-surfaced endoplasmic reticulum (SER). A few, short microvilli (MV) can be seen projecting into the glandular lumen (lu). Several large granules, presumed to be zymogen (z), and mitochondria (m) are also evident in the micrograph. × 16,000. An enlargement of the SER is shown in the insert where close packing is apparent, suggesting a hexagonal array of the tubular membranous elements. Magnification × 48,000.

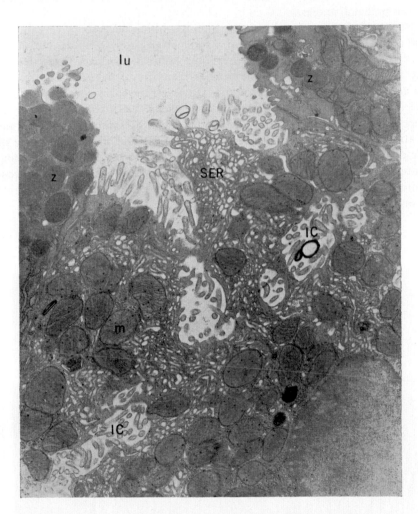

Figure 4. Electron micrograph from rat stomach showing an oxyntic cell between chief cells. Intracellular canaliculi (IC) are present within the oxyntic cell. The smooth-surfaced endoplasmic reticulum (SER) and abundant mitochondria are also evident. Zymogenic granules (z), presumably containing pepsinogen, are seen in the adjacent chief cells. Magnification × 12,000. (Courtesy of A. W. Sedar).

for the possibility of such membrane transformations has come from more extensive fine structural analysis, phospholipid turnover studies, and developmental studies.

Sedar (1965) has pointed out that high resolution microscopy reveals that the apical plasma membrane is not a symmetrical double leaflet, but that a fuzzy amorphous coat exists at the outer luminal interface. Inspection of the fine structural pattern of the tubular members of the SER also reveals an asymmetry of membrane structure. However, the fuzzy coat is present at the internal aspect of the SER membrane, facing the tubular lumen. Cyto-chemical staining techniques indicate the chemical similarity between the coating substance at the inner aspect of the SER and the outer surface of the apical plasmalemma (Sedar, 1969a; Forte and Forte, 1969). The coating material is a glycoprotein virtually devoid of sialic acid, with hexose, glucos-amine and fucose representing the primary carbohydrate components (Forte and Forte, 1970).

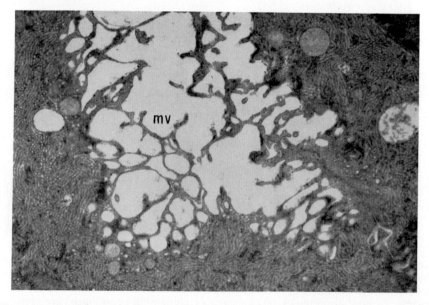

Figure 5. Cross-section of gastric gland from secreting bullfrog gastric mucosa. Oxyntic cells surround the gland lumen with numerous long, filamentous microvilli (mv) projecting from the apical surface. The mucosa from which this section was taken was stimulated with 10^{-5}M histamine, resulting in an H^+ secretory rate of $3\cdot6\ \mu$ equiv./cm$^2\cdot$hr at the time of fixation. Magnification $\times$ 10,200. (From Kasbekar and coworkers, 1968).

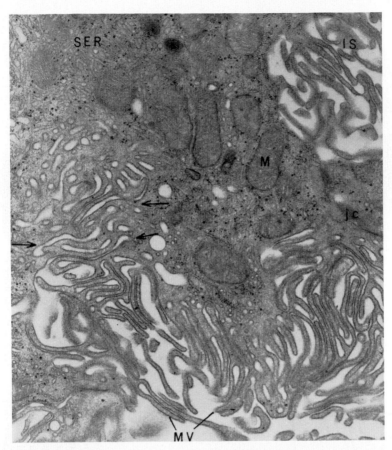

Figure 6. Enlarged view of apical region of oxyntic cells from a bullfrog gastric mucosa secreting H$^+$ at a rate of 5 μ equiv./cm^2·hr. Extensive microvilli (MV) project into the lumen of the gland. Arrows indicate regions where profiles of the membranous SER are seemingly coming in contact with the apical secretory surface of the cells. The intercellular space between adjacent oxyntic cells is indicated by ics. Magnification × 18,000. (From unpublished results of Sedar and Forte).

If the tubules were to fuse with the plasma membrane during HCl secretion, as depicted in the scheme shown in figure 7, the orientation of the membrane asymmetry would be consistent with the observed ultrastructural changes. Using a variety of chemical fixatives, Lillibridge (1968) has verified the finding that the thickness of the plasma membrane of the oxyntic cell is the same as the SER membrane thickness, in contrast to other cytoplasmic membranes, such as mitochondria and rough-surfaced endoplasmic reticulum. In a very recent and significant study, Sedar (1969b) has shown that

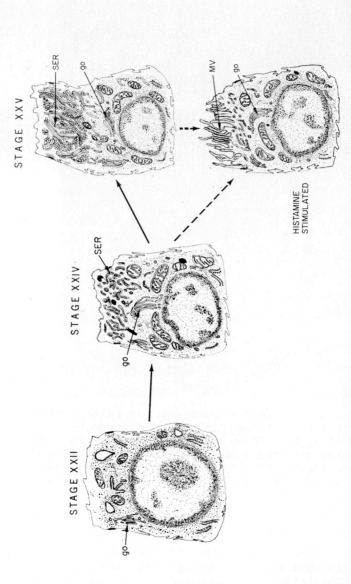

Figure 7. Drawing of developing bullfrog tadpole oxyntic cells showing the pertinent ultrastructural details which occur when H^+ secretion first appears. The various metamorphic stages indicated are i) stage XXII, when tadpole forelimbs have emerged, ii) stage XXIV, when tail reabsorption is nearly complete, and iii) stage XXV, the newly metamorphosed froglet. Also indicated by the dotted arrows are the structural changes associated with stimulation of H^+ secretion by applying histamine to stage XXIV or XXV tadpole stomachs. Virtually the same ultrastructural changes are manifest for histamine stimulation in adult frog gastric mucosa. The most striking structural changes which occur as the preoxyntic cells (stage XXII) mature into functional acid secreting units are the i) elaboration of Golgi structures (go), with blebbing or pinching off of their peripheral lamellae, and ii) synthesis and elaboration of smooth-surfaced endoplasmic reticulum (SER) which accumulates in the apical region of the cells and becomes especially abundant in stage XXV. The most striking change which is seen when a resting cell is stimulated to secrete H^+ is the extensive elaboration of microvilli (mv) at the apical secretory surface, presumably derived from the SER.

the SER may interconnect with the apical secretory surface under certain conditions. He found that a protein as large as the enzyme peroxidase readily penetrates from the free surface into the lumina of the tubular membrane system of bullfrog oxyntic cells. Thus there appears to be ample structural evidence to support the possibility of membrane transformations between the SER and the plasma membrane.

Membrane migration from one region of the cytoplasm to the apical cell surface is a possibility which would render large changes in *de novo* membrane synthesis unnecessary. However, membrane transformations as a function of secretory activity might be associated with changes in the turnover of specific constituents. Recent investigations from this laboratory concerning phospholipid turnover in gastric mucosa have yielded results consistent with the membrane alterations associated with acid secretion (Kasbekar and coworkers, 1968). The incorporation of $^{32}P_i$ into the phospholipid fraction of histamine-stimulated bullfrog gastric mucosae was increased several-fold compared to paired halves which were not secreting HCl. Not only was there an increased turnover of phosphatidic acid and phosphatidyl inositol, as seen in other tissues (Hokin and Hokin, 1958, 1960; Karnovsky, 1964), but in addition the major structural phosphatides of choline and ethanolamine were increased two to five-fold compared to control mucosae. The fact that incorporation of ^{14}C-acetate into mucosal lipids was not significantly different between resting and secreting mucosae suggests that the increased turnover is primarily associated with polar groups of the phospholipid molecules. Such changes in phospholipid metabolism as these may be primary events underlying the membrane transformations known to accompany HCl secretion.

Supporting evidence for the interrelationship between the SER and gastric HCl secretion is available from developmental studies. Several authors have correlated the morphogenesis of oxyntic cells, and associated organelles, with the functional capacity to secrete acid (Menzies, 1958; Wright, 1962; Toner, 1965; Forte and coworkers, 1969). For the metamorphosing tadpole, as well as for rabbit and chick embryos, the biogenesis of a significant quantity of SER in developing oxyntic cells appears to be a necessary prerequisite for the secretion of HCl (Forte and coworkers, 1969; Toner, 1963, 1965; Hayward, 1967). Electron micrographs of glandular cells from bullfrog tadpole gastric mucosa prior to and after the development of the capacity to secrete HCl are shown in figures 8 and 9, respectively. The 'preoxyntic' cell demonstrates a characteristically undifferentiated appearance with numerous ribosomes, free as well as bound, scattered throughout the cytoplasm and relatively little or no SER. An active and elaborate Golgi apparatus, as well as the appearance of numerous tubular elements of the SER (possibly derived from the Golgi) are obvious structural developments

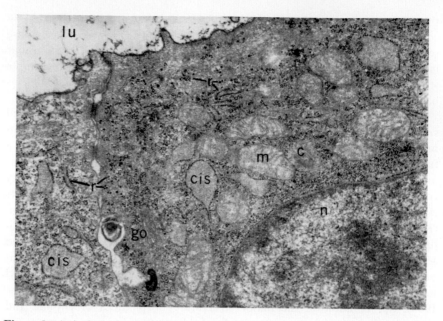

Figure 8. Apical portion of pre-oxyntic cell from stomach of bullfrog tadpole prior to the development of the capacity to secrete H⁺ (stage XXII). Numerous ribosomal particles (r) are seen throughout the cytoplasm, both free and in association with membranes. Several expanded cisternae (cis), probably containing proteinaceous material, are evident. The Golgi (go) complex is present, but not so structurally elaborate and profuse as in later stages. Mitochondria (m), centriole (c), and a nucleus are also evident in the micrograph. Magnification × 20,000. (From Forte and coworkers, 1969).

which coincide with the onset of H^+ secretion (Forte and coworkers, 1969). A schematic representation of the morphogenesis of oxyntic cells, along with the apparent structural alterations associated with stimulation of HCl secretion, is shown in figure 7.

These experimental observations reinforce the evidence already available that the SER of oxyntic cells plays a fundamental role in the membrane transformations which accompany the secretory process. However, from these structural studies alone, the time course of ion translocation events cannot be ascertained. That is to say, it is impossible to distinguish between a mechanism whereby HCl is 'presecreted' into the tubules of the SER and deposited after surface contact during an effective stimulus, or whether these tubular components simply represent membranous reserves to expand the apical secretory surface when HCl secretion is stimulated. Furthermore, it has not been possible to specify sites or loci for H^+ and Cl^- movements.

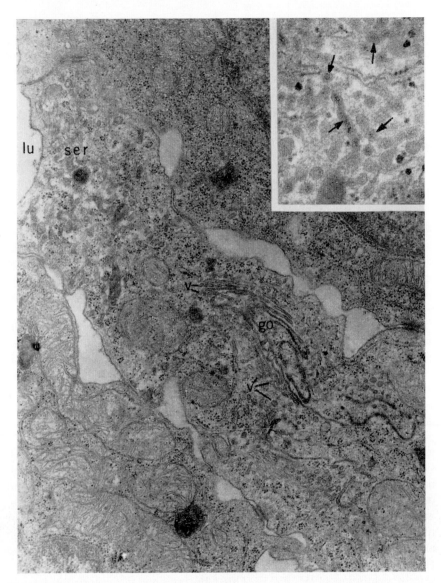

Figure 9. Apex of young oxyntic cell from a gastric gland of stage XXIV tadpole stomach. The prominent Golgi (go) consists of several stacked cisternae which often appear to pinch off vesicular structures (v) at the periphery. Both longitudinal and transverse profiles of the smooth-surfaced endoplasmic reticulum (SER) appear to accumulate in the apical portion of the cell near the glandular lumen (lu). Magnification × 23,000. The inset is an enlargement of the profiles of the SER taken from a similar cell. These structures often contain an electron dense material under the conditions of fixation and staining used. Magnification × 69,500. (From Forte and coworkers, 1969).

Nonetheless, the ultrastructural studies have been valuable in providing morphological markers upon which physiological investigations may be undertaken and interpreted. Clearly, physical and chemical events associated with secretory or metabolic reactions within cells and cell fragments must ultimately coincide with the morphological data in order to provide a complete and coherent theory of the mechanism of HCl secretion.

III. ELECTRICAL PROPERTIES AND IONIC FLUXES ACROSS GASTRIC MUCOSA

A. Transport of H^+ and Cl^- by Gastric Mucosa

Observations on the electrical activity of the stomach extend back to 1834 when Donné demonstrated an electrical potential difference (p.d.) across the stomach wall; the mucosal or secretory side was negative with respect to the serosal or nutrient side. A similarly oriented electrical p.d. has been observed in stomachs of virtually all vertebrate species. [An exception to this may occur in certain elasmobranchs where the p.d. is either too small to be measured or non-existent (Hogben, 1959a; Rehm, 1962a)].

In terms of the major ionic constituents of the secretory product the gastric p.d. is oriented so that it favours H^+ secretion but it represents an electrical barrier against which Cl^- must be transported. From an analysis of the electrical p.d. across the mammalian stomach and the concentrations in the blood and gastric juice, respectively, Rehm (1950) concluded that both H^+ and Cl^- were actively transported across the gastric mucosa; that is to say, secretion of these ions occurs against the respective electrochemical potential gradients. Hogben (1951) applied Ussing's short circuit technique to the study of ion transport processes by isolated frog gastric mucosa. Both sides of the tissue were bathed by identical Ringer's solutions, thus eliminating the chemical gradients which normally exist between the blood and the gastric juice of an *in vivo* secreting stomach. The spontaneous transmembrane potential was reduced to zero by passing an external current through the mucosa (short-circuit current). Under these conditions with no electrochemical potential difference between the two bathing solutions, Hogben demonstrated a net transport of Cl^- from the nutrient to the secretory solution (1951, 1955). Thus, by fulfilling the criteria of Ussing, Cl^- can be said to be actively transported by gastric mucosa. However, the measured short-circuit current was found to be less than the net equivalent of Cl^- flux and this difference was found to be exactly equal to the H^+ transport into the secretory solution, as shown in table 1. The interpretation of these results was that active Cl^- transport occurred both as a current of Cl^- (short-circuit current) and as a component of HCl, the latter being essentially equivalent to the movement in the same direction of two equal and oppositely charged ions (Hogben, 1955). Hogben's data thus provided a more quantitative aspect to what had previously been proposed from the analysis of ion concentrations in the blood and the secretory product in view of the observed electrical field.

B. Cl⁻ Exchange Diffusion

In addition, Hogben (1955) was able to infer from his data that a significant component of Cl⁻ flux across gastric tissue occurred as exchange diffusion. The partial passive conductance of an ion moving between identical solutions when the p.d. is zero may be derived from the flux ratio equation (see Ussing, 1949; Hodgkin, 1951) which can be reduced to the following form:

$$G_i = \frac{z^2 F^2}{RT} \cdot J_i \tag{2}$$

where G_i is the partial ionic conductance, J_i the unidirectional flux of the ion, and z, F, R and T have their usual physicochemical meaning. Hogben (1955) showed that the conductance due to Cl⁻, calculated from the backflux of Cl⁻ from secretory to nutrient solution, was larger than the total mucosal conductance; consequently, simple ionic diffusion could not account for the total Cl⁻ backflux. He reasoned that a significant portion of Cl⁻ movement through the gastric epithelium must occur as an uncharged moiety in

Table 1. Chloride transport by short-circuited bullfrog gastric mucosa

Chloride transfer		Net charge transfer	
$J_{ns}^{Cl^-}$	10·65	Short-circuit current	3·05
$J_{sn}^{Cl^-}$	6·38	H⁺ secretion	1·20
Net Cl⁻ flux	4·27	Sum	4·25
		Mean conductivity	
		Start: 3·21 mmhos/cm²	
		End: 2·28 mmhos/cm²	

Values for $J_{ns}^{Cl^-}$ and $J_{sn}^{Cl^-}$ are the unidirectional fluxes of Cl⁻ from nutrient to secretory (ns) and secretory to nutrient (sn), respectively. All values of flux and current are given in μ equiv./cm² hr. Data from Hogben (1959b).

combination with some component of the membrane, that is to say, an exchange diffusion process.

Several experiments have been designed to specify the membrane interface where transport and exchange of Cl$^-$ occur. Cotlove and Hogben (1956) analysed the steady state intracellular specific activity of chloride when one surface of the gastric mucosa was bathed with ^{36}Cl. They concluded that under normal conditions the relative flux across the secretory, or luminal, interface was about sixteen-fold greater than the flux across the nutrient surface. Furthermore, Cotlove, Green and Hogben (1959) showed that the Cl$^-$ flux at the secretory surface was severely depressed by 2,4-dinitrophenol, while the movement of Cl$^-$ across the nutrient surface was relatively little affected. Inhibition of the active component of Cl$^-$ transport was to be expected, but in addition, it appeared that the uncoupling agent reduces (or abolishes) the exchange diffusion of Cl$^-$ which they localized at the secretory interface.

An alternative proposal was offered by Villegas (1965) on the basis of a study of the desorption kinetics from gastric mucosae previously loaded with ^{36}Cl. He found that the 'slow phase' of ^{36}Cl efflux from the nutrient surface was enhanced when the tissue was bathed with solutions containing exchangeable Cl$^-$, and on that basis he postulated a Cl$^-$ exchange barrier at the nutrient interface. However, there are several problems to be considered before the nutrient surface is considered as the primary locus of Cl$^-$ exchange diffusion. For instance, Villegas (1965) has shown that tracer ^{36}Cl completely equilibrates with tissue Cl$^-$ in less than twenty minutes, which virtually eliminates the utility of the small, slowly-effluent compartment ($t_{1/2}=30$–40 min) as a useful measure of the major component of Cl$^-$ exchange diffusion. Also, Villegas' very interesting data indicate that Cl$^-$ content of whole mucosae was maintained near control *in vivo* levels only when Cl$^-$ was present in the nutrient bathing solution. This would suggest a relatively high free permeability of Cl$^-$ at the nutrient side, as compared to the secretory side. Yet, Cotlove and Hogben (1956) demonstrated rapid equilibration of ^{36}Cl with tissue Cl$^-$ when the isotope was added to the secretory solution. Thus the high flux of Cl$^-$ across the secretory membrane barrier (from secretory solution to mucosal cells) appears to be mainly an isotopic exchange and equilibration process, which is the basis for exchange diffusion, and constitutes the best available evidence to localize a significant exchange process at the secretory surface of the gastric epithelium. This does not eliminate the possibility that some form of anionic exchange or exchange diffusion occurs at the nutrient interface. It should be pointed out that both of the experimental approaches discussed above, namely equilibration (Cotlove and Hogben, 1956) and desaturation (Villegas, 1965) studies, have not sufficiently dealt with problems arising from the large bulk of smooth muscle (muscularis mucosae) and connective tissue relative to the epithelial cell layer nor with the problems of parallel compartmentalization of various cell types within the gastric epithelium.

Heinz and Durbin (1957) and their coworkers (Durbin and coworkers, 1964) attempted to study the specificity of the Cl^- exchange diffusion component in gastric tissue. From their observation that the flux of Cl^- from the nutrient to secretory solution depended upon the concentration of Cl^- on the secretory or trans side, Heinz and Durbin proposed the following carrier-mediated model to explain the transfer of Cl^-.

In the above system Cl^- associates with a diffusible, membrane-bound, carrier (X) with an equilibrium constant, K_{XCl}. Assuming the latter to be equivalent at both the inner (I) and outer (O) facing solutions, i.e. a passive exchange process, Heinz and Durbin examined the kinetics of Cl^- flux as a function of trans concentration and in terms of the relative mobilities of the free (P_X) and bound (P_{XCl}) carrier forms (Heinz and Durbin, 1957). In order to account for the experimental observations (the trans effect) their analysis of the model revealed that the complex carrier form has the shorter translocation time, i.e. $P_{XCl} \gg P_X$. The exchange process appears to have a relatively high degree of specificity (as does the active transport). Bromide will substitute for chloride quite well and iodide to some extent, in the exchange diffusion process, but most other anions are relatively poor substitutes (Heinz and Durbin, 1958; Durbin and coworkers, 1964; Hogben, 1965).

C. Three Components of Cl^- Flux across Gastric Mucosa: the Dependency upon H^+ Secretion

In order to assess the interrelationship between H^+ and Cl^- transport mechanisms it is useful to consider total Cl^- flux from nutrient to secretory side of gastric mucosa in terms of separate flux components consisting of first a pump, or an actively transported component, secondly an exchange diffusion component, and thirdly a passive diffusional leak component. To approach such an analysis, exchange diffusion is simply taken as that component of Cl⁻ flux which is dependent upon the concentration of Cl⁻ on the trans (secretory) side. The active component is taken as that fraction of Cl⁻ flux from nutrient to secretory solution which is dependent upon metabolism and must be measured in the absence of exchange diffusion. The passive leak of Cl⁻ is that flux remaining after the elimination of the active and exchange components (table 2).

Table 2. The dependence of Cl⁻ flux from nutrient to secretory solution ($J_{ns}^{Cl^-}$) and net H⁺ secretion ($J_{ns}^{H^+}$) on the concentration of Cl⁻ in the secretory solution and oxidative metabolism

Secretory Solution	Gas Phase	$J_{ns}^{Cl^-}$	$J_{ns}^{H^+}$
		(μ equiv./cm² hr	
Normal [Cl⁻]$_s$	95%O$_2$–5%CO$_2$	8·6±0·3	3·6±0·3
Low [Cl⁻]$_s$	95%O$_2$–5%CO$_2$	4·4±0·2	3·4±0·2
Low [Cl⁻]$_s$	95%N$_2$–5%CO$_2$	1·0±0·1	<0·05

During the periods indicated by low [Cl⁻]$_s$, Cl⁻ in the secretory solution was completely replaced by an equivalent amount of isethionate. Flux values are given as the mean ± S.E.M. for 22 bullfrog gastric mucosae.

Examination of these flux components under a variety of conditions can generate some predictions regarding the mode of transport and possible interrelationships between H⁺ and Cl⁻ (Forte, 1969). For instance, conditions leading to an increase (e.g. histamine stimulation) or decrease (e.g. anoxia) in H⁺ secretion produced an equivalent change in the active component of Cl⁻ flux. This suggests the possibility of coupling, or at least a rather close relationship, between the transport of these two ionic species and the metabolic events which control them. I also reported that variations in [Cl⁻] on the secretory side of the mucosa (replacement with isethionate) do not affect H⁺ secretory rate, thereby indicating that H⁺ secretion is not dependent upon the operation of Cl⁻ exchange diffusion (Forte, 1969). However, conditions of reduced H⁺ secretion (e.g. anoxia) resulted in decreased Cl⁻ exchange, in agreement with observations made by others (Heinz and Durbin, 1957; Cotlove and coworkers, 1959).

The close association in active transport of both H⁺ and Cl⁻ could be explained by either a tight electrical or a biochemical coupling of the two processes. For electrical coupling one might propose that the transport of H⁺ at the secretory interface of oxyntic cells (deposition of protons or sequestration of OH⁻) is an electrogenic process, that is, one capable of generating an E.M.F. by separation of charge. An obligatory Cl⁻ transfer, so that an equivalent of H⁺ secretion results in an equivalent of Cl⁻ secretion,

Secretory solution

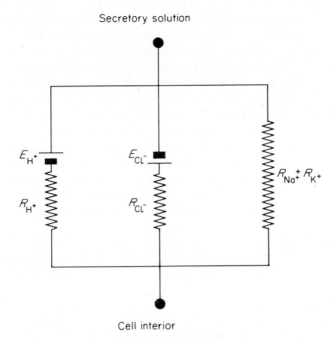

Cell interior

Figure 10. Equivalent circuit for electrical coupling between H^+ and Cl^- transport (after Rehm, 1962). As long as the external resistance is high (open circuit), and no effective shunts are due to other ions ($R_{Na^+} + R_{K^+}$ is high), transfer of H^+ and Cl^- through their respective limbs are stoichiometrically coupled.

is provided by the circuit analogy shown in figure 10 representing the secretory interface of the gastric mucosa. The feasibility of a nearly stoichiometric coupling between an electrogenic H^+ pump and Cl^- transport is based on the fact that the permeability of the secretory interface to other major ionic constituents (Na^+ and K^+) is exceedingly low compared to that reported for Cl^- transfer at that locus (Cotlove and Hogben, 1956; Harris and Edelman, 1964; Davenport and coworkers, 1964). It would thus be conceivable that H^+ and Cl^- carry the bulk of the current through their respective limbs of the circuit. One advantage of this model of electrical coupling is that it is not necessarily confined to a particular site within a cell, or even to a single cell. It is established simply on the basis of pathways, or loops, for specific ionic current flow.

The coupling of H^+ and Cl^- transport through a biochemical mode is more restrictive, in that one is almost forced to propose a common site for both ionic species. However, an extension of the carrier model may be successfully applied to much of the data, and one such system is depicted in figure 11. Cl⁻ flux across the secretory interface of oxyntic cells could occur primarily in association with a protonated carrier which, in turn, is a direct function of the metabolic and secretory state of the tissue. That is, under optimal conditions for secretion, the titre of protonated carrier would be at its highest, hence Cl⁻ transport and exchange are at their maxima. The carrier form is limited by the addition of inhibitors of H⁺ secretion and consequently transport and exchange diffusion of Cl⁻ are diminished. The restriction of Cl⁻ exchange diffusion, by removal of Cl⁻ from the trans solution, would not appreciably affect reactions involving the formation and decay of the protonated carrier form, and hence would not alter net H⁺ secretion. According to the scheme shown in figure 11 there are several degrees of freedom in the carrier model. For instance, one might alter the direct coupling between Cl⁻ transport and exchange diffusion by the values

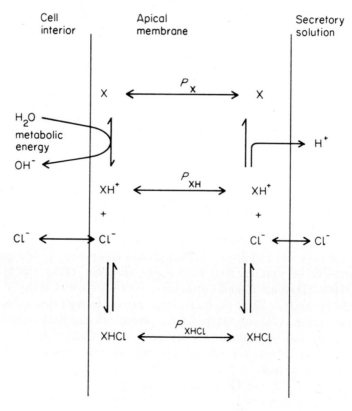

Figure 11. Schematic carrier model to provide tight coupling between H⁺ and Cl⁻ transport by gastric mucosa.

assigned to the relative affinities of Cl⁻ for the free or protonated carrier forms (or the mobilities thereof). One could also establish electrical relationships, such as those discussed above, with appropriate notations for polarity and transfer of charge (i.e. the net deposition of H^+ at the secretory interface could be a metabolically driven source of E.M.F. with an equivalent of Cl⁻ charge transfer maintaining electroneutrality and carrying the current through the appropriate limb).

Clearly, this translocating model system is highly idealized along the classical lines of the carrier hypothesis; however, the observed kinetics would also apply to rotational carrier forms or those mechanisms based on configurational or allosteric changes (Patlak, 1957; Heckmann, 1965; Vidaver, 1966). The main reason for constructing simple model systems is to provide a conceptual framework with which to examine data; besides such models are useful in the design of further experiments. The complex carrier model system was developed to explain the apparent close coupling between H^+ and Cl⁻ transport, the dependence of Cl⁻ exchange diffusion on H^+ secretion, and H^+ secretion independent of [Cl⁻] on the secretory side. There are many other parameters and phenomena which need to be accounted for in the idealized model. Some of these, such as the source of the gastric electrical p.d., the interrelations between ion transport and electrical activity, the nature of the coupling between ion pumps and metabolic energy, etc. will be considered in subsequent sections.

D. Site and Source of the Gastric P.D.

It was pointed out earlier that a net current of Cl⁻ ions (net Cl⁻ transport minus HCl secreted) exactly matched the amount of current required to reduce the transmucosal p.d. to zero mV (Hogben, 1951, 1955). This quantitatively established active transport of Cl⁻ and the participation of the anion in the electrical activity of the amphibian gastric mucosa. Heinz and Durbin (1959) extended these observations by their experimental approach of bathing gastric tissue in Cl⁻-free solutions. Under these conditions the transmucosal p.d., which was ordinarily about −30 mV (secretory side negative), reversed in sign to about +8 mV in sulphate-Ringer's solution. They further demonstrated that the short-circuit current in sulphate-Ringer's solution was directly proportional to the rate of H^+ secretion. In short, these experiments demonstrated that there is an electrical potential difference associated with H^+ transport, and that this potential is normally masked by the electrical activity associated with Cl⁻ transport.

Using a micropuncture technique Villegas (1962) provided some very interesting experimental results on the cellular localization of the mucosal potential difference. Intracellular potentials of the frog gastric mucosa were measured, marked by deposition of a dye, and later identified and localized histologically. Villegas concluded that in chloride-Ringer's solution the normal transmucosal p.d. of −30 mV could be described as a two-step jump across the oxyntic cell. That is to say, the interior of the oxyntic cell was about −18 mV with respect to the nutrient side, with the remainder of the p.d. drop (−12 mV) at the secretory membrane. In addition, he proposed that the surface epithelial cells, which are electrically in parallel with oxyntic cells, are not polarized in the transepithelial sense, but that an equal and opposite p.d. drop of about 49 mV (cell interior negative) occurred at both the secretory and nutrient interfaces. Canosa and Rehm (1968) have criticized such an explicit representation of gastric intracellular potentials on the basis of their observation that the mucosa is in constant motion due to activity of muscularis mucosae. Although such technical difficulties most certainly do exist, Villegas' data appear reasonable and are the best available on the localization of the gastric p.d.

In considering the origin of the gastric electrical potential difference, the two most popular theoretical descriptions for a membrane potential are first, one based upon the equilibrium potential due to ion concentration differences at, or across, the membrane interface and secondly, one based on the electrogenic pump, so-called because its activity is due to a metabolically dependent separation of charge. However, a precise, detailed description of an electrogenic pump in a biological system has not as yet been given, although several models for generating such activity have been proposed.

E. Ion Gradients and Gastric P.D.

Several groups have studied the transmucosal p.d. as a function of variations in the concentration of ionic constituents of the two bathing solutions. Forte and coworkers (1963) replaced the Cl^- in both solutions bathing isolated bullfrog gastric mucosa with 2-hydroxyethylsulphonate (isethionate) and tested the effects of incremental changes in the concentration of Cl^- or H^+ in the solutions. In agreement with the observations of Heinz and Durbin (1959) they found that in the absence of Cl^- the transmucosal p.d. reversed in sign; a steady state value of about 25 mV (secretory side positive) was achieved after several washes. The electrical conductance decreased and the rate of H^+ secretion declined to about one-fifth that observed in chloride-Ringer's solution. Isethionate has several advantages over sulphate as a replacement anion in the bathing solutions. First, it is univalent, thus eliminating the concentration and osmotic problems encountered with sulphate-Ringer's solution; secondly, isethionate is somewhat larger than sulphate; and thirdly, the calcium salt is quite soluble in contrast to calcium sulphate.

Increases in the concentration of Cl^- in the nutrient solution, ($[Cl^-]_n$), produced changes generally consistent with a Cl^- diffusion potential across that interface, but the relationship between p.d. and log $[Cl^-]_n$ was not linear. The slope appeared to vary inversely with the stimulation of H^+ secretion induced by increasing the $[Cl^-]_n$, i.e. an increase in the $[Cl^-]_n$ from 1 to 10 mM produced a 17 mV change in p.d. and was accompanied by a three-fold stimulation of H^+ output, whereas further increases from 10 to 100 mM produced a 50 mV change in p.d. with very little additional enhancement of H^+ secretion (see figure 12). We suggested that the non-linearity was related to an effect of $[Cl^-]_n$ (and specific for the nutrient side) in stimulating the H^+ secreting machinery (Forte and coworkers, 1963). Support for this explanation comes from experimental manipulation of

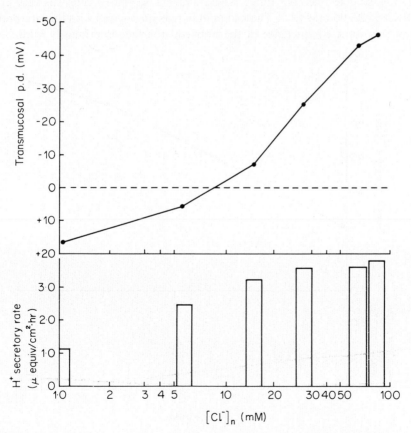

Figure 12. Transmucosal p.d. and rate of H^+ secretion by bullfrog gastric mucosa as a function of the concentration of Cl^- in the nutrient solution ($[Cl^-]_n$). The secretory solution contained no Cl^- throughout the course of the experiment.

$[Cl^-]_n$ under conditions where H^+ secretion is suppressed by a relatively potent inhibitor, e.g. SCN^- (Davenport, 1940; Forte, 1968). One may compare the results shown in figure 12, for a mucosa where H^+ secretion was allowed to develop fully, with those of figure 13, where 5 mM SCN^- was applied to the secretory side throughout the experiment. In the latter case a linear relationship existed between p.d. and log $[Cl^-]_n$; however, the slope of p.d. was only about 25 mV for a ten-fold change in $[Cl^-]_n$.

Harris and Edelman (1964) approached the problem by considering the participation of both Cl^- and K^+ in a diffusion potential at the nutrient interface, and evaluated this possibility by the method of Hodgkin and Horowicz (1959). When Harris and Edelman varied $[Cl^-]_n$ and $[K^+]_n$ at a constant product, they obtained slopes of 50–55 mV for a ten-fold concentration change over the range where the H^+ secretory rate was very little affected (20–90 mM Cl^-). These data fit in reasonably well with the interpretation that ionic conductance of the nutrient interface is primarily a function

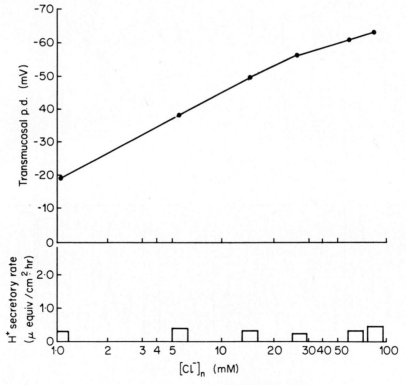

Figure 13. The dependence of p.d. and H^+ secretion upon Cl^- concentration in the nutrient solution ($[Cl^-]_n$) in the presence of SCN^-. Before the $[Cl^-]_n$ was increased, 5 mM NaSCN was added to the secretory solution.

of K^+ and Cl^-, and that these ions contribute to a diffusion potential between the nutrient solution and cell interior. Due to problems inherent in the specification of ionic concentrations in the 'free cytoplasm' of the epithelial cells it is not possible to precisely test the agreement between the predicted equilibrium potential for Cl^- and K^+ and the voltage drop at the nutrient membrane reported by Villegas (1962; see discussion by Durbin, 1967). However, the available data for the ionic composition of gastric tissue are generally consistent with the interpretation that a Cl^- gradient (from nutrient solution to cell interior) and a K^+ gradient (cell interior to nutrient solution) contribute to the potential at the nutrient interface.

There are some interesting and curious results which occur when K^+ and Cl^- concentrations are varied in the secretory solution ($[K^+]_s$ and $[Cl^-]_s$). When frog gastric mucosa is bathed in isethionate-Ringer's solution, variations in $[Cl^-]_s$ are essentially without effect on the gastric p.d. (Forte and coworkers, 1963). The p.d. is also insensitive to variations in $[K^+]_s$ for mucosae bathed in Cl^--solutions (Harris and Edelman, 1964). The simplest conclusion would predict a relatively low specificity or conductance to Cl^- and K^+ at the secretory membrane, and that their concentration gradients at that locus do not appreciably contribute to the gastric p.d. It may be shown, however, that the p.d. does respond to changes in $[Cl^-]_s$ when normal chloride-Ringer's solution is present on the nutrient side (Heinz and Durbin, 1957; Forte, 1969). Thus the 'diffusion potential' for Cl^- at the secretory interface is dependent upon the presence of Cl^- on the nutrient side (or within the epithelial cells). Clearly this relationship is not directly analogous to a simple inanimate membrane system describing an anionic diffusion potential. Although quite speculative, it is possible that these observations are related to the Cl^- exchange process at the secretory interface which in turn appears to be a function of the H^+ secretory capacity of the tissue. That is, the p.d. response due to changes in $[Cl^-]_s$ may be the net result of two boundary potentials (Donnan equilibrium) between Cl^- and a proposed carrier at the inner and outer surfaces of the secretory membrane. If the effective carrier form were produced at the cytoplasmic (inner) interface of the secretory membrane (for example, the protonated carrier mentioned earlier) and if the major access of Cl^- to the epithelial cell cytoplasm was via the nutrient membrane, then severe restrictions might obtain for carrier distribution through the membrane phase. In the absence of a source of cellular Cl^- the amount of carrier at the luminal interface of the secretory membrane would be limited and variations in $[Cl^-]_s$ might not establish an interfacial potential.

Davis and coworkers (1963) found that in Cl^--free solutions the p.d. of bullfrog gastric mucosa responded to changes in $[K^+]_n$ or $[K^+]_s$, and that the resulting changes in p.d. are essentially equal and opposite in sign. It was pointed out earlier that the p.d. was not a function of $[K^+]_s$ in Cl^--Ringer's solution, thus it appears that the permeability of the secretory membrane to K^+ is altered when Cl^- is removed from the bathing solutions. One must therefore be careful to consider the exact conditions and the nature of the replacement ions before interpreting the role of specific ion gradients in the total transmembrane potential.

In addition to its participation in junctional diffusion potentials at the tissue interfaces, K^+ has a role in the maintenance of H^+ secretion by gastric mucosa. That is, after several washes in K^+-free solutions, H^+ secretion is abolished and the p.d. and total mucosal conductance decrease (Davis and coworkers, 1965). Normal conditions may be reestablished by the addition of K^+ to either the nutrient or secretory bathing solution. Rb^+ or Cs^+ will substitute for K^+ in supporting H^+ secretion and the electrical charac-teristics of the tissue (Forte and coworkers, 1967). The entire pattern of response, in terms of time-course and reversibility, may be construed to be simply the removal of an essential participant, K^+, from the series of reactions leading to H^+ secretion (Davis and coworkers, 1965).

The relative effects of varying the concentration of several other ions is also of interest. Changes in $[H^+]$ in the nutrient solution between pH $8\cdot0$ and $6\cdot5$, and in the secretory solution between pH $7\cdot5$ and $2\cdot0$, are without significant effect on the transmucosal p.d. of isolated frog mucosa (Forte and coworkers, 1963; Harris and Edelman, 1964). An increase in $[H^+]_s$ below pH $1\cdot0$ affects the p.d. in both Cl^- and sulphate-Ringer's solution (Harris and Edelman, 1964).

The role of Na^+ in gastric secretory function is a curious one, especially when the results obtained with amphibian and mammalian preparations are compared. Taking the isolated frog gastric mucosa first, variations in $[Na^+]_s$ are without effect on p.d. or H^+ secretion (Rehm, 1962b). When $[Na^+]_n$ is reduced to as low as 10 mM, again there is no appreciable effect on secretion or p.d. (Harris and Edelman, 1964), but both of these parameters are severely depressed when Na^+ in the bathing solution is replaced by choline (Sachs and coworkers, 1966). This suggests that Na^+ has an essential role in gastric secretory function; however, the effects of Na^+ removal appear to be modified by the nature of the buffers and replacement ions used (Davenport, 1963).

Experiments performed on *intact* cat or dog stomach have led to a similar conclusion that Na^+ is not an essential constituent of the mucosal fluid, either for electrical activity or for maintenance of H^+ secretion (Linde and coworkers, 1947; Rehm, 1953; Moody and Durbin, 1965). A somewhat different result has been observed for several of the *isolated* mammalian preparations which have been investigated. For *in vitro* rat stomach the transmucosal p.d. is dependent upon $[Na^+]_s$, whereas variations in Cl^- are without effect (Cummins and Vaughan, 1963, 1965). For isolated stomach of cat, dog, and monkey the p.d. is sensitive to complete removal from the bathing solution of either Na^+ or Cl^- (Kitahara, 1967; Kitahara and co-workers, 1967). For all of these preparations an active transport of Na^+, from secretory to nutrient solution, has been proposed. Since the rate of H^+ secretion appears to be independent of $[Na^+]_s$, the possibility of a forced $Na^+–H^+$ exchange at the secretory membrane may be virtually ruled out (Linde and coworkers, 1947; Cummins and Vaughan, 1965; Moody and Durbin, 1965).

Rehm has suggested that a Na^+ transport system directed from cell interior to nutrient side may be operating in all gastric mucosal preparations, amphibian and mammal, but that under normal physiological conditions this represents only a small fraction of the total H^+ secreted toward the gastric lumen (personal communication). Thus, for the actively secreting, intact mammalian preparation or for the isolated amphibian gastric mucosa (which does very well *in vitro*) the Na^+ pump may be difficult to detect or measure. For the *in vitro* mammalian stomach the observed H^+ secretory rate is considerably below *in vivo* rates, and the operation of the Na^+ pump may be more apparent. This line of reasoning seeks a common basis with which to explain the effects and role of Na^+ in the secretory activity of various gastric preparations. Other explanations stressing inherent species differences or variable progressive deterioration of the mucosae must also be taken into account.

A further line of evidence for a Na^+ pump in gastric mucosa comes from developmental studies. In stomachs isolated from both the rabbit foetus (Wright, 1962) and the metamorphosing bullfrog tadpole (Forte, Limlom-wongse and Kasbekar, 1969), a Na^+ transport system, directed toward the blood or nutrient side, has been identified prior to the development of H^+ secretory capacity. In the early stages of development this activity would most likely be confined to the columnar, mucous-secreting, epithelial cell since it is the only cell type which apparently forms a 'continuous' barrier (via tight junctions) between bathing solutions. Whether the Na^+ pump persists in the structurally (and functionally) similar surface epithelial cells of the adult mucosa, or whether such activity also develops in the newly differentiated oxyntic cells, cannot be ascertained at present.

We may now attempt to summarize the experimental results which specify the role of ion concentration gradients in the transmucosal p.d. The isolated frog gastric mucosa will be the major case in point, but similarities and analogies apply to almost all stomach preparations. A schematic representation of gastric mucosa bathed in normal chloride-Ringer's solution with the diffusion potentials located at the nutrient and secretory interfaces is shown in figure 14. Diffusion potentials for K^+ and Cl^- are located at the nutrient side. A

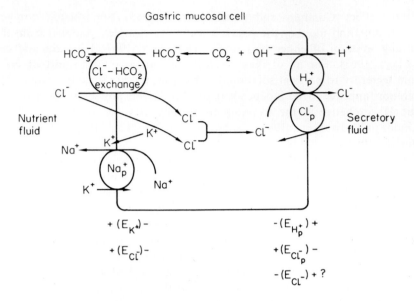

Figure 14. Schematic diagram of gastric epithelial cell showing possible contributions of diffusion potentials due to Cl⁻ and K⁺ (E_{Cl^-} and E_{K^+}). Also shown are the location of a Na⁺ pump (Na_p^+) and Cl⁻–HCO₃⁻ exchange diffusion mechanism. A Cl⁻ pump ($E_{Cl_p^-}$) and H⁺ pump ($E_{H_p^+}$) located at the secretory interface may contribute to the total transmucosal potential under certain conditions. Compare with figures 16 and 19. (Adapted from Forte and coworkers, 1963, and Harris and Edelman, 1964).

diffusion potential for Cl⁻ is located at the secretory side but as pointed out earlier, this is specific for conditions with an exchangeable anion in the nutrient solution and may well be related to a carrier-mediated exchange mechanism. A K⁺ concentration cell exists at the secretory membrane but alterations in [K⁺]ₛ might produce little change in transmucosal p.d. if rapid flux adjustments of Cl⁻ (and H⁺) at the secretory membrane represented an effective shorting loop in Cl⁻-Ringer's solution. The appearance of a typical K⁺-diffusion potential in the absence of Cl⁻ is consistent with such a possibility. Further support is provided by Hogben's finding (1968) that Cl⁻ backflux (S→N) and H⁺ secretion are enhanced by more than 50 per cent of control values when [K⁺]ₛ is raised from 2·5 to 70·0 mM. Significantly, the total short-circuit current flowing through the tissue is very little altered by the change. If K⁺ is simply an impermeable cation at the secretory interface, as is thought to be the case for Na⁺ (Davenport and coworkers, 1964), such large changes in Cl⁻ and H⁺ flux might not be expected.

F. The Maintenance of Ion Gradients

Most of the discussion thus far has centered on the contribution of ion concentration cells and diffusion potentials to the total gastric p.d. during optimal metabolic conditions. If the gastric tissue is deprived of an oxidative energy supply the p.d. falls to zero, the resistance rises (Crane and coworkers, 1948a; Rehm and coworkers, 1953), and tissue ion concentrations tend toward equilibrium, i.e. $[Cl^-]$ increases and $[K^+]$ decreases (Davenport and Alzamora, 1962). It will be of interest to consider the O_2-dependent phenomena which operate to establish and maintain tissue ion balance and to determine whether electrogenicity is associated with these events.

The p.d. response of isolated frog mucosa to changes in $[Cl^-]_n$ and $[K^+]_n$ has been shown to be reduced in the absence of O_2 (Sanders and Rehm, 1967; Rehm, 1968). That is, an apparent decrease in the permeability of the nutrient membrane to these ions occurs, so that the respective slopes of the p.d. response are severely and reversibly depressed. Rehm (1968) has interpreted these results in terms of an interesting equivalent circuit for the gastric epithelium. He considers the resistance pathway between adjacent epithelial cells as high, but not infinite. (It is known that this intercellular pathway is dependent upon Ca^{2+}—Forte and Nauss, 1963; Sedar and Forte, 1964). The abolition of oxidative metabolism produces an increase in total mucosal resistance (Crane and coworkers, 1948a; Rehm and coworkers, 1953). Rehm (1968) proposed that the secretory membrane is the site of the large, O_2-dependent change in resistance so that during anoxia the ratio of the resistance of intercellular channels to the resistance of secretory membrane, is decreased. The net result is to effectively 'short' a portion of the ionic diffusion potential developed at the nutrient membrane and thus on measurement between the two bathing solutions there occurs an apparent diminution of p.d. response during anoxia.

There is an alternative way, with some experimental justification, to explain the data derived from anoxic mucosae. If one considers the details of the p.d. response as Cl^- is added to the nutrient side of an anoxic mucosa, as shown in figure 15, a relatively large transient is observed. That is, the secretory side becomes more negative shortly after the addition of Cl^-, and this is followed by an apparently exponential decay back to near original p.d. levels. Forte and coworkers (1963) suggested that Cl^- secretion was ordinarily closely coupled to metabolism. Moreover, in the anoxic preparation where secretion was inhibited, intracellular $[Cl^-]$ increased shortly following addition of Cl^- to the nutrient side. Thus the characteristics of Cl^- conductance at the nutrient membrane might be the same with or without an O_2 supply but the gradient of Cl^- could not be maintained under conditions of limited metabolic activity. A similar argument may be introduced in the case of $[K^+]_n$. That is, a relatively rapid equilibration of K^+ could occur across the nutrient membrane in the absence of a transport mechanism which might ordinarily stabilize intracellular $[K^+]$. A suggestion for such a phenomenon is apparent in the results given by Rehm (1968, p. 254s) where a transient in the p.d. response may be noted when the nutrient bathing solution of an anoxic mucosa is changed from 79 mM K^+ to 4 mM K^+.

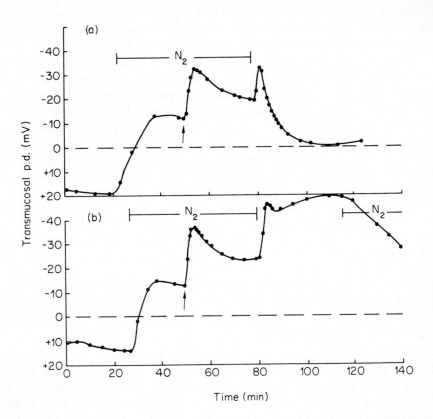

Figure 15. Effect on p.d. induced by addition of Cl^- to the nutrient side of bullfrog gastric mucosa bathed in Cl^--free, isethionate Ringer's solution in the absence of oxygen. During periods indicated by N_2, bathing solutions were equilibrated with 95 per cent N_2 and 5 per cent CO_2. a: Addition of 10 mM NaCl to nutrient fluid indicated by arrow. b: Addition of 5 mM NaCl + 5 mM NaSCN to nutrient fluid indicated by arrow. Note the difference between a (Cl^- alone) and b (Cl^- + SCN^-) for the p.d. response after the readmission of oxygen to the solutions. This difference may be accounted for by the fact that the H^+ secretory rate was restored in a to $2 \cdot 0 \mu$ equiv./cm^2.hr (measured between 100–110 min) compared to a rate of $0 \cdot 5$ μ equiv./cm^2.hr in b measured during the same time period.

G. Electrogenic H^+ and Cl^- Pumps

In addition to the ion diffusion potentials at the gastric mucosal interfaces there also appears to be electrical activity directly associated with the operation of H^+ and Cl^- transporting mechanisms. The anionic pump is responsible for a net flow of Cl^- across the secretory surface. Since it can operate in the absence of H^+ secretion (Forte, 1968) and with no exchangeable anion in the secretory solution (Forte and coworkers, 1963), and since the cellular $[Cl^-]$ is low under normal metabolic conditions (Davenport and Alzamora, 1962), the Cl^- pump seems to be electrogenic. That is, the pump is capable of generating an E.M.F. by the net transport of electrical charge.

Characteristics of the H^+ pump may best be examined in the absence of readily diffusible anions (for example isethionate or sulphate). Under these conditions the p.d. is sensitive to suitable inhibitors of H^+ secretion and it appears to closely approximate the short-circuit current of the tissue (Heinz and Durbin, 1959; Forte and coworkers, 1963). Net H^+ secretion occurs against a large chemical and electrical gradient. Thus the electrical activity observed in Cl^--free solutions may be taken as largely the result of an electrogenic H^+ pump. A more complete discussion of the electrogenicity of the H^+ and Cl^- transport mechanisms may be found in several articles by Rehm (1964a, 1964b, 1967).

The separation of H^+ and Cl^- pumps under the specialized conditions described above does not necessarily argue against the view of the existence of a close coupling between the transport of these two ionic species during optimal conditions for HCl secretion. Using the equivalent circuit given by Rehm (1964a) to describe his separate secretory site theory obvious implications may be found for an electrical coupling between the separate electrogenic transport mechanisms. As depicted in figure 16, an equal and oppositely oriented flow of current through the pumps would provide a deposition of HCl with no net charge transfer. Also shown in figure 16 is a possible adaptation of the basic carrier mechanism in a form which could provide electrogenic H^+ and Cl^- mechanisms under specialized circumstances. The protonated carrier is 'generated' at one membrane interface via endergonic reactions and in this form will act in any one of the following ways:

i) as a vehicle for an electrogenic Cl^- transport mechanisms (if the proton gradient is dissipated via a specific leak pathway);

ii) as a carrier for net H^+ and Cl^- translocation (if the proton leak were inaccessible, for example by histamine binding);

iii) as a locus for isotopic exchange diffusion of Cl^- (if the mobility, or translocation time, of the XHCl form were higher than the other forms).

In the absence of a suitable anion (e.g. Cl^- or Br^-), the system would also provide the basis for electrogenicity associated with the H^+ pump.

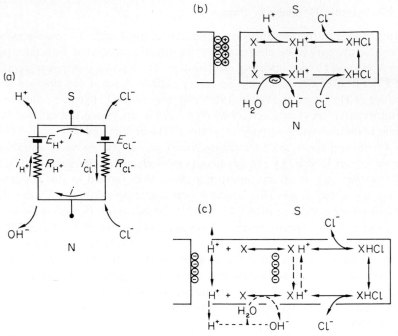

Figure 16. Electrogenic models for H^+ and Cl^- transport. a: Equivalent circuit from Rehm (1963) where current generated by the H^+ pump (E_{H^+}) and Cl^- pump (E_{Cl^-}) flows internally through the pump resistances, R_{H^+} and R_{Cl^-}. In this scheme H^+ current (i_{H^+}) from N→S is balanced by the Cl^- current (i_{Cl^-}) S→N, and net charge transfer does not occur. b: Carrier mechanism analogous to that given in figure 11. The system is driven by cellular metabolic energy. Under optimal conditions of H^+ and Cl^- transport, net transfer of charge does not occur. Deprivation of Cl^- would make manifest an electrogenic H^+ pump. c: Adaptation of system produced in b to demonstrate mechanism for the development of an electrogenic Cl^- pump. The cationic equivalents in the system (XH^+ and/or H^+) would freely leak or exchange back from the S membrane interface to N side. Net movement of H^+ or Cl^- in the case of the respective electrogenic mechanisms would occur only if a shunt leakage (or exchange) were present or if current were passed in the appropriate direction.

H. Current Flow through Gastric Mucosa: the Response of H^+ and Cl^- Flux

A fundamental argument which tends to support electrogenic H^+ and Cl^- transport mechanisms is available from experiments where electrical current from an external source is passed through the gastric mucosa. Passage of current through amphibian or mammalian gastric preparations has been observed to decrease or increase H^+ secretion, depending on the direction of current flow through the tissue (Rehm, 1945, 1959; Crane and coworkers, 1948b).

An experiment showing the relative changes in H⁺ secretion and Cl⁻ flux from N→S following the application of a moderate, constant current in the direction opposing H⁺ secretion (current flow from S→N) is shown in table 3. When the secretory surface of the isolated bullfrog mucosa is bathed with a chloride-free solution, most of the total current flowing through the mucosa may be accounted for by the decrease in H⁺ secretion and the enhanced Cl⁻ flux from N→S, both changes being equivalent to net flux changes. Thus

Table 3. Influence of electric current on H⁺ and Cl⁻ fluxes across isolated bullfrog gastric mucosa with and without Cl⁻ in the secretory solution

E_m (mV)	Secretory solution	$J_{ns}^{Cl^-}$	$J_{ns}^{H^+}$	$\Delta J_{ns}^{Cl^-}$	$\Delta J_{ns}^{H^+}$	$(\Delta J_{ns}^{Cl^-} + \Delta J_{ns}^{H^+})$	ΔI_m
					(μeq/cm²/hr)		
−32	Cl⁻	+8·6	−5·0				
				+1·2	+2·5	+3·7	+6·1
+23	Cl⁻	+9·8	−2·5				
				−1·9	−1·7	−3·6	−6·1
−38	Cl⁻	+7·9	−4·2				
−47	Ise⁻	+4·5	−4·0				
				+4·0	+2·0	+6·0	+6·1
+31	Ise⁻	+8·5	−2·0				
				−3·8	−1·9	−5·7	−6·1
−55	Ise⁻	+4·7	−3·9				
−34	Cl⁻	+8·8	−3·7				

The normal secretory solution containing 90 mM Cl⁻ was replaced by Cl⁻-free isethionate solution during the periods of Ise. The nutrient solution is used as reference for membrane potential, E_m. Positive sign for transfer of charge corresponds to an electron flow from nutrient to secretory side. $J_{ns}^{Cl^-}$ is the unidirectional flux of Cl⁻ from nutrient to secretory solution; $J_{ns}^{H^+}$ is the net H⁺ secretion into the secretory solution; the ΔJ's refer to the algebraic differences between two succeeding periods; and ΔI_m is the net electrical current flow through the gastric mucosa required to maintain the transmucosal potential at the indicated value.

H⁺ and Cl⁻ are the primary constituents which carry current from the tissue to the secretory solution (or in the reverse direction), substantiating a greater conductance of the secretory interface to H⁺ and Cl⁻ than to other components of the solutions. This is also consistent with Hogben's observation (1955) that Na⁺ and K⁺ conductance represented a small part of total trans-tissue conductance. It is interesting that in experiments where Cl⁻ was the major anionic component of the secretory solution (see table 3), a much larger portion of the total current could not be accounted for as the sum of the indicated flux changes. In this case Cl$^-$ flux is not a net flux and for a complete ionic balance sheet one would have to record the decrease in the unidirectional backflux of Cl$^-$ from S→N when current is applied.

I. Transient Changes associated with Voltage or Current Clamp

The electrical and flux changes mentioned above, which occur when current is passed through gastric tissue, may be considered to be steady state changes which can be measured over a relatively long time course. Within the first minutes after application of current, however, there are some rather interesting transient changes which deserve discussion. The observed transient p.d. response has been separated by Rehm (1967) into two distinct phases: a rapid and a slow phase. The former phase can be described as a capacitative response, and is analogous to what is typically observed for most bio-membranes. However, the extremely long time constant of the slow phase ($t_{1/2}=30$–60 sec) virtually precludes a simple membrane capacitance if one assumes a simple RC circuit for a biological membrane approximately 100 Å thick. There are alternative explanations for such slow transient responses (Mauro, 1961). Rehm (1967) for example has suggested that a time-variant polarization E.M.F. at some membrane locus (possibly the K$^+$ equilibrium potential at the nutrient membrane) represents the basis for the long time constant transient.

Slow transients may also be seen for the current passing through the mucosa which is necessary to change and maintain the p.d. at some pre-designated voltage. The typical voltage clamp response of isolated gastric mucosa may be seen in figure 17 during various secretory states (resting, stimulated and inhibited). There appears to be a reasonably good correlation between the magnitude of the current transient and the integrity of the H$^+$ secretory mechanism.

If instead of simply clamping the transmucosal p.d. at zero mV the clamp is applied in increments, first reducing the natural membrane potential and eventually reversing its sign, the typical current transient appears to diminish with each incremental voltage step, as does the rate of H$^+$ secretion (figure 18). Both the secretory rate and current transients can be abolished by high current flows (300–500 μA/cm^2), which further suggests an inter-relationship between H$^+$ secretion and the characteristic transient current response (Forte, unpublished observations). There is at least one treatment, however, which abolishes the long time constant transients without affecting the H$^+$ secretory rate: addition of Ba^{2+} (10^{-4}–10^{-3}M) to the nutrient solution (Jacobson and coworkers, 1965; Rehm, 1967). It is possible to reconcile these various effects, that of an apparent interdependence with H$^+$ secretion

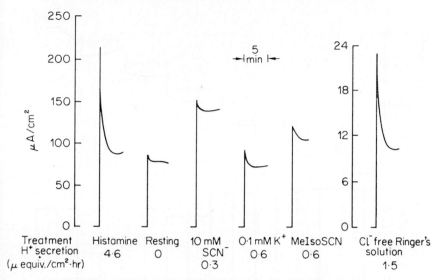

Figure 17. Short-circuit current response during the 5 minutes after reducing the p.d. to zero mV during various secretory states. Note the large 'spike' or current transient which occurs in the first 2–3 min. Under normal secreting conditions (e.g. histamine) the current 'spike' is between 100–200 per cent greater than the steady state current. The various conditions of diminished H^+ secretion were i) resting mucosa, ii) addition of 10 mM NaSCN, iii) lowering $[K^+]$ from normal value of 4 mM to 0·1mM, and iv) addition of 3 mM methyl isothiocyanate (MeIsoSCN). The tracing shown for a mucosa maintained in a Cl^--free isethionate Ringer's solution has virtually an identical pattern to that observed during normal secretory conditions, except that absolute current values are smaller since the tissue resistance is much higher. The reduction of H^+ secretion in Cl^--free solutions does not necessarily indicate direct inhibition, but is rather more likely due to the absence of a diffusible ion to accompany H^+ in the secretory process.

and the Ba^{2+} effect, within the framework of the general model for polarization E.M.F. offered by Rehm.

It was shown above that in normal chloride-Ringer's solution the nutrient surface of the mucosa is more permeable to K^+ than the secretory interface. One might thus expect that a positive current passing from secretory to nutrient surface might disturb the K^+ balance within the cells. Very simply stated, the equivalents of charge carried as K^+ across the nutrient membrane, and hence removed from the cell, would be greater than that replaced from the secretory solution. (It will also be recalled that the bulk of current flowing across the secretory membrane could be accounted for by an increased Cl^- flux into the secretory fluid and a depression of H^+ secretion). The possibility of a time-dependent change in intracellular $[K^+]$ becomes

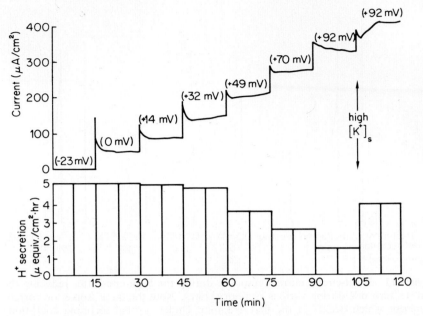

Figure 18. H$^+$ secretion and current required to maintain a series of incremental voltage clamps. The transmucosal p.d. for each 15-min interval is given in parentheses. The electrical sign refers to the secretory side, with the nutrient side being a reference. It is to be noted that the initial current transient, which occurs immediately after the voltage step, becomes more dimished and spread out in time as the secretory side becomes more positive with a concomitant decrease in H$^+$ secretion. At the point indicated by arrows the [K$^+$] on the secretory side was increased from 4 to 100mM, producing an immediate restoration of H$^+$ secretion to near the original level.

even more likely in view of the evidence showing the presence of a K$^+$ equilibrium potential at the nutrient interface (Harris and Edelman, 1964). It is well known that K$^+$ is an essential constituent for normal H$^+$ secretory rates (Gray and Adkison, 1941; Davenport, 1963; Davis and coworkers, 1965), and it may also be shown that a reduction in [K$^+$] in the bathing solutions reduces and eventually abolishes electrical transients of gastric tissue (figure 17; Davis, Rutledge, Noyes and Rehm, unpublished observations, quoted by Rehm, 1967).

From the above line of reasoning it follows that one of the major possible effects of passing current S→N is a progressive decrease in intracellular K$^+$, thus limiting the availability of this essential constituent to the secretory process and, at the same time, establishing a polarization E.M.F. at the nutrient membrane. Experimental tests of this hypothesis might involve the manipulation of conditions affecting the movement of K^+ into and out of the mucosal cells. Rehm (1967) concludes that Ba^{2+} greatly reduces the permeability of the nutrient membrane to K^+, which would be one way of reducing the polarization E.M.F. Another test would be to increase the amount of K^+ entering the cell from the secretory solution. Although the permeability of the secretory surface to K^+ is low, an increase in $[K^+]_s$ would increase the proportion of current flow into the cell which is carried by K^+, thus replacing at least a portion of that which is lost at the nutrient surface. As typified by the results shown in figure 18 for a voltage clamp with large current flow from S→N, raising $[K^+]$ in the secretory solution produced an immediate recovery of H^+ secretion, often up to 80–90 per cent of the rate observed on open circuit. It would be important to know whether K^+ ions carried into the cell by the current were directly participating in a reaction leading to H^+ secretion, or whether there was simply a forced exchange of H^+ for K^+. Rehm (1964a) has produced evidence against this latter possibility. Thus the tentative model whereby K^+ is 'electrolytically' removed from the cell and from the secretory site is consistent with Rehm's (1967) hypothesis of a polarization E.M.F. which produces a net distortion of ion concentration gradients. Feasibility of such model systems using the best estimates of gastric epithelial dimensions and resistance properties is currently being tested by Kidder and coworkers (1968).

From an analysis of electrical data various models or schematic representations of the gastric mucosal cells have been proposed. The scheme shown in figure 19 is a summary of several such existing ones and is taken largely from the model system offered by Rehm (1968). It describes much of the information coming from electrical and flux studies on bullfrog gastric mucosa.

IV. METABOLISM OF GASTRIC MUCOSA

A. Thermodynamic and Stoichiometric Efficiency of Acid Secretion

The expression of the electrical and secretory characteristics of gastric mucosa in terms of equivalent circuits or schematic analogues such as those given above can provide a useful frame of reference with which to examine the physiology of HCl secretion, but ultimately all of the convenient terms and descriptions, such as H^+ or Cl^- pumps must be explained in terms of basic biochemical events. From a study of the metabolism of gastric mucosa, and how the secretory processes relate to metabolism, a conceptual basis for this biochemical translation, as well as some interesting analogies with other transport and secretory phenomena, is available.

It was pointed out earlier in this chapter, and in more general terms earlier in this book, that the minimum thermodynamic work requirement for a

F

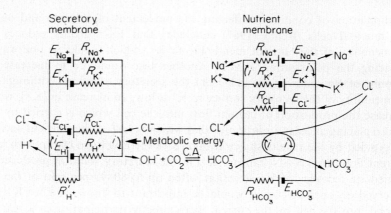

Figure 19. Schematic electrical circuit for the nutrient and secretory gastric membranes (modified from Rehm, 1969). It is assumed that R_{Na^+} and R_{K^+} at the secretory membrane are very high and hence these ions do not ordinarily contribute to the secretory membrane potential (Davenport and coworkers, 1964). Metabolically driven H^+ and Cl^- pumps operate at the secretory membrane to deliver these ions into the secretory solution against their respective electrochemical gradients (Rehm, 1950; Forte and coworkers, 1963). At the basal or nutrient membrane a Na^+ pump (perhaps Na^+/K^+ exchange) is operating to maintain high intracellular K^+ (Bornstein and coworkers, 1959; Cummins and Vaughan, 1965). Conductance of the nutrient membrane is mainly due to K^+ and Cl^-, hence their respective concentration gradients determine the nutrient membrane potential (Harris and Edelman, 1964). A Cl^-–HCO_3^- exchange mechanism of the nutrient membrane serves to promote a rapid egress of HCO_3^- from the cell (Rehm, 1967). Compare with figures 14 and 16.

transport process can be calculated from the chemical and electrical forces against which it is operating. Using the concentration and electrical gradients observed in various gastric preparations, it can be calculated from equation (1) that about 10,000 cal/g mole are required for the secretion of mammalian gastric juice (0·16 N HCl) and about 8,800 cal/g mole are required for the secretion of 0·12 N HCl by amphibian gastric mucosa. These figures clearly represent a minimum free energy requirement if the system is assumed to be 100 per cent efficient in transferring energy, and they ignore such problems as internal friction of the system, solvent drag, etc. It is likely that much more energy is expended by oxyntic cells to produce a secretory product in the range of pH 1·0; however, one could express the ratio of minimum energy required to total energy available from cellular metabolism as the thermodynamic efficiency. It is generally agreed that for frog gastric mucosa about 2 moles of HCl are produced per mole of O_2 consumed by the entire tissue (Teorell, 1949; Crane and Davies, 1951; Davenport, 1952; Forte and Davies,

1963). Taking the complete calorific yield for oxidation of glucose to CO_2 and H_2O, namely a value of 114,000 cal/mole O_2 utilized, it may be realized that the amphibian system is operating at about 15 per cent efficiency. This is clearly an oversimplification of the system since in normal glucose oxidation about half of the available energy is harvested for work processes, but it does demonstrate that ample energy is available within the system to account for HCl secretion, and a reference point is established for comparison.

In addition to the total efficiency of a secretory process, any detailed or molecular model of the system must be able to account for the observed stoichiometry. That is, for gastric secretion a predictable, quantitative relationship must exist between the number of H^+ secreted and the turnover of specific metabolic intermediates. Alternative theories have been offered whereby the stoichiometric relationships are a specific function of the electrochemical potential gradient at the secretory site (Davies, 1957b). In any case the relationship between total equivalents transported and utilization of specific metabolites provides a useful way in which to test various modes or hypotheses for HCl secretion and can be called the stoichiometric efficiency.

Acid secretion is dependent upon oxidative metabolism (Delrue, 1930; Rehm, 1946; Crane and coworkers, 1946; Davenport, 1947). Based upon this observation and the operational simplicity of the oxidation–reduction mechanism developed by Lund (1928) and Lundegardh (1939), three groups of workers independently proposed an analogous system to account for the formation of gastric acid (Robertson and Wilkins, 1948; Conway and Brady, 1948; Crane and Davies, 1948). This mechanism became popularly known as the redox pump hypothesis (Conway, 1951). The basic feature of this scheme is that hydrogen ions (H^+) are produced from substrate hydrogen atoms, with molecular oxygen acting as the final electron acceptor via the cytochrome system. If transport of any monovalent ion were dependent solely upon this process, a quantitative *upper* limit of 4·0 (the electrochemical equivalent of O_2) would exist between the number of ions transported and O_2 consumed.

B. Oxygen Consumption associated with H^+ and Cl^- Secretion

The data mentioned above, showing about 2 moles of HCl produced per mole of O_2 consumed by gastric tissue, are well within both the thermodynamic and stoichiometric limit for a redox pump, but there are some problems which must be considered for a complete analysis of the system. First, respiration measurements discussed thus far involve the entire tissue, and hence include processes and cell types other than those concerned with

HCl secretion. Secondly, direct translation of energy via a redox mechanism would preclude the utilization of the energy for other processes, for example synthesis of high energy intermediates, transport of other ionic species, etc.

The following account is a resume of the experimental results which show that a simple redox mechanism cannot account for the secretion of H^+ or Cl^- by gastric mucosa. When total equivalents of active Cl^- transport across short-circuited bullfrog gastric mucosae are compared to total moles of O_2 consumed a mean ratio of about 3·8 is found, with more than 40 per cent of the individual cases having ratios in excess of 4·0 (Forte and Davies, 1963). Thus for gastric Cl^- transport a simple redox mechanism may be eliminated even without considering some of the more difficult problems.

In the case of H^+ transport, a major problem has been to determine the portion of total tissue O_2 consumption which is directly coupled to H^+ secretion. Crane and Davies (1951) used gastric mucosa from a species of frog (*Rana temporaria*) which did not secrete acid spontaneously. They measured the increase in O_2 uptake (ΔQ_{O_2}) and the net H^+ secretion (ΔQ HCl) upon stimulation of the preparations by the addition of histamine. They found that more than two thirds of the ratios of $\Delta Q_{HCl}/\Delta Q_{O_2}$ were above 4·0. The basic assumption made in calculating these ratios was that the increased O_2 consumption can account for the whole of H^+ secretion. Davenport (1952) and Davenport and Chavré (1953) approached the problem from a different angle; they attempted to evaluate the Q_{HCl}/Q_{O_2} from the slope of the relationship between these two parameters for large populations of frog and mouse mucosae, and to estimate the 'resting respiration' from an extrapolation of the data to conditions of zero acid secretion. These authors concluded that mean ratios of Q_{HCl}/Q_{O_2} were below 3·0 for histamine-stimulated as well as unstimulated mucosae. Davies (1957a) has pointed out that many of the individual points from the data of Davenport fall above the 4·0 mark and he has used this and other arguments to make higher ratios more plausible.

In a further attempt to define the respiratory fraction directly associated with H^+ secretion, Forte and Davies (1964) measured the reversible changes in O_2 consumption induced by inhibition of the secretory process by passage of electrical current from S→N, and addition of SCN^- to the bathing solutions. As pointed out earlier H^+ secretion is reduced by an opposing electrical current. If the current were carried by H^+ (or an equivalent intermediate form) directly through the pump, then one might predict a reduction in the energetic equivalents used to drive the pump. A reversible reduction of H^+ secretion and O_2 consumption most certainly occurs when about 0·5 mA/cm² of current flows from S→N (Forte and Davies, 1964). A similar line of reasoning may be used in connection with the use of SCN^- as a specific inhibitor of the H^+ pump. From their measurements, which were carried out under CO_2-free conditions where H^+ secretion would be very low or non-existent (Davies 1957a), Kidder and coworkers, (1966) found no depression of O_2 consumption by gastric mucosa using concentrations of SCN^- up to 50 mM. However, under more physiological conditions, using a HCO_3^-/CO_2 buffer system, concentrations of 10^{-3}–10^{-2}M SCN^- reversibly depressed acid secretion and O_2 consumption with considerably smaller effects apparent in other parameters, such as Cl^- transport (Forte and Davies, 1964). For an inhibition of H^+ secretion both by an opposing current and by SCN^- an analysis of the ratios of $\Delta Q_{HCl}/\Delta Q_{O_2}$ shows values typically above 4·0 (range 5·0–15·0). It must be pointed out that the assumption underlying these interpretations was that the residual respiration after inhibition of H^+ secretion, was not associated with the secretory process, nor could it be channeled into secretory work.

It is apparent from most of the evidence cited above that a simple redox pump cannot account for HCl secretion by gastric mucosa. Other theories have been proposed whereby intermediary metabolites, such as phosphorylated compounds, may participate in the acid generating sequence in a manner that would increase the observed H^+/O_2 ratios. One of the earliest theories was that advanced by Davies and Ogston (1950). According to their Mechanism II high-energy phosphate bond energy is utilized in parallel with a redox mechanism. It is interesting that the essential elements of the Davies and Ogston scheme, operating in reverse (i.e. a H^+ gradient created by a redox mechanism generating high-energy phosphate compounds), forms the basis for one of the current theories of mitochondrial oxidative phosphorylation (Mitchell, 1961, 1967).

C. Relationships between Acid Secretion and Phosphate Metabolism

It is now worthwhile to examine the available evidence for the involvement of phosphagens, such as ATP or creatine phosphate, in the secretory process of the stomach. One of the earliest arguments for such a proposal was that uncoupling agents, such as 2,4-dinitrophenol, are well-known to inhibit HCl secretion by gastric mucosa whereas respiration is stimulated (Davies, 1951; Heinz and Durbin, 1957). It may be argued that in the case of gastric tissue the uncoupling agents have inhibitory sites other than those typically associated with oxidative phosphorylation. However, studies by Forte and coworkers (1968) of the properties of mitochondria isolated from rabbit gastric mucosa show that these organelles undergo typical phosphorylating reactions which are uncoupled by reasonable concentrations of dinitrophenol. Thus there does not appear to be a specialized system peculiar to gastric mitochondria.

In our laboratory we sought evidence for a role for ATP in the gastric HCl secretion by measuring the changes in mucosal levels of several nucleotides under varied conditions of secretion (Forte and coworkers, 1965). We found a positive correlation between the rate of acid secretion and the concentration of nucleoside triphosphate (NTP) in the tissue*. Normal, oxygenated tissue was found to contain about 1·2 μmole NTP/g wet tissue. Along with the decrease in H^+ secretory rate, anoxia decreased the concentration of NTP and increased levels of NDP, AMP and inorganic phosphate within gastric tissue; however, a small but measureable rate of acid secretion was maintained for at least sixty minutes after exposure to the O_2-free atmosphere, or more appropriately, until tissue levels of NTP fell below 0·4–0·5 μmoles/g tissue. The time course for such changes in a typical experiment is shown in figure 20. Furthermore, the restoration of H^+ secretion upon reoxygenation was only effective when levels of NTP returned to near normal values. The inclusion of glycolytic inhibitors, such as iodoacetamide and sodium fluoride, completely abolished H^+ secretion and reduced NTP to very low values (0·07 μmoles/g tissue). The general conclusion from these results was that the H^+ secretory rate was dependent upon the ATP content of the tissue and that ATP generated by anaerobic metabolism can drive the H^+ secretion machinery, albeit at a greatly reduced rate. Earlier suggestions that energy from glycolysis may be used to support H^+ secretion arose from the work of Davenport and Chavré (1950, 1952).

In addition to the general question regarding ATP as an intermediate in HCl secretion we were interested in obtaining some estimate of the stoichiometry between H^+ secreted and ATP turnover (Forte and coworkers, 1965). To determine this relationship for anoxic mucosae, the rate of lactate production was measured to provide an estimate of ATP production (assuming 1·5 moles of ATP are formed per mole of lactate formed from glycogen) and, correcting for changes in tissue NTP levels, the rate of ATP utilization for the total tissue was calculated as 2·8 μmoles/g tissue for the thirty minute period over which it was measured. When compared to the 4·2 μequiv. of H^+ secreted over the same time period a value of 1·5 was obtained for the H^+/ATP ratio. This value is clearly only an estimate of the actual ratio since several important variables could not be ascertained, such as the participation of ADP in the secretory process and the relative proportion of lactate being formed from precursors other than glycogen. (This latter problem may be of great significance since glucose was present in the bathing media.) However, it is of interest that the ratio calculated under these initial anaerobic

---

*In most of our earlier studies the assays were not specific for ATP and ADP; thus we actually measured total nucleoside tri- and diphosphate. In subsequent studies using the more specific firefly assay method we determined that in oxygenated mucosa ATP represents about 80 per cent of the total tissue nucleoside triphosphate (Forte, 1966).

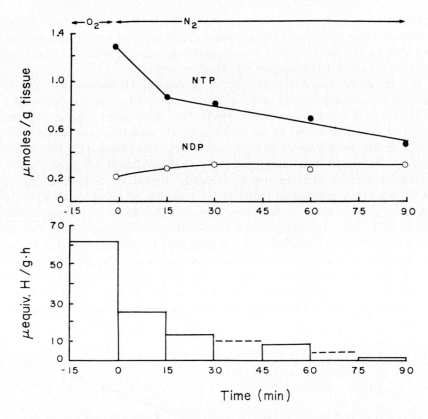

Figure 20. Change in acid secretion and nucleotide concentrations with time after switching to 95 per cent N_2 and 5 per cent CO_2. Oxygen electrode measurements showed that the system was virtually anoxic within 15–20 min after switching to the oxygen-free gas. Acid secretion was not measured during the periods indicated by the dashed line. (From Forte and coworkers, 1963).

conditions is consistent with the range of H^+/ATP ratios which may be calculated from the $\Delta Q_{HCl}/\Delta Q_{O_2}$ values (assuming P/O ratio=3) given by Crane and Davies (1951) and by Forte and Davies (1964) for mucosae where H^+ secretion was either stimulated or inhibited by appropriate reagents or conditions.

More recently, Durbin (1968) has investigated this same problem of the relationship between H^+ secretion and high-energy phosphate compounds. Along with ATP, he measured mucosal levels of phosphoryl creatine and in general his results are in agreement with the data of Forte and coworkers. Durbin reported slightly lower values of 1·04 to 1·48 for the ratio of ions transported to high energy phosphate consumed. His mucosae were maintained on short circuit and he calculated the number of ions transported as the sum of H^+ secreted plus the short-circuit current, which is really an estimate of active Cl^- transport (cf. Hogben, 1951).

It is significant that SCN^- is effective in reducing H^+ secretion by gastric mucosa without appreciably altering the concentration of various nucleotides of the tissue (Forte and coworkers, 1965). This observation suggests that the mode of action of SCN^- as an inhibitor of H^+ secretion is at some point distal to the *production* of energy metabolites. Consistent with this is the finding that respiration and oxidative phosphorylation of gastric mitochondrial preparations are not significantly altered by concentrations of SCN^- which are effective in reducing H^+ secretion by intact mucosa (Forte and coworkers, 1967a). It had been proposed that SCN^- was acting at the site of ATP utilization and to test this possibility the change in ATP level as a function of time of anaerobiosis was measured in two experimental groups of mucosae: a control group which was previously secreting H^+ at a normal rate of about 4 μequiv./cm²/hr and a group treated with 10 mM NaSCN to inhibit H^+ secretion, and presumably decrease the utilization of ATP (Forte and coworkers, 1965). The results of these experiments are shown in figure 21,

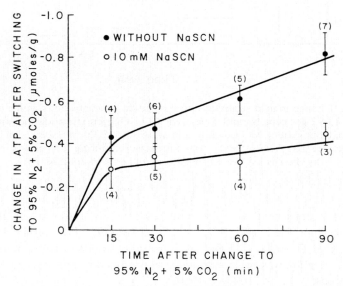

Figure 21. Change in concentration of ATP after onset of anoxia in secreting gastric mucosae and in those treated with 10mM SCN^-. At the point indicated by zero time the bathing solutions were bubbled with 95 per cent N_2 and 5 per cent CO_2 and paired mucosal halves were assayed for ATP periodically. The height of the line through each point indicates the standard error; the number of mucosae is shown in parentheses. (From Forte and coworkers, 1965).

where it may clearly be seen, in agreement with our hypothesis, that the decrement of tissue ATP (and hence rate of utilization) was considerably lower in the SCN^--treated preparations.

D. Relationships between Cytochromes and Gastric H^+ Secretion

Several interesting observations and speculations have been made which suggest a role for various cytochrome components (or other redox compounds) in the process of H^+ secretion. Kidder and coworkers (1966) analyzed the kinetics of oxidation and reduction patterns of several bullfrog gastric cytochromes as a function of H^+ secretion. They observed that the onset of H^+ secretion, which occurred upon the reintroduction of O_2 to anoxic mucosa, was associated with a shift in cytochrome c (peak at 550mμ) to a more reduced steady-state level. This, coupled with their findings that inhibition of H^+ secretion by SCN^- produced an increase in the oxidized state of cytochrome c (and in some cases cytochrome b), led them to suggest a direct role for cytochrome c in the translocation of gastric H^+. In terms of what is known for isolated mitochondria they pointed out that the SCN^- effect was not likely to be due to a depletion of tissue ADP levels, i.e. transition from respiratory state 3 to state 4. However, they did emphasize the difficulty in directly comparing results on isolated mitochondria with those on intact tissue. Variations in concentration and compartmentalization of several intermediary metabolites are a well-known feature for whole cells, and the details for intercommunication between such barriers can provide more 'bizarre' redox patterns than those typically observed with homogeneous mitochondrial fractions (Peters, 1956; Chance and Hess, 1959; Chance, 1963). Nonetheless, the general observations by Kidder and coworkers (1966) concerning the redox state of gastric cytochromes are valid and constitute an important new approach to the interpretation of H^+ secretory phenomena.

Davies (1957b, 1961) has proposed several hypothetical electron-cycle schemes which provide an asymmetric deposition of H^+ through a cytochrome system driven by ATP. He suggested that a membrane-bound cytochrome, such as cytochrome b_5, could be linked to the oxidation and reduction of a hydrogen carrier, such as ubiquinone 50, in order to produce H^+ at the low pH found in gastric juice. He cited evidence that the oxidation of reduced ubiquinone 50 by ferric ions could liberate H^+ in amounts equivalent to the production of a concentration gradient greater than 10^6 (Conrad and Davies, 1961). The findings that microsomes isolated from gastric mucosa contain abundant quantities of cytochrome b_5 (about 0.5 μmoles b_5/g microsomal protein) supports the candidacy of this material in such a reaction (Forte and coworkers, 1967). Also the evidence to be presented below, that gastric microsomes contain enzymes concerned with phosphate metabolism and appear to be derived from the SER of intact oxyntic cells, adds additional evidence to the projected arguments for an electron-cycle mechanism closely associated with phosphate-bond energy. A condensed and slightly modified version of the scheme proposed by Davies (1961) and earlier by Davies and Ogston (1950) is shown in figure 22. An equilibrating membrane-bound carrier is capable of interacting with an electron acceptor

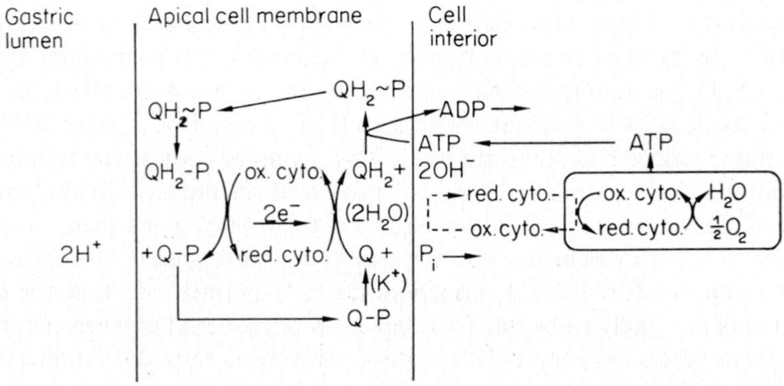

Figure 22. Scheme to account for H^+ translocation by a cyclic oxidation–reduction mechanism in which the asymmetry is provided by a phosphorylated carrier form (after Davies and Ogston, 1950, and Davies, 1961). The oxidation and reduction of a membrane bound carrier, Q, is catalysed by cytochrome components of the membrane. Transfer of phosphate from ATP to reduced carrier at the cytoplasmic interface provides the chemical gradient for net proton flow in the direction of the gastric lumen. As indicated in the scheme, an oxidized form of the carrier may also be dephosphorylated at the interior aspect of the membrane where subsequent reduction and recycling occur. Net flow of electrons is from lumen to cell interior. Cl^- transfer may either take place in close association with the protonated carrier, or as an electrical equivalent through a shunt or alternative carrier path (see figure 16). Also shown are mitochonrdial reactions which might serve to alter the general redox levels of diffusible components, which in turn, could influence membrane events. As indicated there is a stoichiometric limit of $2H^+/ATP$; however, alternative mechanisms operating on the same principles can yield even higher ratios.

(or donor). The asymmetry of the system is provided by the phosphorylation of the carrier. The sequential arrangement of a carrier-phosphate in two energy states is useful to allow both the phosphorylation and dephosphorylation to occur at the same membrane site, but it is not essential to the system. Available evidence demonstrating quinone and quinol phosphates in biological systems (with a proposed participation in oxidative phosphorylation) is consistent with the view that these redox compounds can act both as hydrogen carrier and phosphate acceptor (Vilkas and Lederer, 1962; Clark and coworkers, 1961; Brodie and Watanabe, 1966). The operation of the system as shown in figure 22 would also to some extent be regulated by (or, in turn, regulate) the total redox state of the cell. Cytoplasmic oxido-reduction components, such as cytochrome c (Kidder and coworkers, 1966), might interact with the electron or hydrogen carriers associated with proton translocation. The stoichiometry between H^+ secreted and ATP utilized may be fixed, as indicated in figure 22, or the system could easily be expanded to include a series of coupled oxidoreduction steps capable of delivering greater numbers of ions, depending on the electrochemical gradient (Davies, 1957b, 1961).

E. ATPase of Gastric Mucosal Cell Fractions

Much of the preceding evidence supports the concept that ATP is involved as an intermediate in reactions leading to the production of H^+ by gastric mucosa. It is pertinent to review some of the recent work seeking to identify and localize various enzymatic reactions which may participate in the secretory process, in particular, those reactions concerned with transfer and utilization of phosphate bond energy.

Kasbekar and Durbin (1965) reported an ATPase in the microsomal fraction from homogenates of frog gastric mucosa. This enzymatic activity did not have the characteristic $Na^+ + K^+$ stimulation and ouabain inhibition which is typical of similar cell fractions isolated from tissues which actively transport Na^+. However, the gastric microsomal ATPase activity is depressed by SCN^- and stimulated by HCO_3^-, and on the basis of their findings Kasbekar and Durbin (1965) suggested a role for this enzyme in the H^+ secretory process. Characteristics of the frog microsomal preparation have, in general, been confirmed in our laboratory for microsomes isolated from mammalian gastric tissue. On visualization by electron microscopy with negative staining techniques (as typified in figure 23) the gastric micro-somal fraction appears to be a relatively homogeneous suspension of tubular structures, either extended or turned on themselves to assume the shape of a doughnut (Forte and coworkers, 1966). The dimensions and morphological characteristics of these membranous structures are similar to elements of the SER seen in intact oxyntic cells (compare with figures 3, 6 and 9).

Inhibition of gastric ATPase by SCN^- is consistent with the evidence and hypothesis given above, that the effect of the inhibitory agent on gastric H^+ secretion would be at the site of utilization and not of production of ATP. For intact mucosa Durbin (1964) has shown that SCN^- is more effective in reducing H^+ secretion when Cl^- concentration of the medium is reduced. He demonstrated competition kinetics between SCN^- and Cl^-, and suggested

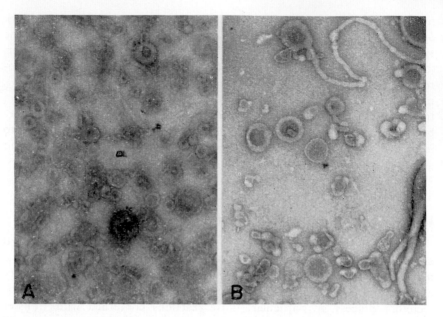

Figure 23. Electron micrograph of microsomes isolated from gastric mucosal homogenates. Preparations are negatively stained with sodium phosphotungstate. Magnification × 24,750. A: Microsomes from rabbit gastric mucosa. B: Microsomes from bullfrog gastric mucosa. (From Forte and coworkers, 1967).

that the competition is for carrier sites. Forte and coworkers (1965) studied the binding affinity of various ions to microsomes isolated from rabbit gastric mucosa and found that SCN^- was bound much more tightly than Cl^-. These results might assume real meaning if the difference in binding affinity could be ascribed to an interaction with specific carrier molecules or enzymic sites responsible for H^+ and Cl^- transport. However, kinetic analysis has shown that SCN^- inhibition of microsomal ATPase activity is not by competition with ATP (Kasbekar and Durbin, 1965). I have recently proposed a model whereby SCN^- competes with Cl^- for membrane sites and the resulting bound form of SCN^- interferes with the velocity constant of the enzyme involved with proton deposition and not the binding of ATP (Forte, 1968). The amount of SCN^- bound, and hence degree of inhibition, would be an inverse function of $[Cl^-]$ and the relative dissociation constants as shown in the following equation

$$[XSCN^-]_m = \frac{K_d^{KCl}}{K_d^{KSCN}} \frac{[SCN^-] [XCl^-]_m}{[Cl^-]} \tag{3}$$

where $[XSCN^-]_m$ and $[XCl^-]_m$ are the amounts of SCN^- and Cl^- bound to membrane 'carrier' sites, respectively; and K_d^{KSCN} and K_d^{XCl} are the respective dissociation constants for the membrane-anionic complexes. Such a relationship would suggest that Cl^- and SCN^- might appear as competitive substrates for the production of H^+ in intact gastric preparations (Durbin, 1964), and it could also be consistent with the inhibition kinetics of SCN^- for microsomal ATPase. Further support for such an interpretation is available from the observation that an increase of Cl^- concentration in the enzyme incubation medium reduced the effective inhibition of SCN^- on the ATPase of frog gastric microsomes (Kasbekar and Durbin, 1965), which is precisely what would be predicted from equation (3).

F. K+-activated Phosphatase of Gastric Microsomes

Another very interesting enzyme activity found in isolated gastric microsomes is a K^+-stimulated, Mg^{2+}-requiring phosphatase, or more appropriately a p-nitrophenyl phosphatase, since this was the measured activity (Forte and coworkers, 1967a). A comparison of the effect of K^+ on various nucleotidases and phosphatases of rabbit and bullfrog microsomes is given in table 4. Neither Na^+ nor ouabain altered the K^+-stimulated phosphatase activity of gastric microsomes, although the latter agent was extremely effective in abolishing K^+-activated phosphatase in kidney and brain

Table 4. Hydrolysis of various phosphate compounds by gastric microsomes

Substrate	0 mM K+	1 mM K+	10 mM K+
Rabbit gastric microsomes			
ATP	56·6	59·5	61·3
ADP	26·7	30·2	32·9
5′AMP	0·3	0·3	0·3
2′AMP	0	0	0
p-nitrophenyl phosphate	2·6	13·0	26·8
Glucose-6-phosphate	0·01	0·01	0·01
O-phosphoryl serine	0	0	0
Bullfrog gastric microsomes			
ATP	17·4		19·7
ADP	3·1		4·0
5′AMP	1·0		1·1
2′AMP	0·6		0·6
p-nitrophenyl phosphate	3·0	4·7	8·3
Glucose-6-phosphate	0·6		0·6

All values are given as μmoles P_i released/mg protein·hr (From Forte and coworkers, 1967a).

microsomes (Ahmed and Judah, 1964; Albers and coworkers, 1965). It is significant that the gastric ATPase also has characteristics quite distinct from ATPase of membranous fractions derived from Na^+-transporting tissues. Another interesting similarity between the gastric microsomal ATPase and the K^+-stimulated phosphatase is that both enzymic activities are reduced or abolished after treatment of the microsomes with phospholipase C (Forte and coworkers, 1967a), thus suggesting a lipoprotein nature for the active enzyme units.

Detailed kinetic analysis of phosphatase activity indicates that a K^+ complex with the phosphate ester is the active substrate for the enzymic reaction. Rb^+, NH_4^+ and Cs^+ will substitute to varying degrees for K^+ as an activator of the microsomal phosphatase. It will be recalled that the H^+ secretory process of intact mucosa is dependent upon the presence of K^+ in the bathing solution. Rb^+ is virtually indistinguishable from K^+ in restoring the electrical and secretory activity of a K^+-depleted mucosa. Cs^+ will also act as a substitute, but restoration of H^+ secretion and p.d. are somewhat slower and high concentrations are required (Forte and coworkers, 1967a). These latter findings may be of special significance in view of the parallel effects of these cations in stimulating microsomal phosphatase activity.

Admittedly p-nitrophenyl phosphate is not the normal physiological substrate for the gastric microsomal phosphatase, but it serves as a useful model for studies dealing with the general characteristics and kinetics of the enzyme reaction. The precise role of a K^+-activated phosphatase in a reaction sequence which would provide H^+ is, at present, purely speculative. It is possible that this latter enzyme activity serves a single step in sequential phosphate exchange and ion translocation reactions, as depicted in figure 22. Translocating mechanisms, involving the turnover of a phosphorylated intermediate, have been proposed for isolated membranous systems thought to be associated with Na^+ and K^+ transport (Ahmed and Judah, 1965; Post and coworkers, 1965; Fahn and coworkers, 1966). Other possible systems for H^+ translocation similar to the asymmetric ATPase of Mitchell (1961) have been suggested, namely systems whereby K^+ serves not only as an enzyme activator, but also as a locally exchangeable cation in the secretion of HCl (Forte and coworkers, 1967a).

Perhaps it is fitting to conclude this section with some recent developmental studies which offer a very interesting correlation between enzyme activities and H^+ secretory capacity. ATPase and K^+-stimulated phosphatase were assayed in this laboratory by Limlomwongse in microsomes isolated from bullfrog tadpole stomach at various metamorphic stages. Not only is the total ATPase activity increased as metamorphosis proceeds, but there is also a marked and perhaps indicative, increase in the SCN^--sensitive ATPase (table 5). In the case of phosphatase activity of gastric microsomes from the

Table 5. ATPase and phosphatase activities of microsomes isolated from tadpole stomach at various stages of metamorphosis in comparison to H[+] secretion measured for the same stages

Stage of metamorphosis[1]	H[+] secretion[2] (μequiv/cm² hr)	ATPase[3] (μmoles P_i/mg protein hr)			Phosphatase[3] (μmoles *p*-nitrophenol/mg protein . hr)		
		Control	+10mM SCN⁻	SCN⁻-sensitive ATPase	Control	+10mM K⁺	K⁺-stimulated phosphatase
XVIII	0	6·3±0·4	5·2±0·5	1·1	1·5	1·5	0
XX	0	9·5±0·5	6·6±0·2	2·9	0·8	0·8	0
XXII	0	11·8±0·2	7·5±0·3	4·3	0·8	0·8	0
XXIV	1·6±0·3	13·0±1·5	7·2±0·8	5·8	0·7	1·0	0·3
XXV	2·5±0·4	17·0±0·2	7·3±0·2	9·7	1·4	3·9	1·6
Full grown adult	3–6	34·7	9·1	25·6	1·1	5·3	4·2

[1]Metamorphic stages according to Taylor and Kollros (1946).
[2]Data are from Forte and coworkers (1969).
[3]ATPase values are given as the mean ±S.E.M. for 4–7 separate preparations (except for adult). Phosphatase data are the mean of two separate experiments. (From unpublished results of Limlomwongse and Forte)

tadpole, K^+ stimulation is not apparent until stage XXIV, a finding which accords well with the appearance of HCl secretion.

This observation of a development of enzyme activity which parallels the capacity of the stomach to secrete H^+ is also like much of the earlier data about ATPase and K^+-stimulated phosphatase of gastric microsomes only circumstantial and not direct evidence. On the other hand, no one has yet succeeded in demonstrating an HCl translocation system in cell-free gastric preparations. One is thus compelled to search for those peripheral and suggestive bits of information on the basis of which a working hypothesis can be formulated and then tested. The observed parallel effects of certain activators and inhibitors on intact preparations and isolated microsomes suggest a possible correlation between acid secretion and enzyme activity. ATPase and K^+-stimulated phosphatase appear tightly bound to microsomal membranes, and chemical treatments which tend to dissociate the protein–phospholipid interactions seem also to destroy enzyme activity. The morphological similarities between isolated microsomes and the SER of oxyntic cells are also suggestive, especially in the light of the observed membrane transformations during various secretory states (Sedar, 1965; Forte and coworkers, 1966). More recently, evidence has been produced which indicates that there is a correlation between the metamorphic appearance of HCl secretion in the bullfrog tadpole stomach and the ultrastructural development of the SER (Forte and coworkers, 1969). The concomitant expression and development of the microsomal ATPase and K^+-stimulated phosphatase thus represents additional, albeit circumstantial, evidence which implicates a direct role for these enzymes in the generation of HCl by the stomach.

Acknowledgement

Part of the work described in this Chapter was supported by a grant from the U.S. Public Health Service, Grant AM 10141.

REFERENCES

Ahmed, K. and J. D. Judah (1964) *Biochim. Biophys. Acta.*, **93**, 603
Ahmed, K. and J. D. Judah (1965) *Biochim. Biophys. Acta.*, **104**, 112
Albers, R. W., G. Rodriguez and E. de Roberts (1965) *Proc. Natl. Acad. Sci. U.S.*, **53**, 557
Brodie, A. F. and T. Watanabe (1966) *Vitamins Hormones*, **24**, 447
Canosa, C. A. and W. S. Rehm (1968) *Biophys. J.*, **8**, 415
Chance, B. (1963) *Ann. N.Y. Acad. Sci.*, **108**, 322
Chance, B. and B. Hess (1959) *J. Biol. Chem.*, **234**, 2404

Clark, V. M., D. W. Hutchinson, G. W. Kirby and A. Todd (1961) *J. Chem. Soc.*, **1961**, 715
Conway, E. J. (1951) *Science*, **113**, 270
Conway, E. J. and T. G. Brady (1948) *Nature*, **162**, 456
Conrad, H. J. and R. E. Davies (1961) Quoted by R. E. Davies in A. Kleinzeller and A. Kotzk (Eds.), *Membrane Transport and Metabolism*. Academic Press, New York p. 320
Cotlove, E. and C. A. M. Hogben (1965) *Federation Proc.*, **15**, 41
Cotlove, E., N. D. Green and C. A. M. Hogben (1959) *Federation Proc.*, **18**, 31
Crane, E. E. and R. E. Davies (1948) *Biochem J.*, **43**, xliii
Crane, E. E. and R. E. Davies (1951) *Biochem. J.*, **49**, 169
Crane, E. E., R. E. Davies and N. M. Longmuir (1946) *Biochem. J.*, **40**, xxxvi
Crane, E. E., R. E. Davies and N. M. Longmuir (1948a) *Biochem. J.*, **43**, 321
Crane, E. E., R. E. Davies and N. M. Longmuir (1948b) *Biochem. J.*, **43**, 336
Cummins, J. T. and B. E. Vaughan (1963) *Nature*, **198**, 1197
Cummins, J. T. and B. E. Vaughan (1965) *Biochim. Biophys. Acta.*, **94**, 280
Davenport, H. W. (1940) *Amer. J. Physiol.*, **129**, 505
Davenport, H. W. (1947) *Federation Proc.*, **6**, 94
Davenport, H. W. (1952) *Federation Proc.*, **11**, 715
Davenport, H. W. (1963) *Amer. J. Physiol.*, **204**, 213
Davenport, H. W. and F. Alzamora (1962) *Amer. J. Physiol.*, **202**, 711
Davenport, H. W. and V. J. Chavré (1950) *Gastroenterology*, **15**, 467
Davenport, H. W. and V. J. Chavré (1952) *Amer. J. Physiol.*, **171**, 1
Davenport, H. W. and V. J. Chavré (1953) *Amer. J. Physiol.*, **174**, 203
Davenport, H. W., H. A. Warner and C. F. Code (1964) *Gastroenterology*, **47**, 142
Davies, R. E. (1948) *Biochem. J.*, **42**, 609
Davies, R. E. (1951) *Biol. Rev.*, **26**, 87
Davies, R. E. (1957a) In Q. R. Murphy (Ed.), *Metabolic Aspects of Transport across Cell Membranes*. University of Wisconsin Press, Madison p. 277
Davies, R. E. (1957b) In Q. R. Murphy (Ed.), *Metabolic Aspects of Transport across Cell Membranes*. University of Wisconsin Press, Madison p. 244
Davies, R. E. (1961) In A. Kleinzeller and A. Kotzk (Eds.), *Membrane Transport and Metabolism*. Academic Press, New York p. 320
Davies, R. E. and A. G. Ogston (1950) *Biochem. J.*, **46**, 324
Davis, T. L., J. R. Rutledge and W. S. Rehm (1963) *Amer. J. Physiol.*, **205**, 873
Davis, T. L., J. R. Rutledge, D. C. Keesee, F. J. Bajandas and W. S. Rehm (1965) *Amer. J. Physiol.*, **209**, 146
Delrue, G. (1930) *Arch. Intern. Physiol.*, **33**, 196
Donné, A. (1834) *Ann. Chim. Phys.*, **57**, 398
Durbin, R. P. (1964) *J. Gen. Physiol.*, **47**, 735
Durbin, R. P. (1968) In *Handbook of Physiology Volume II*. American Physiological Society, Washington, D.C. p. 879
Durbin, R. P. (1968) *J. Gen. Physiol.*, **51**, 233
Durbin, R. P., S. Kitahara, K. Stahlmann and E. Heinz (1964) *Amer. J. Physiol.*, **207**, 1177
Fahn, S., G. J. Koval and R. W. Albers (1966) *J. Biol. Chem.*, **241**, 1882
Forte, G. M., J. G. Forte and R. G. Bils (1965) *Federation Proc.*, **24**, 714
Forte, G. M., L. Limlomwongse and J. G. Forte (1969) *J. Cell. Sci.* (in press)
Forte, J. G. (1966) *J. Indian Med. Profess.*, **12**, 5637
Forte, J. G. (1968) *Biochim. Biophys. Acta.*, **150**, 136
Forte, J. G. (1969) *Amer. J. Physiol.*, **216** (in press)
Forte, J. G. and R. E. Davies (1963) *Amer. J. Physiol.*, **204**, 812
Forte, J. G. and R. E. Davies (1964) *Amer. J. Physiol.*, **206**, 218
Forte, J. G. and A. H. Nauss (1963) *Amer. J. Physiol.*, **205**, 631
Forte, J. G., P. H. Adams and R. E. Davies (1963) *Nature*, **197**, 874
Forte, J. G., P. H. Adams and R. E. Davies (1965) *Biochim. Biophys. Acta.*, **104**, 25
Forte, J. G., G. M. Forte and R. F. Bils (1966) *Exp. Cell. Res.*, **42**, 662
Forte, J. G., G. M. Forte and P. Saltman (1967) *J. Cell. Physiol.*, **69**, 293
Forte, J. G., L. Limlomwongse and G. M. Forte (1967) *Federation Proc.*, **26**, 274

Forte, J. G., G. M. Forte, R. Gee and P. Saltman (1967) *Biochem. Biophys. Res. Commun.*, **28**, 215
Forte, G. M. and J. G. Forte (1969) *The Physiologist*, **12**, 228
Gray, J. S. and J. L. Adkison (1941) *Amer. J. Physiol.*, **134**, 27
Harris, J. B. and I. S. Edelman (1964) *Amer. J. Physiol.*, **206**, 769
Hayward, A. F. (1967) *J. Anat.*, **101**, 69
Heckmann, K. (1965) In P. Karlson (Ed.), *Mechanisms of Hormone Action*. Academic Press, New York p. 41
Heinz, E. and R. P. Durbin (1957) *J. Gen. Physiol.*, **41**, 101
Heinz, E. and R. P. Durbin (1958) *Federation Proc.*, **17**, 945
Heinz, E. and R. P. Durbin (1959) *Biochim. Biophys. Acta.*, **31**, 246
Hodgkin, A. L. (1951) *Biol. Rev.*, **26**, 339
Hodgkin, A. L. and P. Horowicz (1959) *J. Physiol. (London)*, **148**, 127
Hogben, C. A. M. (1951) *Proc. Natl. Acad. Sci. U.S.*, **37**, 393
Hogben, C. A. M. (1955) *Amer. J. Physiol.*, **180**, 641
Hogben, C. A. M. (1959a) *Science*, **129**, 1224
Hogben, C. A. M. (1959b) *Amer. J. Digest. Diseases*, **4**, 184
Hogben, C. A. M. (1965) *Federation Proc.*, **24**, 1353
Hogben, C. A. M. (1968) *J. Gen. Physiol.*, **51**, 240s
Hokin, L. E. and M. R. Hokin (1958) *J. Biol. Chem.*, **233**, 805
Hokin, L. E. and M. R. Hokin (1960) *J. Gen. Physiol.*, **44**, 61
Ito, S. (1961) *J. Biophys. Biochem. Cytol.*, **11**, 333
Jacobson, A., M. Schwartz and W. S. Rehm (1965) *Physiologist*, **8**, 200
Karnovsky, M. L. (1964) In R. M. C. Dawson and D. N. Rhodes (Eds.), *Metabolism and Physiological Significance of Lipids*. John Wiley, London p. 501
Kasbekar, D. K. and R. P. Durbin (1965) *Biochim. Biophys. Acta.*, **105**, 472
Kasbekar, D. K., G. M. Forte and J. G. Forte (1968) *Biochim. Biophys. Acta.*, **163**, 1
Kidder, G. U., T. L. Davis and W. S. Rehm (1968) *Biophys. J.*, **8**, A174
Kidder, G. U., P. F. Curran and W. S. Rehm (1966) *Amer. J. Physiol.*, **211**, 513
Kitahara, S. (1967) *Amer. J. Physiol.*, **213**, 819
Kitahara, S., K. Fox and C. A. M. Hogben (1967) *Abstracts of the Biophysical Society*, Houston, Texas. Abstract No. TB4
Langley, J. N. (1881) *Phil. Trans. Roy. Soc. London*, **172**, 663
Lillibridge, C. B. (1968) *J. Ultrastruct. Res.*, **23**, 243
Linde, S., T. Teorell and K. J. Obrink (1947) *Acta Physiol. Scand.*, **14**, 220
Lund, E. J. (1928) *J. Exp. Zool.*, **51**, 265
Lundegårdh, H. (1939) *Nature*, **143**, 203
Mauro, A. (1961) *Biophys. J.*, **1**, 353
Menzies, G. (1958) *Quart. J. Microscop. Sci.*, **99**, 485
Mitchell, P. (1961) *Nature*, **191**, 144
Mitchell, P. (1967) *Federation Proc.*, **26**, 1370
Moody, F. G., and R. P. Durbin (1965) *Amer. J. Physiol.*, **209**, 122
Patlak, C. S. (1957) *Bull. Math. Biophys.*, **19**, 209
Peters, R. A. (1956) *Nature*, **177**, 426
Post, R. L., A. K. Sen and A. S. Rosenthal (1965) *J. Biol. Chem.*, **240**, 1437
Rehm, W. S. (1945) *Amer. J. Physiol.*, **144**, 115
Rehm, W. S. (1946) *Amer. J. Physiol.*, **147**, 69
Rehm, W. S. (1950) *Gastroenterology*, **14**, 401
Rehm, W. S. (1953) *Amer. J. Physiol.*, **172**, 689
Rehm, W. S. (1959) *Amer. J. Digest. Diseases*, **4**, 194
Rehm, W. S. (1962a) *Amer. J. Physiol.*, **203**, 1091
Rehm, W. S. (1962b) *Amer. J. Physiol.*, **203**, 63
Rehm, W. S. (1964a) In J. Hoffman (Ed.), *The Cellular Function of Membrane Transport*. Prentice–Hall, New Jersey p. 231
Rehm, W. S. (1964b) In F. M. Snell (Ed.) *Transcellular Membrane Potentials and Ion Fluxes*. Gordon and Breach, New York p. 64

Rehm, W. S. (1967) *Federation Proc.*, **26**, 1303
Rehm, W. S. (1968) *J. Gen. Physiol.*, **51**, 250s
Rehm, W. S., H. Schlesinger and W. H. Dennis (1953) *Amer. J. Physiol.*, **175**, 473
Robertson, R. N. and M. J. Wilkins (1948) *Australian J. Sci.*, **1**, 17
Sachs, G., R. L. Shoemaker and B. I. Hirschowitz (1966) *Proc. Soc. Exp. Biol. Med.*, **123**, 47
Sanders, S. S. and W. S. Rehm (1967) *Physiologist*, **10**, 298
Sedar, A. W. (1961a) *J. Biophys. Biochem. Cytol.*, **9**, 1
Sedar, A. W. (1961b) *J. Biophys, Biochem. Cytol.*, **10**, 47
Sedar, A. W. (1965) *Federation Proc.*, **24**, 1360
Sedar, A. W. (1969a) *J. Ultrastruc.* Res., **28**, 112
Sedar, A. W. (1969b) *Anat. Record*, 1969 (in press)
Sedar, A. W. and J. G. Forte (1962) In *Fifth International Congress for Electron Microscopy*. Academic Press, New York p. YY4
Sedar, A. W. and J. G. Forte (1964) *J. Cell. Biol.*, **22**, 133
Teorell, T. (1949) *Experientia*, **5**, 409
Toner, P. G. (1963) *J. Anat.*, **97**, 575
Toner, P. G. (1965) *J. Anat.*, **99**, 389
Ussing, H. H. (1949) *Acta Physiol. Scand.*, **19**, 43
Vial, J. D. and H. Orrego (1960) *J. Biochem. Biophys. Cytol.*, **7**, 367
Vidaver, G. A. (1966) *J. Theoret. Biol.*, **10**, 301
Vilkas, M. and E. Lederer (1962) *Experientia*, **18**, 546
Villegas, L. (1962) *Biochim. Biophys. Acta.*, **64**, 359
Villegas, L, (1965) *Amer. J. Physiol.*, **208**, 380
Wright, G. H. (1962) *J. Physiol. (London)*, **163**, 281

CHAPTER 5

Ion fluxes in the cochlea

Brian M. Johnstone

Department of Physiology,
The University of Western Australia,
Nedlands, Western Australia

I. INTRODUCTION

The inner ear contains two fluid systems, separated by membranes and confined, in the higher animals, within the bony walls of the skull. The two fluids, perilymph and endolymph, fill the bony canals and associated structures. The system is often referred to as the Membranous Labyrinth.

The endolymphatic system is usually surrounded by the perilymph and consists of the fluid-filled semicircular canals, utricle, saccule, endolymphatic sac and the cochlear duct or scala media (figure 1). The ductus reunions which joins the cochlear duct to the rest, (the vestibular apparatus) is small enough in mammals to cause a severe restriction of diffusion and voltage spread, although it is quite large in birds and lower animals.

It has been known for a long time that the endolymph system is a self-contained isolated system (in contrast to the perilymph system which is connected to the cerebral spinal fluid), and recent findings indicate that the fluid composition is grossly different from most other body fluids.

There are many problems in working with this organ and a comparison of the number of papers per year on the cochlea, as compared to say the retina, reflects these difficulties. Many measurements are at best simply an order of magnitude estimate; also those quantities which can be measured

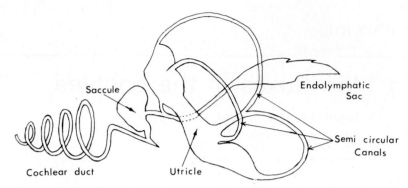

Figure 1. The membraneous labyrinth: a schematic diagram of the endolymph system of the inner ear. The cochlear duct (or scala media) is joined to the rest of the system by a duct (ductus reunions). In birds this is large while in mammals it is quite narrow and represents both a considerable electrical resistance and flow resistance.

with precision (e.g. potential measurements) may show large variability, for example a 'normal' cochlear potential can vary between $+75$ to $+95$ mV. Hence accurate and rigorous mathematical treatment is not yet warranted. In the quantitative treatment in this essay, I have been forced to use many simplifications which would not be permissible in other fields. I regard these as justified as even order of magnitude results help to put ideas in perspective.

II. ELECTROCHEMISTRY

In 1954, Smith and coworkers and later Citron and coworkers (1956) found endolymph high in potassium (144 mM) and low in sodium (16 mM) whereas perilymph was similar to cerebral spinal fluid or plasma. Further work (Johnstone and coworkers, 1963) has revised these early estimates of ionic composition to potassium equal to 150 mM and sodium equal to 2·0 mM and these figures have recently been confirmed by Bosher and Warren (1968). This system is even more unusual as the endolymph in one section, the cochlea, is at a high positive potential of $+80$ mV with respect to perilymph or plasma. The electrochemical situation of the cochlea is set out in figure 2.

Exploration of the potentials in the rest of the labyrinth showed that the high positive potential ($+$EP) in scala media is confined to the cochlea. The endolymph in the rest of the system being only a few millivolts positive. The high potassium concentration and low sodium concentration is however common to all the endolymph.

Furthermore, in reptiles, the low positive potential extends to the cochlea

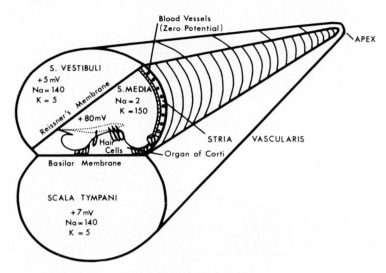

Figure 2. Electrochemistry of the cochlea. The scala vestibuli and scala tympani contain perilymph and the scala media, endolymph. The boundary of the endolymph is along the top of the Organ of Corti, rather than the basilar membrane; hence, the spaces within the Organ of Corti are bathed in perilymph.

although the potassium and sodium compositions are similar to mammals (table 1). Hence, a common denominator in the hearing organ of all animals is the high [K], low [Na] endolymph (compared with plasma concentration) whilst the high +EP is a particular adaptation of the cochlea of mammals.

The +EP is extremely sensitive to oxygen lack, falling to zero in one minute and indeed progressing negatively, sometimes by as much as -50 mV

Table 1. Endolymph concentrations and potentials for various animals
(approximate values)

Species	Structure	K^+ (mM/l)	Na^+ (mM/l)	Potential (mV)
Guinea-pig	Cochlea	150	1·5	+80
Guinea-pig	Utriculus	144	15	+ 5
Turtle	Cochlea	114	2·7	+ 6
Lizard	Cochlea	140	10	+ 9
Elasmobranch	Sacculus	64	273	− 3
Teleost	Sacculus	73	104	0

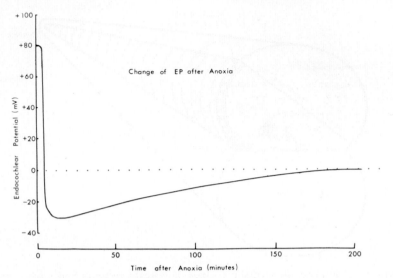

Figure 3. Time course of change of EP after anoxia. On occasions, the EP can become as negative as −50 mV.

some five minutes after blood flow has stopped. This − EP then declines back towards zero over the next two to three hours (figure 3).

The origin of the EP (both positive and negative) and the manner in which the ionic composition is regulated, has been the topic of many papers over the past ten years; however, only sporadic progress has been made.

Many writers ignore completely the concepts of equilibrium potentials and passive ionic distributions and prefer to imagine the endolymph is secreted from some gland as a fluid of the found composition (longitudinal flow theory, Dohlman, 1964). Others take the attitude that the endolymph is best considered as an enclosed fluid system whose ionic composition is regulated by ion leaks from surrounding cells in conjuction with specific active transport systems pumping ions between endolymph and plasma (Naftalin and Harrison, 1958). Only this latter concept will be considered here, as the evidence for it at the moment is overwhelming.

In order to determine which ions, if any, may be passively distributed, it is necessary to compare their equilibrium potentials with the +EP. However, a major difficulty is the accuracy of analysis of endolymph. It should be emphasized that the sample volumes available are extremely small, and the endolymph is mostly surrounded by bone and the high [Na], low [K] of perilymph. There are many problems in obtaining and analysing samples of 10 to 100 nl uncontaminated by outside sodium, and a partial list of sodium analysis obtained by various authors is shown in table 2.

Table 2.

	Smith, Lowry and Wu (1954)	Citron, Exely and Hallpike (1956)	Johnstone, Schmidt and Johnstone (1963)	Rodgers and Chou (1966)	Davies (1966)	Bosher and Warren (1968)
Endolymph sodium (mM/l)	16	26	1·4	79	25	0·91

For a variety of reasons (discussed by Bosher and Warren) the analysis of Johnstone and coworkers and Bosher and Warren must be considered as closest to the true value. Very recent work in our laboratory using sodium-selective microelectrodes (Johnstone and Sellick, unpublished) have confirmed endolymph [Na] to be below 2 mM/l.

The relationship between the ions and the equilibrium potentials is set out in table 3. This table indicates the importance of accurate analysis of endolymph sodium. Many workers' results would indicate an inward active transport of sodium (e.g. for an endolymph sodium concentration of 25 mM/l, the equilibrium potential is equal to 45 mV); whereas, there would appear to be in fact a small outward active transport. The table reveals that all the ions must be pumped: K^+ inwards, Cl^- outwards, Na^+ outwards and probably H^+ inwards.

III. ACTIVE FLUX ESTIMATIONS

The measurement of fluxes is necessary for a full understanding of the system. Ideally, unidirectional fluxes should be measured under short circuit conditions. Such techniques applied to the cochlea have not been developed yet and indeed the enormous difficulties in any isotope experiment on this organ have precluded more than preliminary attempts at the problem.

The basic difficulty is that one animal provides only one point on a rate

Table 3. Equilibrium potentials of various ions—to be compared to the endocochlear potential of $+80$ mV

Ion	Endolymph concentration (mM/l)	Perilymph concentration (mM/l)	Equilibrium potential with respect to perilymph or plasma (mV)
Na^+	1	140	$+128$
K^+	154	7	-80
Cl^-	110	120	-6
H^+	7·8 (pH)	7·5 (pH)	$+18$

curve and contamination of the endolymph is highly probable. The most successful attempts on these problems have been by Rauch and coworkers (1963, 1966). They injected ^{24}Na and ^{42}K into perilymph and sampled the endolymph some time later. Instead of using a micropuncture technique, they froze the cochlea in liquid nitrogen, dissected it out and then allowed it to warm up slowly. The perilymph and endolymph melted at different temperatures, and hence the two fluids could be collected separately. The technique is certainly interesting, but the results show that considerable contamination can occur. For instance, endolymph sodium was reported to be more than 20 mM/l and recovery experiments of ^{42}K injected into perilymph show specific activity measurements of up to 160 per cent. None the less, some valuable information has been gained by these workers. Their results indicate that Reissner's membrane seems highly permeable to sodium and potassium (with a half time of one minute) and this permeability falls during anoxia.

Choo and Tabowitz (1964, 1965) injected ^{22}Na and ^{42}K into blood of cats and sampled endolymph by micropuncture at times up to forty-eight hours later. They report a slow rise in ^{42}K activity with time, reaching a peak some forty-eight hours after injection. Unfortunately, their ^{22}Na results are so contaminated with perilymph (they quote their sampled fluid as having more than 100 mM/l Na) as to be useless. Thus the few studies to date indicate a rapid transfer of K$^+$ across Reissner's membrane and a slow exchange across the stria vascularis. These results appear to be at variance with non-isotopic and conductance studies which will be discussed later.

Another way of estimating flux, or more correctly permeability, is from membrane conductance data. There are three anatomical boundaries to the scala media, so it is necessary to have at least three separate values of membrane resistance. The equivalent circuit of a cochlear cross-section has been proposed (figure 4). Using some simplifying assumptions, principally that the length constant in scala media is 2 mm and that of scala tympani is 5 mm, Johnstone and coworkers (1966) derived figures for three lumped boundary resistances; Reissner's membrane 46 kΩ, basilar membrane–organ of Corti complex 24 kΩ and stria vascularis 13 kΩ (figure 4). All these values refer to 2 mm length of cochlea.

From histological data, the areas of the appropriate length of Reissner's membrane and stria vascularis can be found. Using these together with the resistances found, a specific resistivity of 368 Ω cm^2 for Reissner's membrane and 48 Ω cm^2 for the stria vascularis may be calculated. Hence, the stria vascularis appears more permeable than Reissner's membrane. These figures of course tell us nothing about the specific ionic conductance, although the high potassium concentration in scala media might suggest that most of the current is carried by K$^+$.

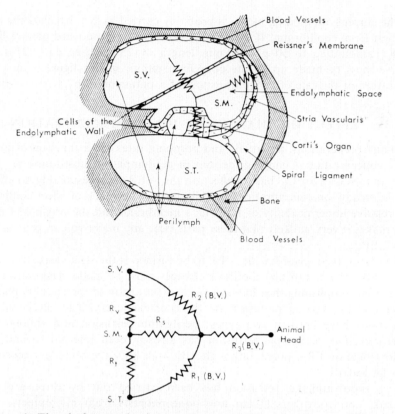

Figure 4. Electrical network of the cochlea. R_v is the resistance from scala media to scala vestibuli and corresponds to Reissner's membrane. R_t is the resistance from scala media to scala tympani and corresponds to the resistance of the Organ of Corti and other cells resting on the basilar membrane. R_s is the resistance through the stria vascularis from scala media to the blood vessels in the spiral ligament. R_1 and R_2 represent the resistances from scala tympani and scala vestibuli to the blood vessels surrounding them within the spiral ligament. R_3 represents the resistance of the cochlear blood vessels to the rest of the animal. These resistances are roughly 1/10th to 1/50th of the membrane boundary resistances.

It is possible to use such data to calculate a minimum total net ionic flux from the scala media. This calculation rests on the assumption that the stria vascularis is the source of the $+EP$ and the major ion leaks occur across Reissner's membrane and the basilar membrane, and back leaks occur around the stria vascularis. The $+EP$ will drive a current through the membrane resistors (figure 4). The current works out to be about $5 \mu A/mm$ length. This is equivalent to $2 \cdot 5 \text{ mA/cm}^2$ of stria vascularis.

The current out of scala media is probably carried by K^+, which is the ion farthest from equilibrium. If this is so, then there is an outward passive flux of K^+ amounting to 50 pM/sec/mm length or 25 mM/sec cm^2. This of course must be made up by the potassium pump, so this figure is also an indication of the rate of active transport of a potassium ion.

IV. EFFECT OF CHANGES OF IONIC CONCENTRATION

In many preparations, potentials and pumping rates are a function of local ionic concentration. Changing the ions in perilymph by substituting K^+ for Na^+ and (SO_4^{2-}) for Cl^- has no effect on the $+EP$ and presumably no effect on the steady ion concentrations in scala media. This lack of effect confirms the relative impermeability of Reissner's membrane and the organ of Corti and makes it very unlikely that these parts bear any major role in generating the $+EP$.

The most likely place for the $+EP$ to be formed is the stria vascularis, and it is across this organ that the ionic changes must be made. There are two ways of accomplishing this, by intracochlear perfusion of the endolymphatic space and by vascular perfusion via the cochlear artery. The intracochlear perfusions have been of two types: perilymph perfusion and endolymph perfusion. The perilymph perfusion has already been referred to and no major effect on EP is noted, although with high KCl, the boundary cells are depolarized.

The endolymphatic perfusion has been carried out by three groups. Katsuki and coworkers (1966) used iontophoresis from micropipettes at currents of about 10^{-6} A. They reported no change when K^+ was injected or when tetrodotoxin was introduced, but a fifty per cent fall in $+EP$ when tetraethylammonium chloride was injected. Tanaka (1963) and later Konishi and coworkers (1966, 1968) perfused by microinjection through two pipettes placed into scala media. This method leads to the complete replacement of endolymph and represents a big advance in cochlea study techniques. Konishi and coworkers perfused with tetrodotoxin and procaine and confirmed Katsuki's finding of lack of effect of tetrodotoxin but found a small (10 per cent) increase of $+EP$ when 0·1 per cent procaine was perfused. They also perfused scala media with a variety of ionic solutions. Perfusing with isotonic KCl caused little change in the $+EP$. In contrast, perfusion with isotonic NaCl caused a gradual decline in $+EP$ which however was not reversible upon reperfusing with KCl.

Vascular perfusion via the cochlear artery has been carried out in cats by Costa and coworkers (1966) and in guinea-pigs by Morizono and Johnstone (1968). Systematic variation in haematocrit of the artificial blood showed that the $+EP$ was sensitive to oxygen supply and needed 15 per cent haematocrit to sustain it (Morizono and Johnstone, 1968). The +EP was only moderately sensitive to ouabain (10^{-4} M) but highly sensitive to dinitrophenol (10^{-6} M) (figure 5). Recently we have perfused with solutions of

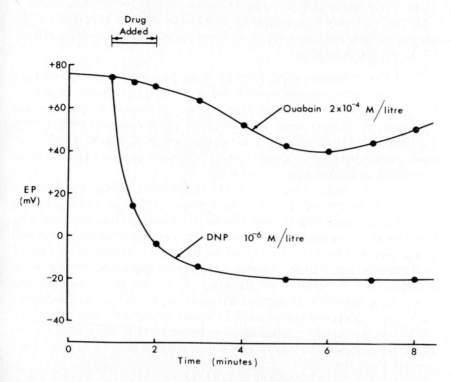

Figure 5. Change of EP when drugs were added to the perfusate of an isolated perfused head–cochlea preparation. A 1 minute bolus of the appropriate concentration of drug was used in both cases.

varying ionic content: perfusion with sodium-free (sucrose or choline substituted) artificial blood had no effect on the +EP, nor did variations of chloride (substituted by sulphate). Raising [K] to more than 15 mM proved difficult as the arteries tended to go into contracture, so shutting off the blood but increasing [K] to 15 mM had no obvious effect. Reducing [K] to zero also had no effect; however, analysis of the perfusate after passing through the animal showed a potassium content of 0·5 mM/l, presumably from extracellular and intracellular leakage. Thus true-zero [K] solutions were not in fact perfused. This problem has not been solved to date, so the effect of reducing the blood [K] to zero cannot be effectively tested.

The results of these experiments imply the following:

i) The $+$EP is not a sodium diffusion potential.

ii) The $+$EP is not generated by an outwardly directed electrogenic sodium pump and is unlikely to be an outwardly directed chloride pump.

iii) The $+$EP may be generated by an inwardly directed K pump; and the only way to fully test this would be to lower the plasma K to zero, a difficult task as already pointed out.

The lack of immediate effect on $+$EP by perfusing scala media with high sodium, low potassium solutions would tend to argue against an inward potassium pump, as one would expect such a procedure to stimulate the pump and thus even increase $+$EP; whereas, $+$EP decreased after a short time. However, this latter effect was not reversible and may be due to other causes; also it may not be essential for a potassium pump to be stimulated by working into a low potassium solution.

A further possibility is for the $+$EP to be generated directly by some sort of redox mechanism as postulated by Conway (1955), in which pumping rates, etc. are controlled by some substrate concentration. An interesting development of this theory is the proton pump envisaged by Mitchell (1966) for the mitochondrion. In his scheme the potential is generated directly by the electron transport chain, the proton (H^+) being ejected on one side of the membrane and the electron on the other. These last two schemes would show a great sensitivity to electron transport inhibitors such as 2,4-dinitrophenol and cyanide and indeed the $+$EP is very sensitive to these substances.

Whatever the specific mechanism of $+$EP generation, there must also be specific ion pumps which may or may not be coupled with it and each other.

V. PUMPING SYSTEMS

Some possible types of systems may now be considered:

i) A neutral sodium–potassium pump with one potassium being pumped in and one sodium out. Such pumps appear to be the principal ionic regulators in many cellular systems such as red blood cells and muscle.

ii) Chloride may be pumped out together with sodium in another neutral pump. Under these conditions (all pumps neutral), one would expect the $+$EP and ionic relationships to be described by the Goldman constant field equation. However, as we have shown (perfusion experiments), this is not so and therefore we may expect one of the pumps to be electrogenic.

iii) The most likely possibility here is for the potassium to be an electrogenic and inwardly-directed pump.

iv) Sodium and chloride may be pumped out as some neutral complex.

This scheme would account for the lack of effect on EP of alterations in sodium and chloride, both in scala media (Konishi and coworkers, 1966, 1968) or in plasma (Morizono and Johnstone, 1968). The net inward flux of potassium and positive charge, being balanced by a net outward leakage of K^+.

Many sodium and potassium pumps have been shown to rely on ATP, and indeed, ATPase appears to be a component of many membrane systems. ATPase has been shown to be high in the stria vascularis and furthermore, Matschinsky and Thalmann (1967) have shown histochemically that the ATPase content of the stria vascularis, but not of the organ of Corti, falls precipitously within thirty seconds of ischaemia.

These ATPases, however, may not be specific and their relation to potassium ions and pumping is obscure at the present time. In addition Morizono and coworkers (1968) report that the concentration of ouabain to just affect $+EP$ is about 2×10^{-4} M (figure 5). This is higher than is usually considered to be a specific blocking function and is approaching concentrations suggesting general metabolic inhibition. A coupled or common mechanism of transport of sodium and chloride has been identified in the gall bladder (Diamond, 1964), and it appears to be uninfluenced by electrical gradients and so is a neutral system. Unfortunately, further information about this interesting scheme is lacking, so any further discussion of this parallelism must await further developments.

The locus of the $+EP$ is almost certainly the stria vascularis, so it is also probably the site of the various pumps. Some reasons for this have already been discussed and other arguments appear to be:

i) A rich supply of blood vessels to this organ, by far the richest to any part of the cochlea.

ii) The marginal cells resemble secreting cells from other tissues, being flat-faced on one side with long finger processes surrounding blood vessels on the other side and with the cytoplasm rich in mitochondria.

iii) Extreme sensitivity to blood flow and oxygen tension.

iv) The presence of ATPase which is very oxygen sensitive.

v) The lack of effect of ions, etc. on the potential when perfusing scala vestibulae, i.e. the lack of effect of ion changes on the outside of the only other candidate, namely Reissner's membrane.

Considering (v) it is to be noted that a similar lack of effect is found when perfusing via the blood vessels and hence perfusing stria vascularis. However, the perfusion fluid in this case cannot be controlled as completely as can a scala vestibulae perfusion, and some effect may become evident when techniques of reducing the potassium concentration to very low values are worked out.

VI. PASSIVE FLUX ESTIMATION

The measurement of the active fluxes have not been very successful to date, but the estimation of passive fluxes have met with some success.

Passive fluxes are usually studied after inhibiting pump action. Anoxia is the usual method of producing this in the cochlea and the results should be applied with caution to the normal cochlea as changes in membrane permeability may have occurred.

The +EP is very sensitive to anoxia as illustrated in figure 3. Several authors have followed the changes in ionic composition of endolymph during prolonged anoxia. Most have used a sampling technique necessitating sacrificing animals at various times after anoxia, each animal providing only one value. Many animals were studied in this way by D. Garfield Davies (1966). Unfortunately his results show considerable scatter and evidence of contamination (lowest [Na] = 25 mM/l), but in general, a roughly linear increase of endolymph sodium amounting to 70 mM/l after one hundred minutes. The potassium results may be more reliable, and show a fall of potassium down to 120 mM. The endolymph sodium increased by approximately 40 mM/l and the potassium decreased by approximately 50 mM/l after one hundred minutes of anoxia. Bosher and Warren (1968) determined endolymph sodium and potassium at various times after anoxia (one animal for each interval) and their results are shown in figure 6.

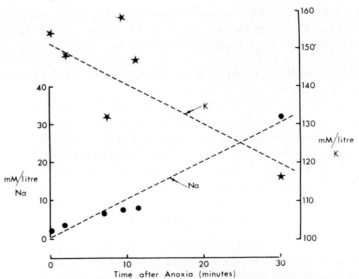

Figure 6. Change of endolymph Na(● ●) and K(★★★) after anoxia. Plotted from the data of Bosher and Warren. The Na line was fitted by eye and has a slope of 1 mM/l/min. The K line was drawn with the same (inverted) slope and starting from the mean value for normal endolymph K of 148 mM/l.

There was little discernible change in potassium for the first ten minutes, probably because the change was too slight in association with the relatively high potassium concentration, but a fall of potassium by about 20 to 30 mM was noted at thirty minutes. These results (K : Na ratio = 3·5 at thirty minutes) are in agreement with the earlier measurements of Johnstone (1965) who found a K : Na ratio of about 4 after thirty minutes of anoxia. However, to relate ion changes and potential changes requires a more exact determination.

Johnstone (1965) attempted a more exact time course estimation when he measured the conductance changes in the endolymph during anoxia. He found virtually no change for the first ten minutes and then a roughly exponential fall in conductance as the KCl ions were gradually replaced by NaCl. Whilst providing a good index of time changes of K^+, the technique is not suitable to the calculation of absolute changes in concentration. However, the time change of conductance closely followed the $-EP$ as it declined back to zero over some two to three hours, and the potassium had fallen to a half after eighty to ninety minutes in good agreement with the directly measured ratio. There was no correlation of conductance and $-EP$ for the first five to ten minutes after the EP first became negative (during this time $-EP$ is in fact increasing), and this suggests that there is some slight positive electrogenic mechanism still operating during this time. The good correlation of conductance change and $-EP$ at later times strongly suggests that the $-EP$ is a diffusion potential, primarily due to K^+.

Recently in this department we have adapted sodium-sensitive microelectrodes to the measurement of changes of endolymph sodium during anoxia. These electrodes gave an excellent measurement of the time course of change of sodium and one such experiment is shown in figure 7. Unfortunately, drift problems have prevented long time recordings, but the short (ten minute) recordings enable an accurate measurement of the net sodium influx to be made.

If the sodium influx is passive, then the change in sodium concentration should be given by the time integral of EP, at least for early times until the sodium concentration has risen appreciably. The time integral of EP is plotted in figure 7 and shows an excellent comparison with the change of sodium.

If we assume that the sodium permeability of the membranes bounding the scala media has not changed during the first few minutes of anoxia, then it is possible to derive the passive sodium flux into the endolymph.

The rise in sodium concentration is approximately 1·5 mM/l per minute. The scala media contains about 50 μl/mm length, so we calculate a sodium influx of 75 pM/min/mm length. Taking $-EP$ as -30 mV and E_{Na} at an average of 100 mV during the first few minutes of anoxia, then the driving

G

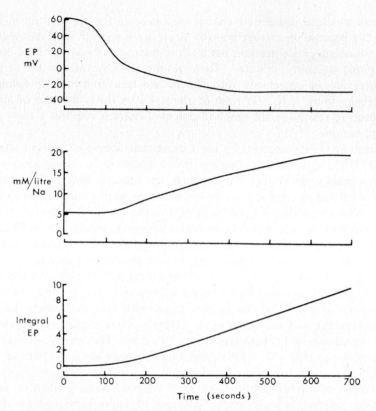

Figure 7. Change of EP, endolymph Na, and integral EP after anoxia. The endolymph Na was measured with a Na-sensitive microelectrode. Times greater than
600 seconds are unreliable due to drift in the electrode system.

force is represented by 130 mV directed inwards. Hence, the influx is approximately 0·6 pM/min/mm length/mV electrochemical driving force.

Dividing by the concentration of sodium in the side from which the
efflux is occurring gives a type of conductance factor (but not a true one)
and this is equal to ·0043 pM/min/mV of driving force/mM of sodium, all
referred to 1 mm length of cochlea.

Bosher and Warren point out that in the normal animal the sodium
equilibrium potential is some 40 mV above +EP. Under these circumstances,
and assuming endolymph is equal to 1 mM/l, a passive efflux of ·0043 ×40 ×1
= ·17 pM/min/mm length would be expected. This is about 1/500 of the
influx during anoxia and so after a short period of anoxia, some sodium
should enter; upon recovery from anoxia, this sodium would take 500 times
longer to passively efflux out. Figure 8 shows the result obtained with a

sodium microelectrode after anoxia. Note the increase of sodium which then remains constant after recovery of EP to normal. This is exactly in accord with the previous calculation and furthermore suggests that the sodium pump (outwards) is very slow. We may apply a similar calculation to endolymph potassium, provided we assume that as sodium increases so potassium decreases, a reasonable assumption, particularly in view of Bosher and Warren's figures of approximately 30 mM sodium increase after thirty minutes anoxia and about 30 mM potassium decrease at the same time.

That is, we may assume a potassium efflux of 75 pM/min/mm length. After a few minutes anoxia, taking EP as − 30 mV and the potassium equilibrium potential as −70, the driving force is 40 mV outwards. Taking the endolymph potassium as 140 mM, then the apparent conductance factor = ·02 pM/min/mV driving force per mM of potassium, i.e. about four times the sodium conductance.

In the normal animal the potassium outwards driving force is 160 mV; assuming the permeability does not change with anoxia, the outward passive flux would be 300 pM/min/mm length.

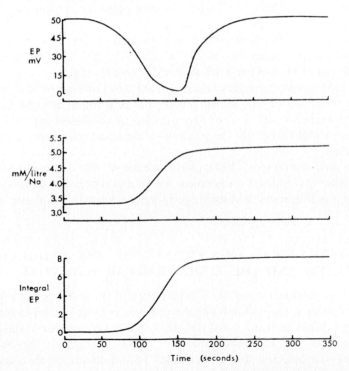

Figure 8. Change of EP, endolymph Na, and integral EP during short term anoxia and subsequent recovery. Recovery was initiated as soon as EP reached zero.

Previously we had calculated from membrane resistance data that the efflux of potassium (the current caused to flow by $+EP$) corresponded to 3000 pM/min (50 pM/sec). This order of magnitude discrepancy strongly indicates that the permeability of the membranes did change with anoxia, and so our efflux and permeabilities are valid for the recently anoxic cochlea and probably valid for sodium in the normal cochlea. It is likely that potassium permeability undergoes drastic changes during the initial stages of anoxia, and indeed we have measured resistance changes in the boundary walls during this time (Johnstone and coworkers, 1966).

With an estimate for the potassium efflux and consequently the pumping requirements for potassium into scala media, it is pertinent to inquire into the energy demands of this system. A very simple way, which will give a lower limit of requirement, is to calculate the power dissipation caused by the potassium efflux and maintenance of $+EP$. There is a potassium efflux corresponding to 5 μA, and this is effectively driven through the membrane resistances (figure 4) with a corresponding power dissipation. This gives a value of about 0·4 μwatt/mm of cochlea, corresponding to an energy expenditure of 6 μcal/min/mm length and the pump must supply at least this amount.

Assuming that five hundred calories are liberated per litre of oxygen, then the minimal pump requirements would be about 1×10^{-3} μl of O_2/min/mm or $1·5 \times 10^{-2}$ μl of O_2/min for a whole cochlea (guinea-pig).

Recently Morizono and coworkers (1968) have measured the blood volume in the inside of the guinea pig cochlea and found it to be 0·2 μl. This would contain 4×10^{-2} μl of O_2; therefore if the blood supply to the cochlea were suddenly cut off, the pump would exhaust the oxygen in under three minutes.

Perlman and coworkers (1959) have measured the various cochlear potentials after occlusion of the cochlear artery and reported a survival time of one and a half minutes, a value in good agreement with the above calculation.

VII.   PHYSIOLOGICAL SIGNIFICANCE OF THE ENDOLYMPH SYSTEM AND THE ENDOCOCHLEAR POTENTIAL

Whilst the endocochlear potential is highly dependent on the ion pumps of the stria vascularis, the endolymph composition is largely independent of the potential. It has been suggested (Davis, 1961) that movement of the hairs of the hair cells due to sound vibration, causes changes in the current through the hair cell, and this in turn leads to stimulation of the auditory nerve. For there to be a standing current through the hair cell, there must be an asymmetry in the system. The top of the hair is bathed in high potassium-concentration endolymph, whereas, its base is in perilymph, i.e. one may consider that the top of the hair cell is 'depolarized' and so a standing current will flow through this region. Another way of looking at it is to simply consider that the high potassium-concentration endolymph is separated from the low potassium-concentration perilymph by a potassium-permeable hair cell, and then the current is simply due to the thermodynamic potential difference of endolymph and perilymph. Von Bekesy (1951) has shown that the hair cell needs an external source of energy for the generation of the microphonic potentials (receptor potentials), and the high potassium-concentration of endolymph would seem to provide this. The development of $+$EP would be in such a direction as to add to this pool of energy and so increase the sensitivity and possibly the linearity of the system.

Some diseases and some drugs apparently affect $+$EP and the endolymph pumps and so cause deafness. Ménière's disease may affect the pumps because it is characterized by a rise in endolymph pressure, a sign of osmotic imbalance probably brought about by some failure of active transport. Both salicylate and quinine appear to cause temporary deafness by interfering with the $+$EP or the endolymph pumps, in this case without osmotic disturbances.

It is possible that these drugs may prove useful in elucidating the active transport mechanism in the cochlea.

Acknowledgement

This investigation was supported in part by grants from the A.R.G.C. and M.S.R.G. of the University of Western Australia. The author thanks Mr. P. Sellick for supplying the data for figures 7 and 8. They are from an honours thesis by Mr. Sellick and a paper on this topic will be published shortly.

REFERENCES

Bekesy, G. Von (1951) *J. Acoust. Soc. Amer.*, **23**, 576
Bosher, S. K. and R. L. Warren (1968) *Proc. Roy. Soc. (London), Ser. B.*, **171**, 227
Choo, Y. B. and D. Tabowitz (1964) *Ann. Otol. Rhinol. Laryngol.*, **73**, 92
Choo, Y. B. and D. Tabowitz (1965) *Ann. Otol. Rhinol. Laryngol.*, **74**, 140
Citron, L., D. Exley and L. S. Hallpike (1956) *Brit. Med. Bull.*, **12**, 101
Conway, E. J. (1955) *Intern. Rev. Cytol.*, **4**, 377
Costa, O. A., R. Thalmann and W. P. Covell (1966) *Laryngoscope*, **76**, 1874
Davies, D. G. (1966) *J. Laryngol. Otol.*, **82**/4, 301
Davis, H. (1961) *Physiol. Rev.*, **41**, 390
Dohlman, G. F. (1964) *Ann. Otol. Rhinol. Laryngol.*, **73**, 708
Johnstone, B. M. (1965) *Acta Oto. Laryngol.*, **60**, 113
Johnstone, C. G., R. S. Schmidt and B. M. Johnstone (1963) *Comp. Biochem. Physiol.*, **9**, 335
Johnstone, B. M., J. R. Johnstone and I. D. Pugsley (1966) *J. Acoust. Soc. Amer.*, **40**, 398
Katsuki, Y., K. Yanagisawa and J. Kanzai (1966) *Science*, **157**, 1544
Konishi, T., E. Kelsey and G. T. Singleton (1966) *Acta Oto-Laryngol.*, **62**/4–5, 393

Konishi, T. and E. Kelsey (1968) *J. Acoust. Soc. Amer.*, **43**, 471
Matschinsky, F. M. and R. Thalmann (1967) *Ann. Otol. Rhinol. Laryngol.*, **76**/3, 638
Mitchell, P. (1966) *Biol. Rev.*, **41**, 445
Morizono, T. and B. M. Johnstone (1968) *J. Oto-Laryngol. Soc. Australia*, **2**/3, 34
Morizono, T., B. M. Johnstone and I. Kaldor (1968) *Otologia Fukuoka*, **14**, 82
Naftalin, L. and M. S. Harrison (1958) *J. Larygol. Otol.*, **72**, 118
Perlman, H. B., R. Kimura and C. Fernandez (1959) *Laryngoscope*, **69**, 591
Ruach, S. (1966) *J. Laryngol. Otol.*, **80**/11, 1144
Rauch, S., A. Kostlin, E. A. Schneider and K. Schindler (1963) *Laryngoscope*, **73**, 135
Rodgers, K. and J. T-Y. Chou (1966) *J. Laryngol. Otol.*, **80**, 778
Smith, C. A., O. H. Lowry and M. C. Wu (1956) *Laryngoscope*, **64**, 141
Tanaka, Y. (1963) *Japan J. Otol. (Tokyo)*, **66**, 999

CHAPTER 6

Ion movement in ciliary processes

V. Everett Kinsey

Institute of Biological Sciences,
Oakland University,
Rochester, Michigan 48063, U.S.A.

I. INTRODUCTION

Movement of ions between blood vessels of the eye and intraocular fluids takes place across membranes which offer varying degrees of resistance from ion to ion and from substance to substance, and which collectively constitute the blood aqueous barriers (Kinsey, 1960). The main areas of resistance are: ciliary epithelium; capillary walls and epithelium of the iris; and the external limiting membrane and Bruch's membrane of the retina. Although active transport may occur across all these barriers, cell-mediated influx of ions across the ciliary epithelium, accompanied by osmotic movement of water into the posterior chamber, constitutes the prime mover in the formation of aqueous humor. It is, therefore, a subject of special interest to the physiologist and clinician alike.

The ciliary epithelium consists of a double layer and is thus unique as a

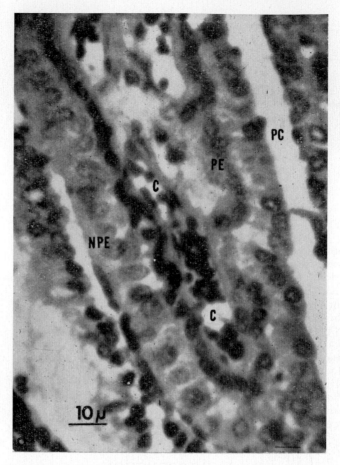

Figure 1. Light micrograph of a ciliary process, showing non-pigmented epithelium (NPE), capillary (C), pigmented epithelium (PE), and posterior chamber (PC). Magnification × 750. (From Holmberg, 1964; Courtesy of the author and publishers).

functioning unit in active transport. The layer closest to the highly vascularized stroma of the process (figure 1) consists of pigmented cells and is a continuation of the pigmented layer of the retina. The inner layer, next to the posterior chamber, is an extension of the whole sensory and neuronal part of the retina, and consists of non-pigmented cells characterized by numerous and extensive cell infoldings (figure 2). Cells nearest the ora serata are cylindrical and have elongated nuclei, and those over the crests of ciliary processes are smaller and cuboidal. Because of their embryonic origin, the apical surfaces of cells from both layers face each other. Overlying the epithelium is the internal limiting membrane, a close-fitting fibrillar structure separate from the cell membranes. Like the analogous basement membrane on the other side of the epithelium, it offers little resistance to passage of even large molecules.

Ion movement across ciliary epithelial cells occurs in both directions by passive diffusion, and unidirectionally by flow from stroma to posterior chamber, partly because of ultrafiltration and partly because of active transport of one or more ions. The multicompartmental nature of the eye complicates the task of determining the relative contribution of each of these processes to ion movement, and the sensitivity of blood aqueous

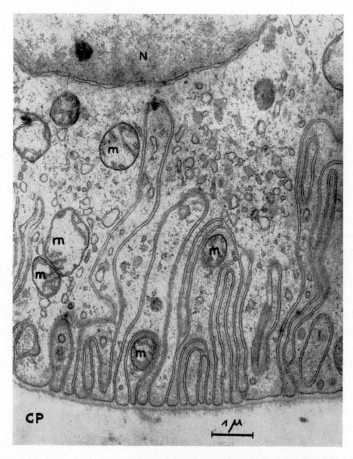

Figure 2. Electron micrograph of non-pigmented ciliary epithelium showing the nucleus (N), mitochondria (m), posterior chamber (CP), and numerous interdigitations. (From Michiels 1968; Missotten, 1964).

barriers to trauma not only limits procedures that can be employed to study ion movement but also adds to the complexity of distinguishing between primary and artifactual effects of metabolic inhibitors on ion flux.

The most direct way to determine net influx of ions into the eye is to block the drainage angle and measure the quantity of ions in the aqueous humor that flows out of normal eyes or in the fluid that emanates from perfused eyes. Rapid changes in aqueous formation induced either by changing the composition of the plasma or the perfusate, or inhibiting transport with metabolic poisons can be determined in this manner, but procedures for obtaining continuous samples of aqueous humor or perfusate are traumatic and may alter the parameters being studied. Moreover, the method does not distinguish between net influx of ions from ciliary processes and that from vitreous humor, anterior surface of the iris, or the cornea, and it provides little information concerning total flux across the ciliary epithelium.

The movement of ions between ciliary body and posterior chamber can also be inferred from time-course studies of accumulation of isotopes in the ocular chambers following systemic administration, and from the ratio of concentration in aqueous humor and plasma at steady state. Because only one sample of aqueous can be removed from each eye, this method makes it necessary to pool data from a number of animals; thus it is not well suited for measuring rapid changes in flux. However, the procedures involved in obtaining samples of aqueous, although possibly traumatic, do not affect the rate of accumulation of isotopes since the eye is not touched during the period when ion movement takes place.

The method for analyzing experimental data distinguishes between uni-directional movement and that which occurs by diffusional exchange across ciliary processes, as well as movement between posterior aqueous and vitreous humor. It does not provide the information necessary to determine whether unidirectional transport occurs by ultrafiltration or by processes that derive energy from metabolism of the ciliary epithelium.

Ion flux can be measured directly across ciliary body–iris preparations placed between plastic or glass cells containing solutions of various com-position, or inferred from the rate at which isolated processes pump them-selves dry when maintained in different media. Whereas information from experiments performed *in vitro* may apply only to a limited extent to the situation in the living eye, use of isolated ciliary bodies affords an opportunity to investigate flux under conditions in which concentration of solutes in fluid bathing the inside and outside of the processes may be varied over a wider range than is possible *in vivo*, and to study electric phenomena associ-ated with transport under conditions not complicated by potentials arising from structures of the eye other than the ciliary body.

All these approaches contribute to present knowledge of ion movement in ciliary processes. Results of some investigations that provide evidence for current concepts of the nature of this movement and mechanisms by which aqueous humor is formed will be described. Unless otherwise indicated, all data pertain to rabbit eyes.

II. DIFFUSION AND FLOW OF SODIUM AND CHLORIDE

The movement of ions between plasma and intraocular fluids has been determined chiefly with radioisotopes, sodium and chloride being studied most extensively. From an analysis of the rates at which radioactive isotopes of these ions accumulate in the posterior chamber, and the steady-state ratios of the naturally occurring species, Kinsey and Reddy (1959) estimate the quantity of each ion that enters and leaves the posterior aqueous by passive diffusion and unidirectionally by flow. Such an analysis requires knowledge of the following factors that control accumulation and steady-state distribution of substances in the posterior chamber: rate of flow of fluid into and out of the posterior chamber from the ciliary processes;* concentration in the fluid that enters by flow; and rate of diffusional exchange between posterior aqueous and blood, lens, and vitreous body. The contribution of each of these factors to the rate of accumulation of substances in the posterior chamber is described by equation (1), the transfer coefficients of which can be evaluated, using the following methods:

$$\frac{dC_h}{dt} = k_{fh}\,(C_s - C_h) + k_{d.ph}\,(C_p - \alpha C_h) - \frac{DA}{\Delta x V_h}\,[C_h - C_v]_{x=0} \tag{1}$$

where C is concentration, k_{fh} and $k_{d.ph}$ are the transfer coefficients by flow and diffusion, respectively, from plasma to the posterior chamber. The subscripts have the following meanings: h, posterior chamber; s, secreted fluid; p, plasma; and v, vitreous humor. The meaning of the other symbols and those used in subsequent equations, and the units in which they are expressed, are shown in table 1. The transfer coefficients are expressed as a fraction of the volume of the chamber to which they refer, and so they vary in magnitude, although volume of flow or quantity of ion moved by diffusion may be identical.

The first term on the right-hand side of the equation is valid if the rate of flow into and out of the posterior chamber is identical; if ions enter

---

*To describe movement of ions across the ciliary epithelium it is unimportant whether this fluid originates in epithelial cells and then enters the posterior chamber as a result of pinocytosis, as a consequence of hydrostatic pressure, or both, or whether ions are actively transported across ciliary epithelium accompanied by an osmotic flow of water. Fluid transport, if not indeed its formation, is energy-dependent and has been referred to as "secreted fluid" (Kinsey and Palm, 1955). Cole (1960a) uses the same designation to describe the portion of influx that is cell mediated.

Table 1. Mathematical nomenclature and numerical values used in solving text equations. All transfer coefficients are expressed as a fraction per minute of the volume of the chamber indicated by the subscript

Symbol	Designation	Units
A_{hv}	Area of posterior chamber-vitreous interface	$1 \cdot 1$ cm^2
V_h	Volume of posterior chamber	$0 \cdot 055$ cm^3
V_v	Volume of vitreous	$1 \cdot 5$ cm^3
V_a	Volume of anterior chamber	$0 \cdot 25$ cm^3
Δx	Increment of space variable in vitreous	$0 \cdot 136$ cm
E	Potential difference across blood aqueous barriers	$2 \cdot 4$ mV
α_{Cl}[1]	Electrochemical equilibrium coefficient from Nernst equation	$0 \cdot 912$
α_{Na}[1]	Electrochemical equilibrium coefficient from Nernst equation	$1 \cdot 096$
D_{Cl}	Diffusion coefficient	$1 \cdot 38 \times 10^{-3}$ cm^2/min
D_{Na}	Diffusion coefficient	$0 \cdot 92 \times 10^{-3}$ cm^2/min
k_{fh}	Transfer coefficient by flow into and out of posterior chamber	$0 \cdot 06$ min^{-1}
$k_{d.hvCl}$	Transfer coefficient by diffusion from posterior chamber to vitreous with respect to volume posterior chamber	$0 \cdot 055$ min^{-1}
$k_{d.vhCl}$	Transfer coefficient by diffusion from vitreous to posterior chamber with respect to volume vitreous	$0 \cdot 002$ min^{-1}
$k_{d.hvNa}$[2]	Transfer coefficient by diffusion from posterior chamber to vitreous with respect to volume posterior chamber	$0 \cdot 037$ min^{-1}
$k_{d.vhNa}$[2]	Transfer coefficient by diffusion from vitreous to posterior chamber with respect to volume vitreous	$0 \cdot 0013$ min^{-1}

[1] $\alpha = e^{nFE/RT}$

[2] Based on the assumption that sodium flux across posterior chamber-vitreous interface relates to that of chloride as their respective rates of diffusion.

secreted fluid from plasma rapidly, relative to their rate of penetration into the posterior aqueous; and if ionic concentration in secreted fluid is a constant linear function of the concentration in plasma. The second term is valid if mixing of constituents in posterior aqueous is sufficiently rapid that Fick's law is applicable to the barrier.

The coefficient α is used to compensate for the retardation or acceleration that differences in electrochemical potential exert on diffusional movement of ions across membranes (Ussing, 1950). Values for α are calculated from the relation $\alpha = e^{nFE/RT}$, and are based on an indirect determination of the electric potential difference of $2 \cdot 4$ mV found to exist between plasma and posterior aqueous; they are $1 \cdot 096$ and $0 \cdot 912$ for cations and anions, respectively (Kinsey, 1967).

Neither vitreous humor nor the lens is well stirred; thus the rate of diffusional loss from the posterior chamber [last term in equation (1)] not only depends on differences in concentration between these structures and posterior aqueous, but on the rate of diffusion into and through them. Since loss of ions to the lens is relatively small (10 percent) compared with that to the vitreous, and since the mathematical model (figure 3) employed to calculate influx of ions into the lens and vitreous only approximates the geometry of these two structures, they are considered together as a lumped environment and referred to as vitreous.

Kinsey and Reddy (1959) consider the model of the vitreous to be shaped like a cylinder having a volume equal to that of the vitreous body and a cross-sectional area equal to that of the interface between posterior chamber and vitreous. Ten discrete stations are chosen at equally spaced points through-out the vitreous model (figure 3); the shading illustrates a concentration gradient that might exist at a particular time after systemic administration

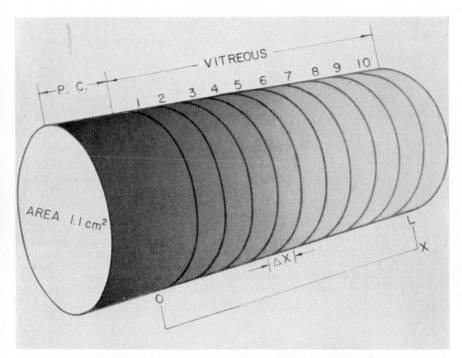

Figure 3. Model of posterior chamber and vitreous considered as a cylinder with ten equally spaced stations or compartments in which the cross-sectional area is equal to that of the posterior chamber–vitreous interface. The depth of shading illustrates a concentration gradient at a particular time as a function of the distance away from the posterior chamber. (From Kinsey, 1960).

of a tracer substance. A space derivative $\delta^2 C_v / \delta x^2$ is approximated by finite differences, thus an ordinary differential equation, with time as the independent variable, is established at each of these stations [equation (2)]:

$$\left[\frac{dC_v}{dt} \right] n \approx \frac{D}{\Delta x^2} \left[C_{n+1} - 2C_n + C_{n-1} \right] \tag{2}$$

where n is the number of the station.

The exchange between posterior chamber and vitreous at the interface between these compartments, the first station only, can be established by using an analog computer to calculate the resulting system of simultaneous differential equations for the vitreous. The numerical values obtained are then continually substituted for the last term of equation (1) where the value for C_v denotes concentration at the interface, rather than average concentration in vitreous, and is referred to by Kinsey and Reddy (1964) as the 'effective' concentration in vitreous.

Equation (1) contains two unknowns: concentration in secreted fluid (C_s) and the value for the coefficient of diffusion ($k_{d.ph}$). The equation can be solved, however, because these terms bear a fixed relation to one another [equation (3)]:

$$C_s = C_h - \frac{k_{d.ph}}{k_{fh}}(C_p - \alpha C_h) + \frac{1}{k_{fh}} \frac{DA}{\Delta x V_h}(C_h - C_v)_{x=0} - \frac{1}{k_{fh}} \frac{dC_h}{dt} \tag{3}$$

This relation can be established from knowledge of the concentration in plasma and intraocular fluids at steady state where $dC_h/dt = 0$, when C_h and C_v are so nearly alike that flux across the posterior chamber–vitreous interface can be either ignored or established for some particular difference in concentration in posterior aqueous and vitreous,* usually at steady state.

---

*The method for determining flux between posterior chamber and vitreous, based on that proposed by Maurice (1957), consists of injecting a tracer into the center of the vitreous and measuring its concentration in this body and the posterior chamber as a function of time. Since all ions that leave vitreous anteriorly must pass through the posterior chamber, their amount is equal to the sum of the quantity flowing out between the lens–iris space to the anterior chamber and that diffusing into blood by way of the ciliary process. The flux rate anteriorly, therefore, is equivalent to a transfer coefficient, $k_{d.vh}$, as shown in equation (4):

$$k_{d.vh} = (k_{fh} + k_{d.ph}) \frac{C_h}{C_v - C_h} \frac{V_h}{V_v} - \frac{dC_h}{dt} \frac{V_h}{V_v} \frac{1}{c_v - c_h} \tag{4}$$

The magnitude of $dC_h V_h/dt$ was calculated by Kinsey and McLean (1969) and found to be small under the conditions of measurement; therefore, this term may be neglected when computing $k_{d.vh}$ without introducing appreciable error. The flux rate posteriorly, $k_{d.hv}$ in terms of that in the reverse direction, is described by equation (5):

$$k_{d.hv} = k_{d.vh} \frac{V_v}{V_h} \tag{5}$$

Table 2. Concentration of sodium and chloride in plasma and ocular humors of rabbit eyes at steady state

	Concentration (relative units)[1]	mmoles/kg H_2O
C_{pNa}	100	150
C_{aNa}	98	146·5
C_{hNa}	97	145
C_{vNa}	94	140
C_{pCl}	100	101
C_{aCl}	104	105
C_{hCl}	99	100
C_{vCl}	108	109

[1]A relative unit is defined as a percentage of the concentration in the plasma once the concentration of a labeled ion in the plasma becomes essentially constant. (Kinsey, 1967.)

The relation between concentration in the secreted fluid and the coefficient of diffusion from plasma to the posterior chamber for sodium and chloride is calculated for steady-state conditions from equation (3) using the appropriate values for the coefficients involved, including those for flux between vitreous and posterior chamber. The concentration of these ions in plasma and ocular fluids is shown in table 2. The results, in relative units, are described by equations (7) and (8):

$$C_{sNa} = 99 + 105 \, k_{d.ph} \tag{7}$$

$$C_{sCl} = 90 - 167 \, k_{d.ph} \tag{8}$$

The relation between C_s and $k_{d.ph}$ is established from steady-state data, and equation (1) is solved using an analog computer. Values for these

---

Therefore, the net movement between posterior chamber and vitreous, $k_{d.hv} (C_h - C_v)$, is equal to the last term of equation (1),

$$k_{d.hv}(C_h - C_v) = \frac{DA}{\Delta x V_h}(C_h - C_v)_{x=0} \tag{6}$$

where C_v on the left side of the equation is the average concentration in the vitreous and that on the right side is the effective concentration, i.e. the one existing at the interface. The value $k_{d.hv}$ for chloride is 0·055 min^{-1}, and that for sodium is 0·037 min^{-1}, based on the assumption that sodium flux across posterior chamber–vitreous interface is proportional to the relative rates of diffusion of chloride and sodium, D_{Na}/D_{Cl}, in vitreous humor (Kinsey, 1969).

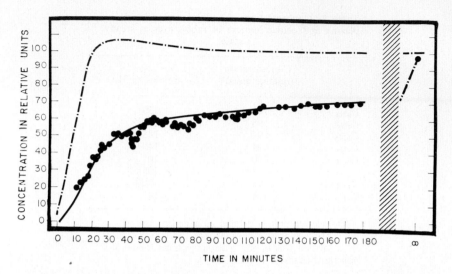

Figure 4. Rate of accumulation of ^{24}Na in the posterior chamber compared with experimental data (filled circles). Dot-dashed line indicates values for plasma. The value for the transfer coefficient by diffusion is 0·02 min^{-1}. (After Kinsey, 1967).

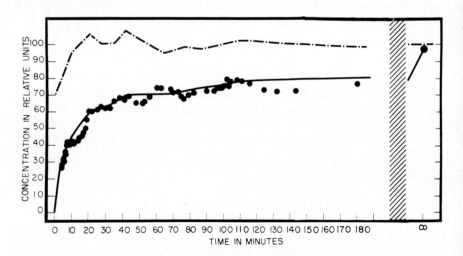

Figure 5. Rate of accumulation of ^{36}Cl in the posterior chamber compared with experimental data (filled circles). Dot-dashed line indicates value for plasma. The value for the transfer coefficient by diffusion is 0·13 min^{-1}. (After Kinsey, 1967)

Table 3. Coefficients of transfer by flow and diffusion into and out of the posterior chambers and the concentration of sodium and chloride in the secreted fluid of rabbit eyes. Average plasma level is 100 relative units.[1]

	Flow k_{fh} (min^{-1})	Diffusion $k_{d.ph}$ (min^{-1})	Concentration C_s (rel. units)
Sodium	0·06	0·02	101
Chloride	0·06	0·13	70

[1]Kinsey, 1967.

parameters are selected until the resulting curves (figures 4 and 5) fit experimental data that show rates of accumulation of sodium and chloride in the posterior chamber of rabbit eyes following parenteral administration of radioactive isotopes of these ions. The numerical values thought to provide the best fits are given in table 3.

Knowing the coefficients of transfer and the absolute units of concentration (table 3), the proportion of sodium and chloride that moves across the ciliary epithelium by flow and diffusion can be readily calculated. At steady state, influx by flow is $(k_{fh})(C_s)(V_h)$ and by diffusion $(k_{d.ph})(C_p)(V_h)$. Since the transfer coefficient for diffusion is the same in both directions, $k_{d.ph} = k_{d.hp}$, efflux to blood by diffusion is $(k_{d.ph})(V_h)(C_h)(\alpha)$.

Table 4 shows that 75 per cent of the sodium enters the posterior chamber by flow and 25 per cent by diffusion, whereas only 25 per cent of the chloride enters by flow and 75 per cent by diffusion.* The relatively large proportion

Table 4. Flux of sodium and chloride by flow and diffusion across the ciliary epithelium[1] of rabbit eyes. Flow rate is assumed to be 3·3 µl/min.

	Influx		Efflux
	by flow (µmoles/min)	by diffusion (µmoles/min)	by diffusion (µmoles/min)
Sodium	0·50	0·165	0·175
Chloride	0·23	0·72	0·65

[1]Influx of sodium and chloride from vitreous to the posterior chamber is 0·28 and 0·33 µmoles/min; efflux is 0·30 and 0·30 µmoles/min, all respectively. (Kinsey, 1969.)

*Although the net influx of sodium can be accounted for exclusively on the basis of flow, approximately 30 per cent of the net influx of chloride occurs by diffusion from the ciliary process and vitreous humor. To this extent the quantity of chloride moving out of the eye is not an accurate indication of unidirectional transport of this ion across ciliary epithelium.

of sodium moving into the posterior chamber unidirectionally suggests that, at least in the rabbit, sodium is the predominant ion actively transported across the ciliary epithelium.

The rates of diffusion of sodium and chloride across the anterior surface of the iris into the anterior chamber, where no unidirectional transport of inorganic ions occurs, are essentially equal; the transfer coefficient, $k_{d.pa}$, for both is 0.014 min^{-1}.

Chloride is the anion present in highest concentration in plasma, but the quantity that enters the posterior chamber unidirectionally is sufficient to neutralize the charge on only about half of the sodium. Bicarbonate, therefore, must exist in the secreted fluid in a concentration several times higher than in plasma and must balance most of the remaining charge (Kinsey and Reddy, 1959). This conclusion is compatible with knowledge of the excess of bicarbonate in aqueous humor of rabbit eyes (Kinsey, 1953). In primates the situation is reversed; bicarbonate is deficient in aqueous humor relative to plasma and the concentration of chloride is correspondingly higher in aqueous than in plasma (Becker, 1957).

III.   NET IONIC INFLUX DEPENDENT ON METABOLISM AND ULTRAFILTRATION

Unidirectional movement of electrolytes across ciliary epithelium may occur by active transport of at least one ion, or by ultrafiltration. Energy for the first process is derived from metabolic activity of cells of the ciliary body and for the second process from differences in hydrostatic pressure between capillaries and intraocular humors, estimated by Davson (1956) to be approximately 10 cm of water. To determine the relative contribution of each process to total net influx, advantage has been taken of the different effects produced by metabolic poisons on the sources of energy. After normal outflow channels are blocked with mineral oil, the ion content of either perfusates or of aqueous humor is determined before and after administration of various metabolic inhibitors (Cole, 1960b).

Dinitrophenol (DNP) and fluoroacetamide (FA), given by way of the lingual artery to minimize systemic effects, reduce the amount of sodium, chloride and potassium that leaves the eye (table 5). The reduction in efflux of sodium and potassium amounts to approximately 75 per cent and that of chloride to 60 per cent of the total, and is presumed to represent the fraction that enters the eye by active transport. If net influx is blocked completely, the remainder, or from 25 to 40 per cent of these ions, must enter the eye by ultrafiltration, assuming, as Cole does, that net influx and efflux are approximately equal. However, a significant fraction of the net movement of chloride into the posterior chamber occurs by diffusion (Kinsey and Reddy,

Table 5. Effect of DNP and FA given by intracarotid infusion on the influx rates of sodium, chloride, potassium and water in rabbit eyes.[1]

Group	(Control Animals)	Animals given DNP	Animals given FA
Sodium (μmoles/min)	$1\cdot01\pm0\cdot10$[2]	$0\cdot28\pm0\cdot08$	$0\cdot31\pm0\cdot09$
Chloride (μmoles/min)	$0\cdot77\pm0\cdot07$	$0\cdot30\pm0\cdot15$	$0\cdot35\pm0\cdot10$
Potassium ($\mu\mu$moles/min)	$29\cdot1\pm1\cdot5$	$9\cdot5\pm3\cdot0$	$8\cdot6\pm2\cdot8$
Water (μl/min)	$6\cdot2\pm0\cdot7$	$2\cdot2\pm0\cdot6$	$2\cdot3\pm0\cdot5$

[1]After Cole, 1960
[2]Standard error

1959), which may account for why the influx of chloride is affected proportionately less than sodium by the metabolic inhibitors, since diffusion, like ultrafiltration, is not subject to inhibition.

Reduction in total ion and water influx by inhibitors of metabolism shows that active transport is involved in formation of aqueous humor, but it is the source of energy available for transport that is affected by poisons rather than a specific transport system. The experiments do not, therefore, indicate the ion or ions that are actively transported.

IV. ACTIVE TRANSPORT OF SODIUM AND CHLORIDE

A. In vivo studies

The most conclusive way to determine which ion is actively transported across a cell membrane is to demonstrate differences in flux across it under zero electrochemical potential. Thus, Ussing (1949) who employed two isotopes of sodium to measure simultaneous flux across short-circuited frog skin, found not only that sodium is actively transported, but also that movement of this ion itself accounts for the 60 to 70 mV difference in potential existing across the skin. The double label technique employed is not feasible in the eye *in situ* and has yet to be performed under the exacting conditions of zero electrochemical potential in isolated ciliary bodies.*
Nevertheless, studies relating changes in electric potential and short-circuit current to ion movement are revealing with regard to mechanisms of

*Holland and Stockwell (1967) determined sodium flux in both directions across ciliary bodies of cat eyes using ^{22}Na and ^{24}Na under what was thought to be zero electrochemical potential. Subsequently, however, Holland and Gipson (1970) found that a small potential (0·55 mV) existed across the membrane because of an unsuspected resistance between the voltage sensing electrodes.

transport across the ciliary epithelium. For instance, systemic administration of DNP and strophanthin G (ouabain) reduces the potential and short-circuit current along with influx of sodium, although change in current increases more in direction than in magnitude when sodium flux is altered (Cole, 1961a). These observations are generally consistent with the idea that sodium is actively transported across ciliary epithelium, especially when it is observed that on the basis of dosage, strophanthin G, a specific inhibitor of Na–K ATPase, is a more effective inhibitor than DNP or FA.

The eye is a complex organ, and isolated observations must, therefore, be interpreted with caution, because structures other than ciliary processes may contribute to transport and electric phenomena. Sodium azide, for instance, increases ion influx, potential difference, and short-circuit current, but probably does so through its action on the retina or pigmented epithelium (Noell, 1953; Cole, 1961a); and iodoacetate, which reduces potential and current but increases influx of sodium, may, through its effect on the blood aqueous barriers, enhance ultrafiltration to a greater extent than it retards active transport.

Other results relating movement of ions and electric manifestations in the living eye are also not explicable on the basis of a model system of a sodium pump operating across a relatively simple membrane. Cole (1961a) observes that imposition of an external EMF in a direction that tends to increase the naturally occurring potential difference by making it more positive, increases influx of sodium and water. This effect appears to be contraindicated on thermodynamic grounds unless it is argued that the electromotive force in some way accelerates active transport across the epithelium to a greater degree than increased electrochemical gradient reduces it by increasing electrostatic repulsion.

B. In vitro studies

Isolated iris–ciliary body preparations placed between microcells can be used to determine which ion is actively transported by the ciliary epithelium. The composition of solutions on either side of the epithelium can be altered separately with this procedure, and the effect of individual ions on potential difference and short-circuit current can be studied independently of each other.

The potential difference increases with the logarithm of the sodium ion concentration on the stromal side of the isolated ciliary body but is unaffected by changes in concentration of potassium (Cole, 1961b). An increase in concentration of sodium on the posterior chamber side of the ciliary body is accompanied by a small reduction in potential difference, but a similar increase in concentration of potassium results in a decided decrease in potential. These results indicate that the stromal surface of the ciliary process is more permeable to sodium by passive diffusion than the epithelial side, whereas the reverse is true for potassium, which permeates from the epithelial side much more readily than from the stromal side.

Short-circuit current and potential increase when chloride in the media on two sides of *in vitro* preparations is replaced with ethyl sulfate or pyroglutamate, whose conductances are lower than chloride, and decrease when concentration of both sodium and chloride is reduced with mannitol (Cole, 1961b, 1969). The greatest reduction occurs when choline is used to replace sodium, in which case both the potential and current become negative (table 6). These results are consistent with the view that chloride, as well as

Table 6. Short-circuit current and transepithelial potential in isolated ciliary epithelium (mean differences ± standard error of means)[1]

	Experimental values				Differences			
	"NaCl"	"Na free"	"Cl free"	"Na,Cl free"	"NaCl–Na free"	"NaCl–Cl free"	"NaCl–NaCl free"	"Na free–Na,Cl free"
Short-circuit current (μA cm^{-2})	24·2 ±0·5	−10·2 ±1·5	34·4 ±0·7	2·5 ±1·1	33·2 ±3·7	−9·5 ±3·2	20·1 ±2·2	−11·4 ±2·4
Potential (mV)	+3·1 ±0·3	−1·1 ±0·4	+6·7 ±0·5	+0·5 ±0·4	+4·2 ±1·2	−3·7 ±0·9	+2·7 ±0·6	−1·6 ±0·5
Number of experiments	30	15	15	17				

[1]After Cole (1969).

sodium, is actively transported across the ciliary epithelium, apparently by independent or loosely coupled processes. This hypothesis is further supported by the demonstration that short-circuit current is temperature-dependent in a medium containing sodium, with or without chloride, and in 'sodium-free' medium, whereas current is small and unaffected by temperature in a 'sodium, chloride-free' medium (Cole, 1969).

The presence of thiocyanate, a competitive inhibitor of chloride transport (Durbin, 1964; Forte, 1968), increases short-circuit current, but only when chloride is present (Cole, 1969). The inhibition, in the presence of 20 mM thiocyanate, corresponds to the calculated chloride contribution to the short-circuit current and indicates almost complete suppression of chloride transport. The relative contribution of sodium and chloride ions to short-circuit current, expressed per cm^2 of projected epithelial surface, is approximately 34 μA for sodium and −10 μA for chloride. Cole (1969) estimates the ratio of metabolic entrainment of chloride to sodium to be in the range of 15 to 23 per cent.

A chloride pump probably exists in the ciliary body of cat eyes also (Holland and Gipson, 1970), where the inner epithelial surfaces of *in vitro* preparations, unlike those in rabbit eyes, are electronegative with respect to stromal surfaces. The net influx of chloride across isolated ciliary bodies occurs against an electrochemical gradient and is about 2.9 μequiv./cm^2/hr. The direction of short-circuit current is from epithelial to stromal surface, as would result from transport of negative (chloride) ions in the opposite direction or positive ions from the epithelium to the stroma. The short-circuit current generated by chloride ions amounts to 36 per cent of the net chloride transport. The remainder is balanced, presumably, by net transport of other anions from epithelium to stroma, or by cations moving inward with chloride. In the absence of evidence that bicarbonate or any other anion is actively transported from epithelium to stroma, the existence of a pump moving sodium into the posterior chamber seems to be the most likely possibility. In 1967, Holland and Stockwell cited evidence for the existence of a sodium pump in cat eyes, but later experiments, again using *in vitro* preparations, suggest that any net influx of sodium must have been small and beyond the sensitivity of methods used to measure it (Holland and Gipson, 1970). The earlier data, however, do not exclude the possibility that sodium is actively transported in cat eyes, and the differences in the nature of ion transport across ciliary epithelia of rabbit and cat eyes must be more of degree than of basic mechanism.

Experiments using an *in vitro* technique of a different kind are also compatible with the idea that sodium is actively transported by the ciliary process of rabbit eyes, but they provide no new data on the question of active movement of chloride. Berggren (1964) has used shrinkage of ciliary processes to measure secretory activity. Eyes *in situ* are first perfused through the carotid artery with cold Krebs solution bubbled with a mixture of nitrogen and carbon dioxide to allow the ciliary processes to fill maximally. The processes are then removed from the eye and attached to a plate mounted in a chamber containing chilled buffer solution. The bathing solution is warmed and supplied with oxygen to restore normal pumping action which causes the processes to shrink. The degree of shrinkage is determined by measuring photographs of optical sections with a planimeter, and is used as an indication of secretory activity.

Shrinkage is dependent on oxidative metabolism and the presence of sodium and potassium. It is inhibited with concentrations of ouabain as low as 10^{-5} M and unaffected by iodoacetate, sodium azide, or by enrichment of the media with potassium (Berggren, 1965). These results indicate that secretory activity occurs across the ciliary epithelium by a mechanism which is dependent on potassium and which consists of active transport of sodium accompanied by diffusional movement of water.

C. Site of Transport

Although active transport of ions and potential difference across ciliary epithelium seem to be related, the double cell layer that comprises the epithelium complicates an explanation of the mechanisms involved and the role played by each layer. Berggren (1960), using microelectrodes to determine the electric potential profile of ciliary processes maintained *in vitro* or transplanted into the anterior chamber of rabbit eyes, shows that the potential gradient across the ciliary process is far from uniform (figure 6). The pigmented and non-pigmented epithelial layers have potentials of -55 and -20 mV, respectively, and the potential of aqueous humor is $+5$ mV, all related to plasma or stroma. The potentials are relatively independent of temperature between 20 and 36° C, and of pH between 7·2 and 8·0, but

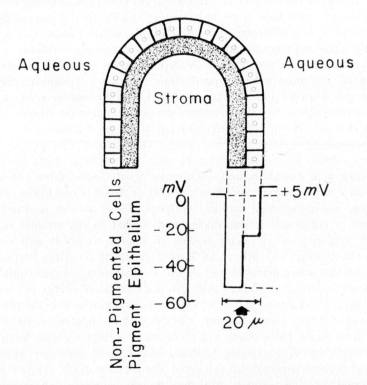

Figure 6. Sketch of ciliary process showing the potential in the two cell layers and aqueous humor compared with stroma. (After Berggren, 1960).

decrease in the presence of cyanide, when the pH is less than 7·2, and when potassium is replaced by sodium in the bathing media.

The significance of variations in electric potential across the ciliary processes with respect to the site of active transport is not obvious, but the dissimilarity of potential probably reflects differences in metabolic patterns of the two cell layers. Histochemical methods for localization of enzymes in the epithelium are revealing in this regard. Cole and coworkers (Cameron and Cole, 1963; Cole, 1963) show that while lactic dehydrogenase is distributed equally between the two epithelial layers, succinic dehydrogenase is much more active in non-pigmented cell layers, as it is also in anterior regions of ciliary processes where transport is greatest. The distribution of enzymes, in addition to other evidence from direct measurements of respiration and glycolysis (Shimizu, Riley and Cole, 1967), seems to indicate that oxidation by way of the citric acid cycle predominates in the inner cell layer of the epithelium and, thus, because of its relatively high yield of energy may provide a more ample supply of high-energy phosphate bonds for ion transport than would be available from glycolysis. Although energy from glycolytic processes is probably either insufficient or unavailable to compensate for loss of energy in the absence of oxidative sources, it is from 10 to 100 times the amount required for active transport of sodium.

Active transport is not necessarily dependent on oxidative metabolism because of the more efficient production of energy. However, there is considerable evidence that sodium transport is highly sensitive to oxygen. The potential across the ciliary epithelium is dependent on oxygen (Cole, 1961b; Holland, 1970), and both potential and rate of aqueous formation are inhibited by cyanide (Miller, 1962; Mishra and coworkers, 1963; Berggren, 1965). Moreover, inhibitors such as malonate or fluoroacetate block citric acid oxidation, and uncoupling agents like cyanine and styryl quinoline dyes (Ballintine and Peters, 1954) or DNP (Cole 1960a, 1960b; Berggren, 1965) decrease sodium transport. The converse is also true; succinate, a citric acid intermediate, when added to the stromal side of isolated preparations of ciliary bodies stimulates transport and oxygen uptake (DeRoetth, 1953; Riley and Baroncelli, 1964), an effect attributable chiefly to the non-pigmented cell layer. Furthermore, oxygen uptake is diminished under circumstances that affect utilization of energy for sodium transport as, for example, in low or sodium-free solutions, or in the presence of ouabain (Riley, 1964). Sodium transport thus appears to be closely linked to oxidative metabolism and to derive its energy through hydrolysis of ATP by a (Na^+-K^+)-activated ATPase which, like the respiratory enzymes, is located in the non-pigmented cell layer. The location of the chloride pump is at present unknown, but it, too, may be associated with the innermost layer of the ciliary epithelium.

V. TRANSPORT OF OTHER INORGANIC IONS

A. Bicarbonate

The concentration of bicarbonate in the posterior aqueous of rabbit eyes exceeds that in plasma and, next to sodium and chloride, is the ion most prevalent in these fluids (Kinsey, 1953). Attempts have been made to measure its rate of movement across ciliary epithelium by determining the rate at which bicarbonate labeled with carbon-14 ($H^{14}CO_3$) accumulates in posterior aqueous following intravenous administration (Kinsey and Reddy, 1959; Green and Sawyer, 1959). Within less time than it takes to perform the experiment (40 seconds), the concentration of labeled bicarbonate in posterior aqueous exceeds that in plasma by an amount indistinguishable from that prevailing at steady state. Kinsey and Reddy (1959) believe that most of the $H^{14}CO_3$ is dehydrated in blood and stroma by the enzyme carbonic anhydrase and forms $^{14}CO_2$ which moves with great rapidity across the ciliary epithelium where it rehydrates and is recovered in the posterior aqueous as $H^{14}CO_3$. The movement of carbon dioxide thus completely masks that of bicarbonate. When acetazoleamide is given prior to labeled bicarbonate to inhibit carbonic anhydrase, the rate of accumulation is reduced, but even under these circumstances it is still so rapid that meaningful measurements of rate of penetration of the bicarbonate ion cannot be obtained.

The importance of bicarbonate in aqueous humor formation originates from observations made by Friedenwald more than 30 years ago. Friedenwald and Stiehler (1938) postulated the existence of an electron pump in ciliary processes which carries electrons across a negatively charged barrier from stroma to epithelium. Energy for the current is supplied by differences in oxidation–reduction potential existing between these structures. The net result of one oxidative cycle involving reduction of one atom of oxygen in the epithelium and oxidation of two atoms of hydrogen in the stroma is the production of two unbalanced anions in the epithelium and two incompletely balanced cations in the stroma. Thus, the system would produce four hydroxyl ions in the stroma from each molecule of oxygen. Through reaction with carbon dioxide, hydroxyl ions would be converted into bicarbonate which would diffuse into the posterior chamber, accompanied by sodium (Friedenwald, 1944). Observations showing that bicarbonate concentration and pH are higher in aqueous humor than in plasma are consonant with the theory (Kinsey, 1950), and were thought to provide a possible explanation for the mechanism by which sodium is transported into the posterior chamber. Subsequently, Berliner (1956) stated that only four sodium ions could be moved for each oxygen molecule utilized and he questioned, because of stoichiometric limitations, whether the redox pump could account for sodium transport in the eye.

A single ciliary process in the rabbit eye utilizes about 0·045 μmole of oxygen per minute (Ballintine, 1956; Cole, 1963). Since net flux of sodium is about 0·5 μmole/min, about 10 sodium ions per molecule of oxygen must be transported, a number slightly less than Berliner (1956) estimates to be transported by the renal tubule and about equal to that moved across frog skin for the same amount of oxygen consumed. A major portion of sodium cannot, therefore, be transported into the posterior chamber by means of a redox pump, although thermodynamically the pump could provide more than enough energy to move all of the sodium.

The amount of oxygen consumed is about twice that needed to maintain the concentration gradient of bicarbonate between plasma and aqueous humor (15 mmoles/l) if it depended upon a redox pump. In the absence of alternative hypotheses to account for the alkaline pH and excess bicarbonate in aqueous humor of rabbit eyes, the explanation offered by Friedenwald is attractive. For a redox pump to operate in primates it would have to act in the reverse direction, since in these species, aqueous humor is acid compared with plasma and is deficient in bicarbonate. Perhaps hydrogen ions are conserved in aqueous humor and hydroxyl ions are lost to the blood (Becker, 1959).

B. Iodide

Movement of iodide across the ciliary epithelium is unique among inorganic ions since it is actively transported out of the ocular fluids into the blood by the ciliary processes (Becker, 1961a). Not only do ciliary body–iris preparations concentrate iodide above the level present in media in which they are incubated, but also intravitreal injections or systemic administration of iodide saturates the system that moves this ion out of the eye. Thiocyanate and perchlorate administered in a similar way also decrease the rate of loss of iodide from the eye, but chloride, bromide, ascorbate, and those ions that reduce loss of iodopyracet (Diodrast) from the eye do not affect its transport.

Movement of labeled iodide from blood into the posterior chamber takes place at an extremely slow rate unless the carrier system responsible for active transport in the outward direction is first saturated by giving non-labeled iodide systemically (Kinsey and Reddy, 1964) or it is inhibited by perchlorate (Becker, 1961a).

The transport mechanism for iodide appears to be located in the ciliary epithelium, and resembles the type of pump responsible for accumulation of iodide by thiouracil-treated thyroid glands, the salivary glands, and the choroid plexus. So far as is known, it serves no useful purpose in the eye.

C. Thiocyanate and Bromide

The rate of accumulation of several inorganic ions not normally present in appreciable concentrations in aqueous humor, notably thiocyanate (Kinsey and Palm, 1955) and bromide (Becker, 1961b) has also been determined, but analyses have not been made to determine the relative proportion of these ions that enter the posterior chamber by different mechanisms.

VI. TRANSPORT OF ORGANIC IONS

Ciliary processes are capable of actively and selectively transporting a number of organic acids both into and out of the posterior chamber by highly specific mechanisms. For instance, ascorbic acid which is present in posterior aqueous humor in a number of species, including man, in a concentration twenty-five times that in plasma, is transported across the ciliary epithelium of rabbits and guinea pigs (Kinsey, 1947; Bárány and Langham, 1955) by a mechanism that becomes saturated at plasma levels of 0·2 m moles/l. The enantiomorph of ascorbic acid, D-isoascorbic acid also concentrates in the aqueous, but D-glucoascorbic acid which differs in structure only in the addition of another CHOH group between carbon atoms 5 and 6 of ascorbic acid, does not (Kinsey, 1947).

In rabbits, the concentration of all but three amino acids is higher in posterior aqueous than in plasma (Reddy and coworkers, 1961), suggesting that these compounds may also be actively transported across the ciliary epithelium. Kinsey and Reddy (1962) observed that the model amino acid α-aminoisobutyric acid (αAIB) enters the posterior chamber by means of a system that is saturated when the concentration in the plasma is raised to about 10 m moles/kg water (figure 7). At this level, $2·4 \times 10^{-3}$ μmoles/min of αAIB, along with unknown amounts of naturally occurring amino acids that share the same carrier system, are actively transported across the ciliary epithelium. With progressively higher concentrations of αAIB the amount entering the posterior chamber is directly proportional to the concentration of non-labeled compound. The slope of the line that shows this relationship is a measure of the rate of accumulation due to diffusion and ultrafiltration. The situation is analogous to influx of inorganic ions occurring in the presence of metabolic inhibitors, as discussed earlier in this chapter. The rate of diffusion and ultrafiltration of αAIB across the ciliary epithelium is $0·006$ min^{-1}, which is less than one-third that of the rate of diffusion alone, of sodium.

From cross-saturation experiments involving transport of αAIB, particularly, Kinsey and Reddy (1962) conclude that at least three carrier systems function in the ciliary processes to move amino acids into the posterior chamber, one each for basic, acidic, and neutral compounds.

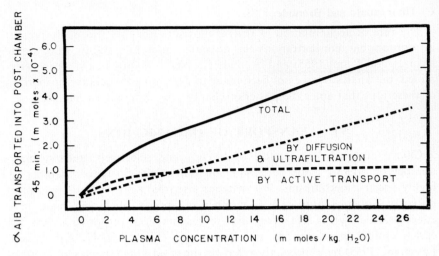

Figure 7. The effect of plasma concentration of α-AIB on the total amount of amino acid which *entered* the posterior chamber during forty-five minutes and that entering by diffusion and active transport. (From Kinsey and Reddy, 1964; courtesy of the author and publishers).

In contrast to the situation in rabbits, amino acids are not present in higher concentrations in the aqueous humor of monkeys, cats and rats (Reddy, 1967). In the rat, the neutral amino acids are transported across the blood aqueous barriers by a process that is carrier-mediated, thus active transport of amino acids by ciliary processes is probably not limited to rabbits.

Becker (1960) and Forbes and Becker (1960) demonstrate the existence of a system capable of actively transporting *p*-aminohippurate (PAH), Diodrast, and a number of other organic anions, including penicillin, out of the eye. Diodrast is accumulated by ciliary body–iris preparations to concentrations 10 to 15 times that in the incubation medium by mechanisms that require glucose, oxygen and potassium. The accumulative process is inhibited by a number of metabolic poisons and saturated competitively by such organic anions as bromcresyl green, probenecid and PAH. Diodrast labeled with radioactive iodine, when injected into vitreous humor, disappears so rapidly that none can be recovered from the anterior chamber. Introduction of non-labeled Diodrast into the vitreous appears to saturate a carrier involved in moving this compound out of the eye, as does also the presence of related organic anions. Non-labeled Diodrast, administered intravenously, does not affect transport out of the vitreous.

The same transport system is present also in guinea pigs and monkeys and is thought to be located mainly in the ciliary process; its physiological and pharmacological significance is unknown, but it closely resembles the one present in the renal tubule. The system is not involved in formation of aqueous humor, since it can be completely inhibited with probenecid without altering the intraocular pressure, rate of flow of aqueous, or changing the concentration of bicarbonate, chloride, or ascorbic acid.

A detailed discussion of movement of non-charged molecules from the ciliary processes into the posterior chamber is beyond the scope of this chapter. Although neither active nor facilitated transport seems to be involved, their penetration is dependent on subtle differences in molecular configuration. Thus the exquisite selectivity of the ciliary epithelial cells in transporting or allowing passage of substances from the blood to the posterior aqueous is not limited to the movement of ions.

VII. SUMMARY

The movement of three quarters of the sodium and one quarter of the chloride across the ciliary processes into the posterior chamber of rabbit eyes takes place unidirectionally by flow, and the remainder enters passively by diffusion. Approximately 70 per cent of the net influx of sodium occurs by reason of cell-mediated processes, probably as a result of active transport involving a carrier system located in the non-pigmented cell layer of the epithelium. The mechanism is activated by Na–K ATPase which derives energy chiefly from oxidative metabolism. Approximately one-fifth as much chloride is similarly transported by an active process but the mechanism and its location are less well defined. The remaining fraction of the unidirectional component of influx of sodium and chloride takes place by ultrafiltration, energy for which is provided by differences in hydrostatic pressure between capillaries and ocular humors.

Bicarbonate also enters by bulk flow and diffusion, but because of the extremely rapid diffusional exchange of carbon dioxide and the subsequent hydration, the proportion that moves by each means cannot be determined. A redox-pump may provide energy for maintaining the concentration gradient of bicarbonate between aqueous humor and plasma.

Iodide is transported *out* of the eye by a carrier system in the ciliary processes which resembles systems for accumulating iodide elsewhere in the body.

Ascorbic acid is actively transported by the ciliary processes into the posterior chamber against a concentration gradient of 25 to 1·0. The transport mechanism is highly selective for ascorbate and is saturated at plasma levels of 0·2 m moles/l.

Amino acids are transported inward across the ciliary epithelium by at least three carrier systems, one each for acidic, basic and neutral molecules. Iodide, and such organic acids as *p*-aminohippuric acid, iodopyracet, and penicillin are actively moved *outward* from aqueous to plasma across the ciliary processes by systems not involved in aqueous formation.

Movement of uncharged substances across the ciliary epithelium does not involve active transport, but like ionic movement is highly selective.

Acknowledgment

This work was supported in part by Research Grant NB 08339 from the National Institute of Neurological Diseases and Blindness of the National Institutes of Health, United States Public Health Service, and by the United States Atomic Energy Commission Contract No. AT(11-1)-2012-3.

REFERENCES

Ballintine, E. J. (1956) in F. W. Newell (Ed.), *Glaucoma, Transactions of the Second Conference*, Josiah Macy Jr. Foundation, Princeton, N.J., p. 109
Ballintine, E. J. and L. Peters (1954) *Amer. J. Ophthal.*, **38**, 153
Bárány, E. and M. Langham (1955) *Acta Physiol. Scand.*, **34**, 99
Becker, B. (1957) *AMA Arch. Ophthal.*, **57**, 793
Becker, B. (1959) *Amer. J. Ophthal.*, **47**, Pt 2, 342
Becker, B. (1960) *Amer. J. Ophthal.*, **50**, Pt 2, 192
Becker, B. (1961a) *Amer. J. Physiol.*, **200**, 804
Becker, B. (1961b) *Arch. Ophthal.*, **65**, 837
Berggren, L. (1960) *Acta Physiol. Scand.*, **48**, 461
Berggren, L. (1964) *Invest. Ophthal.*, **3**, 266
Berggren, L. (1965) *Invest. Ophthal.*, **4**, 83
Berliner, R. W. (1956) in F. W. Newell (Ed.), *Glaucoma, Transactions of the Second Conference*, Josiah Macy Jr. Foundation, Princeton, N.J., p. 105
Cameron, E. and D. F. Cole (1963) *Exp. Eye Res.*, **2**, 25
Cole, D. F. (1960a) *Brit. J. Ophthal.*, **44**, 225
Cole, D. F. (1960b) *Brit. J. Ophthal.*, **44**, 739
Cole, D. F. (1961a) *Brit. J. Ophthal.*, **45**, 202
Cole, D. F. (1961b) *Brit. J. Ophthal.*, **45**, 641
Cole, D. F. (1963) *Exp. Eye Res.*, **2**, 284
Cole, D. F. (1964) *Exp. Eye Res.*, **3**, 72
Cole, D. F. (1965) *Docum. Ophth.*, Separatum 21, 116
Cole, D. F. (1969) *Exp. Eye Res.*, **8**, 5
Davson, H. (1956) *Physiology of the Ocular and Cerebrospinal Fluids*, Little, Brown and Company, Boston, p. 340
DeRoetth, A., Jr. (1953) *AMA Arch. Ophthal.*, **50**, 491
Durbin, R. P. (1964) *J. Gen. Physiol.*, **47**, 735
Forbes, M. and B. Becker (1960) *Amer. J. Ophthal.*, **50**, Pt 2, 197
Forte, J. G. (1968) *Biochim. Biophys. Acta*, **150**, 136
Friedenwald, J. S. (1944) *Brit. J. Ophthal.*, **28**, 503
Friedenwald, J. S. and R. D. Stiehler (1938) *Arch. Ophthal.*, **20**, 761
Green, H. and J. L. Sawyer (1959) *Amer. J. Ophthal.*, **48**, Pt 2, 71
Holland, M. G. and C. C. Gipson (1970) *Invest. Ophthal.*, **9**, 20
Holland, M. G. and M. Stockwell (1967) *Invest. Ophthal.*, **6**, 401
Holmberg, Å. (1964) in J. H. Prince (Ed.), *The Rabbit in Eye Research*, Charles C. Thomas, Springfield, Illinois, p. 183

Kinsey, V. E. (1947) *Amer. J. Ophthal.*, **30**, 1262
Kinsey, V. E. (1950) *Arch. Ophthal.*, **44**, 215
Kinsey, V. E. (1953) *AMA Arch. Ophthal.*, **50**, 401
Kinsey, V. E. (1960) *Circulation*, **21**, 968
Kinsey, V. E. (1967) in W. Leydhecker (Ed.), *Glaucoma, Tutzing Symposium*, S. Karger, Basel, Switzerland, p. 15
Kinsey, V. E. (1969) unpublished
Kinsey, V. E. and I. W. McLean (1969) unpublished
Kinsey, V. E. and E. Palm (1955) *AMA Arch. Ophthal.*, **53**, 330
Kinsey, V. E. and D. V. N. Reddy (1959) *Docum. Ophthal.*, XIII, 7
Kinsey, V. E. and D. V. N. Reddy (1962) *Invest. Ophthal.*, **1**, 355
Kinsey, V. E. and D. V. N. Reddy (1964) in J. H. Prince (Ed.), *The Rabbit in Eye Research*, Charles C Thomas, Springfield, Illinois, p. 218
Maurice D. M. (1957) *J. Physiol.*, **137**, 110
Michiels, J. (1968) *L'humeur Aqueuse et La Barriére Hémato-Camérulaire*, Université Catholique de Louvain, Louvain, Belgium
Miller, J. E. (1962) *Invest. Ophthal.*, **1**, 363
Mishra, R. K., L. P. Agarwal, R. Rao and M. Mohan (1963) *Orient. A. Ophthal.*, **1**, 181
Missotten, L. (1964) *Bull. Soc. Belge. d'ophthal. fasc.*, **136**, 206
Noell, W. K. (1953) U.S.A.F. School of Aviation Medicine, Project No. 21-1201-0004 Report No. 1
Reddy, D. V. N. (1967) *Invest. Ophthal.*, **6**, 478
Reddy, D. V. N., C. Rosenberg and V. E. Kinsey (1961) *Exp. Eye Res.*, **1**, 175
Riley, M. V. (1964) *Exp. Eye Res.*, **3**, 76
Riley, M. V. and V. Baroncelli (1964) *Nature*, **201**, 621
Shimizu, H., M. V. Riley and D. F. Cole (1967) *Exp. Eye Res.*, **6**, 141
Ussing, H. H. (1949) *Acta Physiol. Scand.*, **17**, 1
Ussing, H. H. (1950) *Acta Physiol. Scand.*, **19**, 43

II
Cell Water

CHAPTER 7

Water movements in cells

D. A. T. Dick

Department of Anatomy,
The University of Dundee,
Dundee, Scotland

I. INTRODUCTION

A cell swells or shrinks by taking in or giving out water through the cell membrane. The processes which govern the extent and rate of water movement are therefore also those which control the cell volume and are of biological and clinical significance. However the significance of studies of water movements does not end there. From observations of the movement of water, information can be gained first, as to whether certain intracellular materials are in solution or not, and secondly, as to the structure and composition of the cell membrane through which the water moves.

From an experimental point of view studies of water movement fall into two categories:

i) those in which the total change in cellular volume in response to a given change in the external environment is measured and interpreted, that is, the extent of the change from one position of equilibrium with the environment to another.

ii) those in which what is studied is the rate of response to an environmental change, rather than the extent of the response. The latter are frequently regarded merely as studies of the permeability of the cell membrane to water. However, as we shall see later, the cell membrane is not necessarily the only factor governing cell water movement, and it may be preferable to think of the shrinking or swelling process more generally as kinetics of cell volume change rather than merely as a property of the cell membrane alone. For convenience this section classifies water movements on this experimental basis; but an attempt will always be made to look at the results both as providing practical information regarding cell volume changes and as leading to fundamental conclusions about the ultrastructure of the cell.

II. MECHANISMS OF WATER TRANSFER

There is, as yet, no compelling evidence of the active transport of water into or out of living cells, so that, unlike many ions and some nonelectrolytes, water appears to play a passive role in the life of the cell. Cell water fluxes reflect, not water pumping by the membrane, but some change in the external environment of the cell or in the pumping by the membrane of some other cellular constituent such as a cation or a nonelectrolyte. However, any interpretation of water movements in these terms must be based upon a thorough understanding of the physicochemical processes involved (see Section II A). Approximations and assumptions are needed in deriving mathematical expressions for interpreting water flux data; and the use of approximate formulae without an understanding of the approximations can lead to significant errors in interpretation.

Several different mechanisms have been proposed as playing a role in the control of water fluxes into and out of cells:

   i) osmosis
  ii) pinocytosis
 iii) electroosmosis.
 iv) the action of contractile vacuoles.

Of these by far the most important is osmosis.

While osmotic water transfer is, at first sight, a simple physical concept, several assumptions must be carefully noted when cell water fluxes are interpreted osmotically. An osmotic water flux is defined strictly in terms of a semipermeable membrane, that is, a membrane which is permeable only to water and not to solute. Thus it is assumed (a) that the cell membrane is the effective site of control of water flux; (b) that this cell membrane is, in fact, semipermeable; (c) that the cell whose behaviour is being interpreted has reached an equilibrium with its environment, and (d) that the interior of the cell consists of a solution whose behaviour may be compared with that of solutions *in vitro*. This condition particularly applies to the protein component of the cell interior.

These assumptions may be dealt with in turn.

1. *Water Flux Control by the Cell Membrane*

Perhaps the most convincing reason for attributing to the cell membrane the control of cell water fluxes is the electron microscopical demonstration of a constant physical structure or unit membrane at the boundary of all cells. The lipids in the membrane, especially the hydrocarbon chains of the phospholipids, certainly impose a substantial restriction on water fluxes. Nevertheless there is also some evidence that, although the cell membrane controls the water content at osmotic equilibrium, the cell interior is also important in limiting the rate of water flux after displacement from equilibrium (see Section IX A).

2. *Semipermeability of the Cell Membrane*

Although this implies permeability to water, it does not necessarily exclude fluxes of solutes, and particularly of electrolytes, provided such fluxes are equal in both directions across the membrane so that no net flux of solute takes place. This condition is substantially met by most cell membranes over short periods of time. However, over longer times and particularly under nonphysiological or stress conditions, unbalanced solute fluxes can occur; for example, during cold storage of erythrocytes, an uptake of sodium and chloride takes place which exceeds the simultaneous loss of potassium so that a net uptake of solute occurs. Thus the condition of semipermeability will be strictly fulfilled by the cell membrane only during comparatively short periods of time and under strictly physiological conditions. Where these conditions are not observed deviations from the osmotic pressure laws are often attributable, not to some osmotic abnormality of the cellular constituents, but merely to the failure of the cell to observe the fundamental conditions necessary for the application of osmotic laws.

Where some of the solutes inside or outside the cell can pass through the membrane, modified osmotic laws can be employed to cope with the situation. For example, solutes, to which the cell membrane is readily permeable, may simply be ignored in computing osmotic pressures across the cell membrane. With partially permeable solutes, the more complex treatment of the thermodynamics of irreversible processes must be employed in place of simple but approximate osmotic pressure laws (e.g. Johnson and Wilson, 1967).

Simple osmotic laws are in fact only a particular case of the laws of irreversible thermodynamics where the membrane is permeable only to the solvent.

3. *Condition of Osmotic Equilibrium*

There was at one time some doubt as to whether the cell interior was in osmotic equilibrium with its environment since some measurements of the freezing point depression of cellular homogenates indicated hypertonicity. However, measurements of freezing point depression on tissues which had been rapidly boiled to destroy autolytic enzymes (Appleboom and coworkers, 1958), and measurements of the melting point of frozen tissues (Maffley and Leaf, 1959), have shown that the intracellular osmotic pressure is equal to that of the environment (Opie's (1966) claim that the internal osmotic pressure of liver cells is twice that of plasma cannot be accepted since his melting points were determined with a high thermal capacity Beckman thermometer and not with a thermistor such as that used by Maffley and Leaf). An exception is the kidney medulla in which intracellular hypertonicity appears to be developed as an integral element of kidney function (Bray, 1960). Thus, provided sufficient time has elapsed (see Section IX) and no significant solute flows are taking place (see above), a cell is in osmotic equilibrium with the fluid which surrounds it.

4. *Solubility of Intracellular Constituents*

It will probably never be possible to give a general answer to the question of whether or not the intracellular proteins can be regarded as in solution. Proteins which are readily leached out of the cell, are probably in solution;

the difficult question is whether any significant fraction is so complexed as to be truly insoluble in the living state.

Measurements of intracellular viscosity offer perhaps the best overall indication of intermolecular complexing. According to Staudinger (1935), the increase of viscosity of an asymmetrical polymeric colloid suspension over that of the pure solvent is given by the equation

$$\eta - \eta_0 = K\eta_0 nC \tag{1}$$

where η_0 is the viscosity of the pure solvent, n is the number of subunits in the polymer chain, and C is the concentration in g/l. Thus increase of viscosity is governed by the increase in the size of the molecular aggregate even when the concentration is constant. Various authors have produced evidence that the viscosity of the cytoplasm is relatively low in the range 1–14 centipoise and with little or no elasticity (see studies of Wilson and Heilbrunn (1960) on clam eggs, Ashton (1957) on amoeba, Dintenfass (1968) on erythrocytes and Burns (1969) on yeast and *Euglena*). Such figures, no more than 10 times the viscosity of pure water, do not suggest any high degree of aggregation of the intracellular proteins. On the other hand Crick and Hughes (1950), and Hiramoto (1969), who studied the movement of intracellular iron particles in a magnetic field, concluded that a degree of intracellular elasticity was present. It seems likely that such particles, introduced by phagocytosis or by magnetic force, remain surrounded by some form of intracellular membrane which governs their movements. From X-ray diffraction studies on haemoglobin in erythrocytes, Perutz (1948) concluded that 'despite the existence of short range order, the haemoglobin molecules in the red cell are suspended in true solution'.

Thus it seems unlikely that there exists in the cell a degree of intermolecular aggregation sufficient to prevent much of the cell proteins being regarded for osmotic purposes as effectively in solution.

III. OSMOTIC EQUATIONS APPLIED TO LIVING CELLS

Lucké and McCutcheon's (1932) classical equation is still the best

$$\Pi(V-b) = \Pi_0(V_0-b) \tag{2}$$

where Π_0 and V_0 are the original osmotic pressure and cell volume in an isotonic solution and b is the nonsolvent volume in the cell. There are two consequences of this equation by which its validity may be tested:

i) when V is plotted against Π_0/Π a straight line should result.

ii) from equation (2), the slope of this plot should be

$$\frac{dV}{d(\Pi_0/\Pi)} = V_0 - b \tag{3}$$

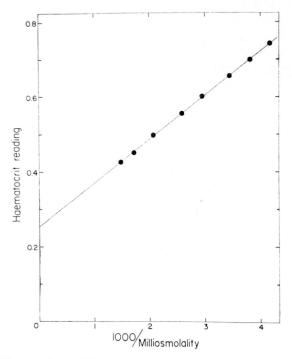

Figure 1. Linear relationship between erythrocyte volume (haematocrit reading) and the reciprocal of external osmotic pressure in accordance with equation (2). (From LeFevre, 1964).

$V_o - b$ is the apparent water content of cell at isotonic osmotic pressure. Condition (i) is obeyed by living cells fairly exactly (figure 1) provided physiological conditions are maintained and no leakage of solute occurs during osmotic water movements. However, $V_o - b$ as measured by equation (2) is never found to be exactly equal to the isotonic water content of the cell (W_m) when measured directly by drying. The discrepancy is conveniently expressed by the ratio

$$\frac{V_o - b}{W_m} = R \tag{4}$$

This ratio and the symbol R given to it were first described by Ponder (1948). R is always found to be less than 1·0 for two main reasons:

i) Under poor physiological conditions or in damaged cells, solute leakage takes place during osmotic adjustments. Such solute movements reduce the volume changes necessary to attain osmotic equilibrium with the external solution. Thus R is much less than 1·0, often 0·5–0·7.

ii) Even under good physiological conditions R is still less than $1\cdot0$. This is because equation (2) is not strictly correct since no account has been taken of changes in the osmotic coefficient of the solute with change of osmotic pressure (or of concentration) of the solution.

As stated by Paterson (see Chapter 5, Volume 1), equation (2) should read

$$\Pi(V-b) = \Pi V_1 = \Phi R T n_2 \tag{5}$$

(This equation has recently been rederived on the basis of classical and irreversible thermodynamics (Nobel, 1969)). Since Φ, the osmotic coefficient varies with concentration, $V_o - b$ in equation (2) is not strictly constant, even when no solute leakage occurs and n_2, the number of grammolecules of solute in the cell, remains the same. If (5) is used in place of (2) then

$$V - b = \frac{\Phi R T n_2}{\Pi} \tag{6}$$

$$\text{and } W_m = \frac{\Phi_o R T n_2}{\Pi_o} \tag{7}$$

Thus $V_o - b = \dfrac{dV}{d\Pi_o/\Pi} = \dfrac{R T n_2}{\Pi_o} \dfrac{d(\Phi/\Pi)}{d(1/\Pi)}$

$$\approx \frac{R T n_2}{\Pi_o} \cdot \frac{\Delta(\Phi/\Pi)}{\Delta(1/\Pi)} \tag{8}$$

for a given change of Π.

Ponder's R is then given by

$$R = \frac{V_o - b}{W_m} = \frac{\Pi_o}{\Phi_o R T n_2} \cdot \frac{R T n_2}{\Pi_o} \frac{\Phi/\Pi - \Phi_o/\Pi_o}{1/\Pi - 1/\Pi_o}$$

$$= \frac{1}{\Phi_o} \frac{\Pi_o\Phi - \Pi\Phi_o}{\Pi_o - \Pi}$$

$$= \frac{1}{\Phi_o} \frac{\Phi_o(\Pi_o - \Pi) + \Pi_o(\Phi - \Phi_o)}{\Pi_o - \Pi}$$

$$= 1 - \frac{\Pi_o}{\Phi_o} \cdot \frac{\Delta\Phi}{\Delta\Pi} \tag{9}$$

Since the intracellular solute contains much protein for which the osmotic coefficient rises steeply with concentration, then $\Delta\Phi/\Delta\Pi$ is in general positive so that $R < 1\cdot0$ as found.

Thus the most fundamental parameter of osmotic equations is Ponder's R. Measurements of b (or $V_o - b$) by osmotic methods are not by themselves of much theoretical value unless simultaneous measurements of W_m are made so as to obtain values of R. Values of R may then be interpreted either in terms of solute leakage i.e. of the integrity of the cell membrane, or of the osmotic coefficient of the intracellular solutes provided solute leakage can be excluded.

IV. CELL VOLUME AND WATER CONTENT: ADJUSTMENT TO THE EXTERNAL OSMOTIC PRESSURE

Owing to the ready availability of the erythrocyte and the ease of obtaining large numbers of uniform cells, its osmotic behaviour has been more studied than that of any other cell. The application of equation (2) to its osmotic behaviour has been confirmed many times (figure 1). Since the intracellular protein concentration of the erythrocyte is high, 33 g haemoglobin per 100 ml of cells, the large osmotic coefficient of haemoglobin will have a significant effect on the value of Ponder's R obtained.

When measurements are made in hypotonic solutions there is good agreement between recent values of Ponder's R and theoretical predictions calculated from equation (9), allowing for the effect of haemoglobin on the average osmotic coefficient of the cell constituents (see table 1 and Dick, 1969). In hypertonic solutions, most recent experimental values are in the range 0·78–0·83 and are considerably lower than those predicted from equation (9), using data for the osmotic coefficient at high haemoglobin concentrations obtained by McConaghey and Maizels (1961) on erythrocytes desalted and shrunken by long exposure to hypertonic lactose solutions. The ionic strength was very low in McConaghey and Maizels' experiments. On the other hand, in erythrocytes, shrunken for a brief period in hypertonic saline solutions as in normal osmotic experiments, the ionic strength is high. Some data of Adair (1967) suggest that the osmotic coefficient rises with the ionic strength. This implies that the osmotic coefficients used to calculate the Ponder's R values shown in table 1 are too low. Theoretical values of Ponder's R at an appropriately higher ionic strength would certainly be lower than those shown in this table but since Adair's data do not extend to sufficiently high concentrations, it is not possible to test whether they would be sufficiently low to account for the low R values found experimentally.

Osmotic responses have, of course, been studied in many other cells and tissues other than red cells. Some results in muscle are summarized in table 2. The R values obtained are low and this may be partly attributable to the high osmotic coefficient of intracellular proteins as in the red cell. Since it is however difficult or impossible to define all the proteins concerned, and even when defined the osmotic properties of muscle proteins have been insufficiently studied, it is not possible to account for the behaviour of

Table 1. Values of Ponder's R in human erythrocytes

Range of relative osmotic pressure	Ponder's R	Reference
	A. Hypotonic Measurements	
1–0·425	0·96–1·05	Guest and Wing (1942)
1–0·5	0·93–0·97	Ponder (1944)
1–0·62	0·99	Ørskov (1946)
1–0·425	0·98	Guest (1948)
1–0·2	0·9	Ponder (1950)
1–0·58	0·97	Hendry (1954)
1–0·62	0·95	Dick and Lowenstein (1958)
1–0·56	0·90	Gaffney (unpublished observations)
1–0·66	0·91[2]	Gary-Bobo and Solomon (1968)
1–0·2	0·963[3]	From Adair's data (1929)
1–0·69	0·953[3]	
	B. Hypertonic Measurements	
1–2·39	0·79	Ørskov (1946)
0·5–1·7	0·78	Ponder and Barreto (1957)
1–2·82	0·89[1]	Olmstead (1960)
1–1·25	1·02[2]	White and Rolf (1962)
0·80–2·33	0·83	Le Fevre (1964)
0·68–1·69	0·80	Savitz, Sidel and Solomon (1964)
0·6–1·8	0·85[2]	Cook (1967)
1–1·65	0·67[2]	Gary-Bobo and Solomon (1968)
1–2·4	0·87[3]	From McConaghey and Maizels' (1961) data
0·6–1·7	0·92[3]	From Adair's (1929) and McConaghey and Maizels' (1961) data

[1]Olmstead's original method of calculation was incorrect and the value shown was recalculated by Le Fevre (1964).
[2]Calculated by present author.
[3]Values computed from Equation 9 for comparison with experimental data.

Table 2. Values of Ponder's R in muscle

Animal	Range of relative osmotic pressure	Ponder's R	Reference
Frog	1·0–2·0	0·83	Dydynska and Wilkie (1963)
Frog	4·0–0·5	0·82[1]	Reuben and coworkers (1963)
Frog	0·39–5·75	0·85[1]	Blinks (1965)
Frog	0·46–2·1	0·83	Gainer (1968)
Crayfish	1–1·78	0·79[1]	Reuben and coworkers (1964)
	1–0·54	0·82[1]	

[1]Assuming $W_m = 0.79$, after Dydynska and Wilkie (1963).

muscle cells with precision. Moreover, in some cases it seems likely that low
R values are due to an apparently physiological mechanism of solute flux
tending to reduce changes of cell volume (e.g. Shaw, 1958; Lang and
Gainer, 1969).

Another factor which may be involved in muscle is the sarcoplasmic
reticulum which appears to behave as an extracellular compartment and
actually *increases* in volume with increase of external osmotic pressure
(Huxley, Page and Wilkie, 1963; Birks and Davey, 1969). Further, McLaughlin
and Hinke (1966) and Hinke (1970), using a variety of techniques, found that
32–42 per cent of fiber water in barnacle muscle was neither osmotically
active nor available as solvent for Na^+, K^+ or Cl^-. Such a large fraction of
osmotically inactive and nonsolvent water cannot be accounted for either as
sarcoplasmic reticulum or by the known osmotic coefficients of muscle
proteins; its physical status remains obscure. It is possible that it may be
related to the large fraction of water in muscle, 72–74 per cent, which has
recently been found by nuclear magnetic resonance studies to be restricted in
its movements (Hazlewood and coworkers, 1969; Cope, 1969).

There has been some controversy as to the cause of the apparent high
osmotic coefficient of the intracellular solutes. There are two main mechanisms which might be responsible for this.

i) Hydration or water binding to protein or other solute molecules. This
mechanism has the attraction of being apparently easy to visualize. The
situation is not, however, as simple as it appears. The mere fact that a protein
binds water which is carried along with it in the ultracentrifuge (*total
solvation* as described by Ogston, 1956) does not necessarily mean that such
water is unavailable to dissolve smaller solutes. Only such water as is
unavailable to small solutes is significant from the osmotic point of view
(this is Ogston's *selective solvation*) and this may be much less than the total
solvation. To attribute the osmotic coefficient of red cell solutes wholly to
'bound water' and then relate this to the selective solvation of haemoglobin
with respect to particular solutes (Savitz, Sidel and Solomon, 1964) is
certainly unsound. As subsequently shown (Gary-Bobo, 1968) the selective
solvation of haemoglobin varies considerably with respect to different small
solutes and for glucose it is zero (in agreement with Miller, 1964). Cook (1967)
found that nonsolvent water for Cl^- was less than 5 per cent although 15
per cent of the water of the red cell was apparently osmotically inactive.

ii) Excluded volume in macromolecular solutions. The volume in a
solution, which molecules of solute have to move around in, is 'free' volume,
while that into which they (or strictly their centres) cannot move is 'excluded'
(see Dick, 1966). The excluded volume consists, of course, of the volume of the
molecules themselves but in addition also includes a space around each
molecule of depth equal to the molecular radius (figure 1). The osmotically

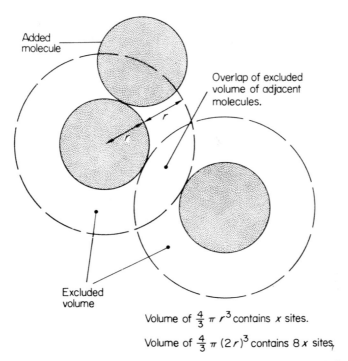

Volume of $\frac{4}{3} \pi r^3$ contains x sites.

Volume of $\frac{4}{3} \pi (2r)^3$ contains $8x$ sites.

Figure 2. The excluded volume is that into which the centre of an added molecule cannot enter; it consists of the volumes of the molecules already present plus a space around each molecule of depth equal to the molecular radius. The total excluded volume for a spherical molecule is thus 8 times the molecular volume (on average slightly less than this owing to the overlap of the excluded volumes of adjacent molecules). (From Dick, 1966).

effective concentration of molecules is given by their number divided by the 'free' volume, not the total volume. Since the excluded volume reduces the 'free' volume it raises the osmotically effective concentration of the molecules; thus the osmotic pressure rises out of proportion to the overall molecular concentration (molecules/total volume); in other words, the osmotic coefficient is greater than 1·0. With small ions and molecules the effect of excluded volume is slight, but it becomes very great with macromolecules at high concentration, such as intracellular proteins, where the osmotic coefficient can exceed 2·0. Thus while water binding to protein solutes certainly occurs, osmotic data must not be interpreted solely in terms of bound water; the excluded volume effect must be taken into account as well.

The theoretical conclusion from osmotic data is thus that in the only case where an intracellular protein is well defined, that of haemoglobin in the erythrocyte, its behaviour is not inconsistent with that of a solution *in vitro;* the status of protein in other cells is unknown. On the other hand, the important empirical fact is that the extent of cell swelling and shrinkage is noticeably less than predicted from simple inverse proportion to the osmotic pressure. The combined effect of the nonsolvent volume and of the high protein osmotic coefficient is such that, for example, in muscle fibres, which comprise three-quarters of the volume of cells of the whole body, shrinkage or swelling is only about two-thirds of the corresponding change of osmotic pressure, i.e. a rise of 25 per cent in serum electrolytes such as can occur in hypernatraemic conditions is followed by only a 16 per cent shrinkage in cell volume. This is a significant mechanism of cellular homeostasis.

V. WATER IN ORGANELLES

A. Water in the Nucleus

The water content of the nucleus is high, 80–88 per cent (v/v). Varying values of the relative solid concentration in nucleus and cytoplasm have been obtained (table 3). In the two egg cells it may be supposed that the low relative solid content of the nucleus is due to the high solid content of cytoplasm due to yolk. Other variations are, however, not so easily accounted for.

The function of the nuclear membrane in controlling nuclear volume has been studied. Although it was early shown that in egg cells the nucleus swells and shrinks along with the cytoplasm during osmotic changes, the nuclear membrane is not semipermeable in the same way as the cell membrane

Table 3. The water content of nuclei

Cell	Fraction of water in nucleus (v/v)	Ratio of nuclear to cytoplasmic solid concentration	Reference
Fibroblast	0·72	3·2	Mellors, Kupfer and Hollender (1953)
Fibroblast	0·88	0·8–1·0	Barer and Dick (1957)
Sea urchin egg	0·87	0·65	Mitchison and Swann (1953)
Rat liver cell	0·88	1·62	Schiemer and coworkers (1967)
Frog oocyte	0·80	0·40	Naora and coworkers (1962)
Thymus	0·83	—	Itoh and Schwartz (1957)

[1]To convert where necessary from (m/m) data, a specific volume of 0·75 ml/g has been assumed for the cell solid.

since it is found to be permeable to salts but impermeable to protein (Battin, 1959; Hunter and Hunter, 1961; MacGregor, 1962). Harding and Feldherr (1959) demonstrated that, while hypertonic salts injected into the cytoplasm of an amphibian egg always cause nuclear swelling, injection of polyvinyl-pyrrolidone or bovine plasma albumin caused shrinkage or swelling according to the concentration of colloid used. The permeability to salts of nuclear membrane of the amphibian oocyte is consistent with the low electrical resistance found by Loewenstein and Kanno (1963), although the resistance of the nuclear membrane of salivary gland cells of *Drosophila* was much higher; considerable variations of salt permeability can thus occur in different nuclear membranes.

B. Water in Mitochondria

Overall the mitochondrion is osmotically rather unresponsive. Values of the apparent osmotically inactive volume b for mitochondria are shown in table 4. Although Tedeschi and Harris (1955) obtained a low value for b of

Table 4. Water compartments in mitochondria

Osmotically inactive volume of mitochondria (b value)	Reference
0·4–0·5	Tedeschi and Harris (1955)
0·75	Amoore and Bartley (1958)
0·80–0·84	Gamble and Tarr (1966)
0·85	Bentzel and Solomon (1967)
0·82	Packer and coworkers (1968)
0·78–0·84[1]	Stoner and Sirak (1969)

[1]Calculated by present author

0·43 and Tedeschi (1961) claimed that this could be wholly accounted for by the volume of solid in mitochondria, recent values are in the range 0·75–0·85, too high to be explained as in whole cells (Section IV). Bentzel and Solomon (1967) described two different compartments in the mitochondrion which offer at least a partial solution to this problem. Thus, the water concentration of isolated mitochondria is 1·9 l/kg dry weight or 0·67 μl/μl wet mitochondria at an osmotic pressure of 272 mOsm (0·20 M sucrose). Of the 0·67 μl of water in each μl of mitochondria 0·46 μl lie in a compartment which is accessible to external sucrose and which does not vary in volume with external osmotic pressure; this compartment thus cannot be surrounded by a semipermeable membrane. The remaining 0·21 μl lie in a compartment not accessible to sucrose but which does respond to external osmotic pressure and is presumably surrounded by a restrictive membrane; the value of Ponder's R for this compartment is 0·49. There are several possible explanations for this finding:

i) as suggested by Bentzel and Solomon, half the water in the sucrose-inaccessible space may be nonsolvent water, presumably bound to solute. Other explanations are, however, possible as discussed above in Sections II and IV, for example

ii) the restrictive membrane may not be perfectly semipermeable, or

iii) large excluded volume effects may occur if the protein concentration in this osmotically responsive volume is very high.

It is not at all clear what possible morphological spaces might correspond to these functional compartments. The space between the inner and outer mitochondrial membranes might be the sucrose-accessible compartment; yet in electron micrographs it is normally smaller in volume than the mitochondrial matrix within the inner mitochondrial membrane, which might represent the sucrose-inaccessible space. These relative morphological volumes are the opposite of those predicted by Bentzel and Solomon's functional studies.

Apart from osmotic changes, mitochondria undergo other changes of volume and water content which are not directly related to the external osmotic pressure. These nonosmotic changes are of two kinds (Lehninger, 1964). In Phase I or low amplitude changes of about 20–40 per cent of mitochondial volume, swelling is produced by changes in respiratory activity, i.e. excess of substrate in absence of ADP causes rapid swelling which is completely reversed by addition of ADP. Phase II or high amplitude changes involve swelling up to 200 per cent of mitochondial volume and are brought about by Ca^{2+} ions, thyroxine, glutathione, and polypeptide hormones containing disulphide bonds such as vasopressin and insulin. The change seems to be related to phosphorylation since it is prevented by dinitrophenol and ATP. It is partially but not completely reversed by ATP.

The mechanism of these nonosmotic changes remains obscure but it has been pointed out by Blondin and Green (1967), Packer and coworkers (1968) and Stoner and Sirak (1969) that they are associated with the internal mitochondrial membrane which during swelling remains intact with reduction of the cristae. Tedeschi and Hegarty (1965) also observed that Ca^{2+} induced swelling of mitochondria could be reversed provided extra NaCl was added to the external medium; some degree of semipermeability thus remains even after high amplitude swelling. Valinomycin and gramicidin, which simultaneously cause mitochondrial swelling and uncoupling of respiration from phosphorylation, are now known to act by increasing the permeability of the mitochondrial membrane to alkali cations, so that in this case at least metabolically linked swelling is fundamentally osmotic in nature (Moore and Pressman, 1964; Chappell and Crofts, 1965).

C. Other Organelles

Evidence of osmotic behaviour has been observed in microsomes (Tedeschi, James and Anthony, 1963), cytoplasmic granules of sea urchin eggs (Harris, 1943), and chloroplast granules (Gross and Packer, 1967).

VI. WATER FLOW NOT ARISING FROM EXTERNAL OSMOTIC CHANGES

A. Pinocytosis

In pinocytosis, a whole droplet of external fluid is engulfed by a cell by means of extensions of its membrane and then gradually absorbed into the cytoplasm. Uptake by this means is therefore, initially at least, completely nonselective. It occurs in amoebae, macrophages, capillary endothelium, smooth muscle, and epithelium of kidney and gall-bladder. It was at first suggested that pinocytosis was a compensatory mechanism of fluid uptake following cell shrinkage; however, pinocytosis has been observed in the amoeba actually following an increase in cell volume (Chapman-Andresen and Dick, 1961) so that this explanation is unlikely. Solute adsorption to the cell membrane precedes or accompanies pinocytosis (Brandt, 1958; Schumaker, 1958; Rustad, 1959; Brandt and Pappas, 1960; Chapman-Andresen, 1962, 1967a; Nachmias, 1968). The process of pinocytosis which follows solute adsorption in *Amoeba* is temperature dependent and prevented by metabolic inhibitors (Chapman-Andresen, 1962, 1967b). Solutes which cause pinocytosis are basic dyes, alcian blue, neutral red and acridine orange, proteins such as bovine plasma albumin, lactoglobulin, lysozyme, and concentrated salt solutions. Proteins only stimulate in acid solution except for those with alkaline isoelectric points which stimulate pinocytosis in neutral solutions; thus only protein bases appear to stimulate. After uptake by the amoeba, alcian blue was found in secondary lysosomes (phagosomes) and was eventually eliminated after five days (Chapman-Andresen, 1967a; Chapman-Andresen and Nilsson, 1967). Gosselin (1967) estimated from uptake of colloidal gold in rabbit macrophages that in each minute between 2 per cent and 20 per cent of the surface membrane was ingested along with the gold by pinocytosis.

Pinocytosis is thus a very rapid process. It is now clear, however, that it is more significant from the point of view of solid rather than fluid uptake into cells. However, it is still possible that it plays a role in transferring solutes and fluids across epithelial membranes (Palade, 1960; Karnovsky, 1968).

B. Electroosmosis

When an electrical potential difference is applied across a membrane which contains fixed electric charges, water is caused to move through the membrane even though there is no difference of osmotic pressure across it; this is called electroosmosis. Although it is easy to detect in artificial membranes, it is only in giant plant cells that it has been possible to detect electroosmosis in a living membrane (table 5). However, an analogous effect, the streaming potential, or voltage caused by water flow through a charged membrane, has been detected in the gall-bladder and in squid axon. From the streaming potential it is possible to calculate the magnitude of the electroosmotic coefficient (Table 5).

Table 5. Electroosmotic coefficients in various tissues

Tissue	Streaming potential (μV/mOsm)	Electroosmotic coefficient (μl/coulomb)	Reference
Chara australis (alga)	—	10	Barry and Hope (1969)
Nitella (alga)	—	20	Fensom and Dainty (1963)
Fish gall-bladder	21	9	Diamond (1962)
Rabbit gall-bladder	36	16	Diamond and Harrison (1966)
Squid axon	6·3	3	Vargas (1968)
Rabbit gall-bladder	47–49·5	21	Wright and Diamond (1969a)

Electroosmotic coefficients are seen to be similar in different tissues and very small. Physiological currents, associated with ion flows, are of the order of 1–10 pM/cm^2/sec, i.e. 10^{-7}–10^{-6}amp/cm^2, so that any associated electroosmotic flow would be 10^{-6}–$10^{-5}\mu$l/cm^2/sec. Such a flow would be small in relation to flows arising from osmotic changes under physiological conditions, for example in an average cell with permeability 10^{-3}cm/sec (see Section IX), the water flow resulting from an osmotic difference of only 10 mOsm would be $2 \times 10^{-4}\mu$l/cm^2/sec. In the same way the streaming potential for an osmotic pressure difference of 10 mOsm is less than 1 mV and is similarly small in relation to cell membrane potentials of around 70 mV. Further, Vargas (1968) and Barry and Hope (1969) have pointed out that approximately half the fluid transport which accompanies an electric current through a cell membrane is due, not to electroosmosis itself, but to the osmotic effect of unequal concentrations in the boundary layers arising from the different transport numbers of the ions in the membrane and adjacent solutions.

C. Action of Contractile Vacuoles

Contractile vacuoles are fluid-containing structures which intermittently expel fluid from the cytoplasm. Their occurrence is limited to certain primitive organisms, protozoa, algae and sponges. Since the work of Kitching (1938), who showed that the rate of contraction of protozoan vacuoles diminished when the osmotic pressure of the medium was raised, it has been concluded that their essential function is to counteract influx of water from a hypotonic environment. Schmidt-Neilsen and Schrauger (1963) and Riddick (1968) confirmed that the fluid expelled by the contractile vacuole in *Amoeba* has only 1/3 to 1/2 of the osmotic pressure of the cytoplasm. In addition to water, however, sodium is also expelled by the contractile vacuole (Chapman-Andresen and Dick, 1962; Riddick, 1968), which appears to form part of the normal sodium pumping mechanism in these organisms.

D. Water Flux Secondary to Ion Transfer

As described in Section III, cells normally maintain impermeability to ions or at least an equality of the fluxes in each direction; when, however, this condition is not maintained and a net flow of ions occurs across the cell membrane, a flow of water accompanies it. Since many cell membranes, and especially that of the erythrocyte, are relatively permeable to anions, it is the cation fluxes which control net salt transfer and thus water movement. The control of erythrocyte volume by changes in the pumping or leakage of cations across the membrane has been worked out in detail by Tosteson and Hoffman (1960). When there is a change of pH, a shift of anions also occurs owing to change in the ionization of haemoglobin; thus occurs the water flux, accompanying the chloride shift, which will be mentioned in Section VII. Net anion transfer can also occur when divalent anions are used in bathing solutions, since to maintain electrical neutrality a one-for-two exchange occurs with internal monovalent anions; the net efflux of anions leads to cell shrinkage (Parpart, 1940; Hempling, 1958).

VII. CELL WATER FLUXES UNDER PHYSIOLOGICAL CONDITIONS

A. Water Flux accompanying the Chloride Shift in Erythrocytes

In the lung the loss of CO_2 and gain of O_2 causes increased acidic ionization of haemoglobin, partly owing to the increased pH, and partly because oxyhaemoglobin is more ionized than reduced haemoglobin. Apart from the exchange of HCO_3^- for Cl^-, which occurs simultaneously, the increase in haemoglobin multianions causes a net displacement of univalent anions from the cell, water is lost, and the erythrocyte shrinks. In the capillaries the reverse changes occur and the cells swell taking up water. The water flux is about 1 per cent of cell volume, i.e. about 2 litres for each complete cycle of all the blood red cells. Tybjaerg-Hansen (1961) has pointed out that, since the red cell normally fits the capillary neatly, some of this water uptake may occur direct across the capillary wall from tissue fluid to red cell and may be an important mechanism of tissue fluid regulation, independent of the colloid osmotic pressure of the plasma. It may be that the oedema which occurs when the red cell does not fit the capillary tightly, i.e. in microcytic anaemia, is to be explained in this way.

B. Cellular Shrinkage during Dehydration of Hypernatraemia

It has already been pointed out in Section IV that, when the plasma osmotic pressure rises, cells do not in general shrink in proportion owing to the nonsolvent volume and to the high osmotic coefficient of the intra-cellular proteins; muscle cells, which make up more than half of the body cells, show only three-quarters of the expected shrinkage. This relative stability of the cell volume may be important in relation to the known deleterious effects of volume change on cell division (Hughes, 1952) (although cell respiration seems to be relatively insensitive (Robinson, 1950, 1952; Olmstead and Granum, 1964)).

Cell water loss on the other hand is proportional to the rise in the external osmotic pressure modified only by the osmotic coefficient of the intracellular solutes (since the nonsolvent volume is unaffected by osmotic changes). McDowell, Wolf and Steer (1955) found that on administering hypertonic solutions of salts intravenously to dogs the apparent volume of distribution was 11 per cent less than the total water volume in the animal. This effect is due to the failure of the nonpenetrating solute to extract as much water as expected from the cells owing to their high osmotic coefficient. Assuming that cell water is 70 per cent of total body water, the average value of Ponder's R for the cells was 0.84, a value comparable to that obtained in muscle (table 2).

C. Cellular Dehydration during Extracellular Freezing

If cooling is gradual, tissues can be lowered to $-20°C$ without intracellular freezing and consequent irreversible damage taking place. When freezing occurs extracellularly, the separation of pure ice causes a rise of external salt concentration and osmotic pressure. This in turn causes a withdrawal of water from the cells, thus lowering the freezing point of the cytoplasm and preventing intracellular freezing. The cells therefore shrink but do not freeze (Meryman, 1956; Mazur, 1963). This effect is clearly important in tissue preservation and in poikilothermic animals during winter.

VIII.   DIFFUSION AND PERMEABILITY

Although other mechanisms of water movement occur, the rate of water movement in and out of cells is usually studied in relation to gradients either of osmotic pressure or of tracer water concentration. As pointed out in Section II, water movement appears to be purely passive. The principal force acting on it is a gradient of concentration or chemical potential which results in a diffusion of water down the gradient (although bulk flow is important in osmosis as discussed in Section IX B, diffusive parameters can be employed initially for both osmotic and tracer movements). Fick's Law states that in a diffusive process the flux of diffusing substance per unit cross-sectional area is directly proportional to the concentration gradient, i.e.

$$\frac{\partial n}{\partial t} = -DA\frac{\partial C}{\partial x} \tag{10}$$

where n is the number of diffusing molecules, A the area available for diffusion, C the concentration, x the distance along the concentration gradient, and D is the constant of proportionality called the diffusion coefficient; it is usually expressed in units of cm^2/sec (The negative sign takes account of the fact that the direction of flow is necessarily the opposite of that of the upward or positive concentration gradient). If diffusion is occurring through a uniform membrane in which the concentration gradient may be assumed to be uniform, then $\partial C/\partial x$ may be replaced by $\Delta C/\Delta x$ where ΔC is the concentration difference across the membrane and Δx is the membrane thickness. Equation (10) thus becomes:

$$\frac{\partial n}{\partial t} = -DA\frac{\Delta C}{\Delta x} \tag{11}$$

In thin biological membranes where there has been some uncertainty as to the thickness, D and Δx are usually combined into one coefficient, k, called the permeability coefficient so that

$$k = \frac{D}{\Delta x} \tag{12}$$

and equation (11) rewritten:

$$\frac{dn}{dt} = -kA\Delta C \tag{13}$$

(k is usually expressed in units of cm/sec).
This is the equation normally used in permeability studies.

Cell water permeability can, as mentioned above, be measured by two completely different techniques, first by causing an actual net flow of water in or out by means of an osmotic gradient, and second, by exposing the cell to tracer water, ^2H_2O, ^3H_2O or $H_2^{18}O$ and measuring its rate of exchange with ordinary water within the cell. In the latter case, there is no net flow of water, only an exchange of labelled for unlabelled molecules. The two techniques involve different mechanisms of water movement, and the two sets of results obtained must be compared carefully and never confused.

The case of tracer water permeability is easily dealt with by means of equation (13). If the cell volume remains constant as it usually does in tracer experiments then,

$$\frac{dn}{dt} = V\frac{dC}{dt} \tag{14}$$

so that equation (13) becomes

$$\frac{dC}{dt} = -k\frac{A}{V}(C-C_e) \tag{15}$$

where C_e is the concentration of tracer in the external medium. If C_e is constant (as it usually is for practical purposes if the volume of external medium is much greater than that of the cells), then equation (15) is easily integrated to give:

$$k = \frac{V}{At} \ln \frac{C_o-C_e}{C-C_e} \tag{16}$$

where C_o is the initial internal tracer concentration (usually zero) and C, the concentration after time, t.

The case of osmotic flow is more difficult since the water concentration difference which causes the flow is usually expressed in terms of the osmotic pressure difference across the membrane, that is according to a modified equation (c.f. equation (13)):

$$\frac{dV}{dt} = kA(\Pi-\Pi_e) \tag{17}$$

(since change of volume is assumed to be due solely to water flux) (Π and Π_e are the osmotic pressures inside and outside the cell respectively). To use this equation, Π is expressed in terms of V by means of equation (2), i.e.

$$\Pi = \frac{\Pi_o(V_o-b)}{(V-b)}$$

Equation (17) thus becomes

$$\frac{dV}{dt} = kA\Pi_o (V_o - b) \left(\frac{1}{V-b} - \frac{1}{V_e - b} \right) \tag{18}$$

In certain cells, for example, the erythrocyte, A remains constant (within limits) as the cell swells or shrinks. Thus (18) may be integrated to give:

$$k = \frac{V_e - b}{V_o - b} \cdot \frac{1}{\Pi_o At} \left[(V_e - b) \ln \frac{V_o - V_e}{V - V_e} + V_o - V \right] \tag{19}$$

If, on the other hand, A changes with cell volume, it has to be expressed in terms of V, for example for spherical cells

$$A = (36\pi)^{\frac{1}{3}} V^{\frac{2}{3}} \tag{20}$$

On inserting this into (18) and integrating, there results

$$k = \frac{V_e - b}{V_o - b} \cdot \frac{V_e^{\frac{1}{3}}}{(36\pi)^{\frac{1}{3}} \Pi_o t} \left[\left(1 - \frac{b}{V_e} \right) \left(\frac{1}{2} \ln \frac{V_e^{\frac{2}{3}} + V_e^{\frac{1}{3}} V^{\frac{1}{3}} + V^{\frac{2}{3}}}{(V_e^{\frac{1}{3}} - V^{\frac{1}{3}})^2} + \right. \right.$$

$$\left. \left. \sqrt{3}\arctan \frac{2V^{\frac{1}{3}} + V_e^{\frac{1}{3}}}{\sqrt{3} V_e^{\frac{1}{3}}} \right) - 3 \left(\frac{V}{V_e} \right)^{\frac{1}{3}} \right]_{V_o}^{V} \tag{21}$$

If the cell volumes are observed initially (V_o), at time t (V), and finally at equilibrium with the external medium (V_e), then the osmotic permeability coefficient, k, can be calculated from (19) or (21). k is thus obtained in units of cm/sec/atm of osmotic pressure. If it is required in the fundamental unit of cm/sec then the term, Π_o, in (19) and (21) must be expressed in terms of the mole-fraction of water in the solution having osmotic pressure Π_o. An equation for doing this is

$$\Pi = \frac{RT}{V_1} (1 - N_1) \tag{22}$$

where N_1 is the mole-fraction of water in a solution. Since $RT/\bar{V}_1$ is equal to 1338 atm at 20°C when there is one atmosphere of osmotic pressure, $(1-N_1)$ equals 1/1338 or N_1 equals 1337/1338. Various units used by different authors may be compared thus. A permeability of 1 cm/sec is equivalent to 449 μ/min, atm at 20°C; $7 \cdot 47 \times 10^{-4}$cm/sec, atm at 20°C; $7 \cdot 23 \times 10^{-7}$cm/sec, cmH$_2$O pressure at 20°C; $0 \cdot 018$ cm/sec, osm.

IX. WATER PERMEABILITY OF CELL MEMBRANES

A. Osmotic Water Permeability

Values of water permeability coefficients of animal cells are shown in figure 3, plotted against the cell surface/volume ratio (for a full table of data

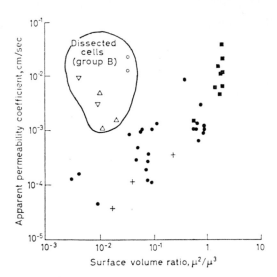

Figure 3. Relationship between apparent osmotic water permeability coefficient of cell membrane and surface/volume ratio in living isolated cells. Most cells are free living but a few were isolated by dissection; except for dissected cells there is a significant correlation. (From Dick, 1966).

see Dick, 1966). The range of permeability coefficients is very large, from 4×10^{-2} to 4×10^{-5}cm/sec; there also appears to be a correlation between water permeability and cell surface/volume ratio for free-living cells (some cells isolated by dissection appear to have higher permeabilities; a number of different reasons for this are discussed below). If taken at its face value, this correlation implies that the membranes of large cells (with small surface/ volume ratios) are either 1000 times thicker or have 1000 times fewer pores than small cells (see Section IX B). However, it is now well established that cell membranes are of relatively uniform thickness, 70–100Å. Variations in pore number are possible, since membrane pores, if they exist at all, have not yet been demonstrated in electron micrographs. However, even if such variations occur, there is no obvious reason why they should be correlated with cell size; *a priori* one might have expected large cells to have more and not less porous membranes as a compensatory mechanism.

Some reasons for the high permeability coefficients of dissected cells appear to be:

i) In ovarian oocytes of amphibia, surface microvilli increase the surface area so that the apparent permeability coefficient is high. If the extra surface is allowed for, a coefficient is obtained about 1/5 to 1/10 of the apparent value (see below).

ii) In giant axons of the squid, definite solute leakage seems to accompany osmotic changes (Hill, 1950).

iii) The calculated values of membrane pores in dissected cells appear much greater than in free living cells (see Section IX B).

It seems unlikely that the correlation between water permeability and surface/volume ratio in free living cells is either accidental or due to changes in membrane properties. A reasonable explanation is that in large cells slow water diffusion in the cytoplasm produces a resistance which is significant in relation to that produced by the membrane, so that permeability calculations which take account only of the membrane give a false impression of low membrane permeability. A recent attempt has been made to assess the relative effects of membrane and cytoplasmic resistance to water flow in the amphibian oocyte (Dick, Dick and Bradbury, 1970). Water permeability appears to be correlated with variation in membrane area, due to development and then regression of surface microvilli, so that the membrane is clearly important in controlling water flow (figure 4a and b). However, even after allowance for microvilli, apparent permeability still decreased with increasing oocyte size, indicating that internal diffusion was also a significant resistive factor (figure 5). It was possible to calculate that the true membrane permeability is in the range $(2–30) \times 10^{-4}$cm/sec and that the diffusion coefficient of water in the cytoplasm is in the range $(6–100) \times 10^{-8}$cm^2/sec. If this type of explanation applies to all the large cells with apparently low membrane permeabilities, then it may be concluded that true membrane permeabilities are of the order of 10^{-3}cm/sec and are relatively uniform. (It must be noted, however, that the calculated true membrane permeability for the oocyte is lower than the apparent permeability of the red cell (table 6), which is a minimum, and probably close, estimate of the true permeability owing to the small amount of cytoplasm.) Two points of comparison may be made.

i) The estimated membrane permeability compares well with the range of values $(1–3) \times 10^{-3}$cm/sec recently obtained for the water permeability of artificial bimolecular phospholipid membranes (Cass and Finkelstein, 1967; Price and Thompson, 1969; Everitt, Redwood, and Haydon, 1969).

ii) The estimated cytoplasmic diffusion coefficient is a mutual diffusion coefficient largely determined by movement of the macromolecular component (see Dick, 1964). It may be compared with the selfdiffusion coefficient of ovalbumin in 24 per cent solution, $8 \cdot 7 \times 10^{-8}$cm^2/sec (Wang and coworkers, 1954). It cannot be compared with the coefficient of diffusion of tracer water in the cytoplasm as has been mistakenly done by some authors (Edelman, 1961; Bunch and Kallsen, 1969; Kushmerick and Podolsky, 1969). Such measurements of tracer water diffusion in cytoplasm are much higher,

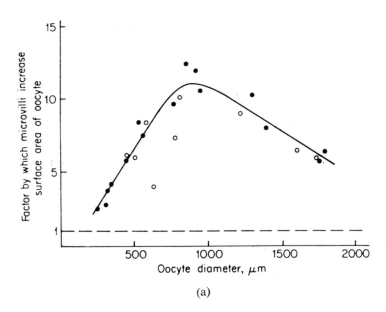

(a)

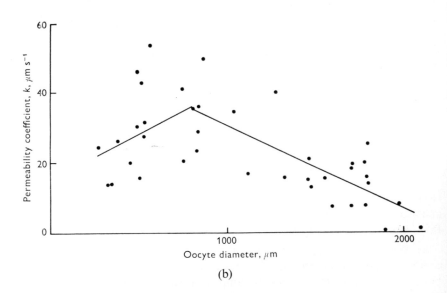

(b)

Figure 4. Relation between size of an amphibian oocyte and (a) increase of surface area produced by microvilli, and (b) apparent osmotic water permeability coefficient (k). Since k varies with microvillar area, the surface membrane is a significant regulator of water permeability. (From Dick and coworkers, 1970).

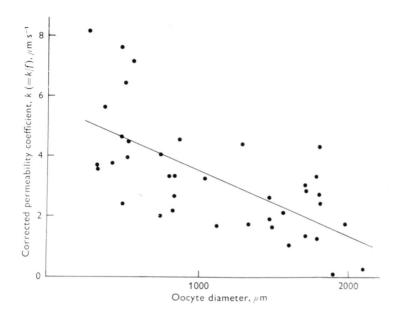

Figure 5. Relation between oocyte size and corrected osmotic water permeability coefficient (*k* divided by the factor by which microvilli increase surface area). The corrected coefficients decrease with increasing oocyte size; since the effect of microvilli has been allowed for, this decrease is probably due to slow water diffusion in the cytoplasm. (From Dick and coworkers, 1970).

$(5–15) \times 10^{-6} \mathrm{cm}^2/\mathrm{sec}$ (Lvøtrup, 1963; Ling and coworkers, 1967; Bunch and Kallsen, 1969), since the exchange of tracer water takes place *between* the macromolecules which do not themselves move.

Thus, except in very small cells, whose diffusion problems are negligible, true water permeability coefficients of cell membranes are probably larger than the apparent ones and similar to those of artificial phospholipid membranes. The diffusional properties of the cytoplasm are not, however, different from those of a colloidal protein solution *in vitro*.

In several cell membranes, it has been found that the rate of osmotic water flow is not linear with the osmotic gradient; high gradients do not produce flows as great as expected or in other words, the osmotic permeability coefficient decreases with rise of osmotic gradient. Diamond (1966) who described this 'nonlinear osmosis' in the gall-bladder, suggested that an osmotic effect on the cell membrane itself might restrict the size of aqueous channels through which flow occurs (see also Rich and coworkers, 1968).

Table 6. Comparison of osmotic and tracer water permeabilities of isolated cells

Cell	Osmotic permeability μ/sec.	Tracer permeability μ/sec.	Pore radius (after Paganelli and Solomon) Å
	A. Free Living Cells		
Frog egg	1·30[1]	0·75[1]	2·1[2]
Xenopus egg	1·59[1]	0·90[1]	2·1[2]
Zebra fish egg	0·45[1]	0·36[1]	0·9[2,3]
Amoeba (*Chaos chaos*)	0·37[1]	0·23[1]	1·8[2]
Amoeba proteus	1·2[4]	0·21[5]	6·9[2]
Beef erythrocytes	156[6]	51[6]	4·1[6]
Dog erythrocytes	200[13]	44[13]	5·9[13]
Human erythrocytes	127[6]	53[6]	3·5[6]
Foetal erythrocytes	117[7]	32[8]	3·9[8]
Human erythrocytes	122[5]	48[8]	4·5[14]
	B. Dissected Cells		
Frog ovarian egg	89·1[1]	1·28[1]	29·9[2]
Zebra fish ovarian egg	29·3[1]	0·68[1]	23·1[2]
Squid axon	11[9]	1·4[9]	8·5[9]
Squid axon	49[10]	1·3[11]	21·5[2]

[1]Prescott and Zeuthen (1953)
[2]Calculated by present author
[3]This value cannot be physically valid
  since it is less than the radius of
  the water molecule, 1-5Å
[4]Mast and Fowler (1935)
[5]Prescott and Mazia (1954)
[6]Villegas *et al.* (1958)

[7]Sjolin (1954)
[8]Barton and Brown (1964)
[9]Villegas and Villegas (1960)
[10]Hill (1950)
[11]Nevis (1958)
[12]Shaafi *et al* (1967)
[13]Rich *et al* (1967)
[14]Solomon (1968)

B. Comparison of Osmotic and Tracer Water Permeability

1. *Evidence for Membrane Pores*

As pointed out in Section VIII, water permeability measured by tracer water indicates the rate of an exchange of water molecules due to diffusion; there is no net flow of water such as takes place during osmotic permeability measurements. During such osmotic flow in a porous membrane the principal mechanism of water movement is not diffusion but bulk flow through the pores. The effect has been well explained by Dainty (1965) (figure 6). Robbins and Mauro (1960) showed that in a porous collodion membrane less than 1/36 of the flow was due to diffusion, the remainder being a bulk flow. Since bulk flow is a much more rapid process than diffusion the osmotic water permeability coefficient in a porous membrane will be higher than the tracer coefficient. This difference has been used as a test to determine whether biological membranes are in fact porous.

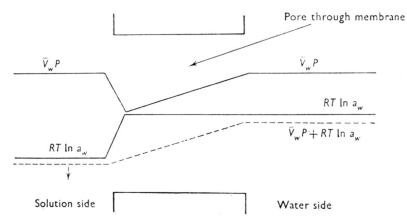

Figure 6. Potential profiles within a pore in a membrane separating a solution from pure water; there is no difference of hydrostatic pressure. The hatched line shows a gradual increase in total chemical potential of water ($\mu = \bar{V}_w P + RT \ln a_w$) from solution to water side of pore. The water activity component of the chemical potential ($RT \ln a_w$) shows a sharp rise at the entrance to the solution end of the pore owing to the exclusion of solute. Since no sudden change of total chemical potential occurs, the pressure component of the chemical potential ($\bar{V}_w P$) must undergo a sharp fall at the same point. A gradient of hydrostatic pressure thus exists within the pore and creates a bulk flow in it. (From Dainty, 1965).

Comparative values of osmotic and tracer water permeabilities for cell membranes are shown in table 6. It is seen that in general the osmotic values are clearly higher. With reservations that will be noted below (see Section X), it has thus been concluded that cell membranes are actually porous. From the osmotic and tracer water permeabilities it is possible to calculate the size of the pores which are believed to exist in the membrane. It is assumed to a first approximation that osmotic flow is wholly bulk flow and tracer flow is wholly diffusive.

Osmotic bulk flow through the membrane pores is given by Poiseuille's Law

$$M = \frac{1}{\Delta \Pi} \frac{dV}{dt} = \frac{n\pi r^4}{8\eta\Delta x} = \frac{Ar^2}{8\eta\Delta x} \tag{23}$$

where M is osmotic flow, n number of pores, r radius of pore, η viscosity, A pore area, and Δx pore length. The validity of this equation as applied to pores of 4–5Å radius is discussed by Mikulecky (1967) and Solomon (1968).

Tracer water flow is given by Fick's Law (equation (11)).

$$m = \frac{1}{\Delta C} \frac{dV}{dt} = \frac{DA}{\Delta x} \tag{24}$$

On dividing (23) by (24) to eliminate A and Δx we obtain

$$r^2 = 8\eta D \frac{M}{m} \qquad (25)$$

In (25) D is taken to be equal to the self-diffusion coefficient for water in bulk, $2\cdot4 \times 10^{-5} \text{cm}^2/\text{sec}$, i.e. it is assumed that the water in the pore is not altered. As will be seen below (see Hays and Leaf, 1962; Derjaguin, 1965) this is not necessarily entirely true. Another difficulty is that (25) has to be modified considerably when the pore size approaches that of the water molecule, as it appears to do in cell membranes (see Paganelli and Solomon, 1957). Values of pore radius calculated from their modified versions of (25) are shown in Column 4 of table 6. For free living cells they vary from $0\cdot9$ to $6\cdot9$Å. In dissected cells they are much larger, and as noted above this may in part account for the very high osmotic permeability of such cells.

Various criticisms have been offered of the interpretation of the ratio of osmotic and tracer permeabilities in terms of membrane pores, especially to take account of unstirred layers on the membrane surface (see Section X).

In at least one cell membrane, that of the marine alga, *Valonia*, no difference has been found between osmotic and tracer water permeability after correction for unstirred layers so that it has been concluded that no membrane pores are present (Gutknecht, 1967, 1968). Gutknecht has pointed out that such an identity also implies that in *Valonia* water flow must be completely surface limited and not affected by internal diffusion. This is, of course, not unexpected since in these cells the protoplasm is a layer only 7–12 μ thick surrounding the vacuole where diffusion must be rapid. However, there remains the problem that the osmotic permeability of *Valonia* is very low $2\cdot4 \times 10^{-4} \text{cm/sec}$. If this is not to be explained by slow cytoplasmic diffusion (Section IX A), then the membrane of *Valonia* must have a real permeability one tenth of that of other cells and of artificial membranes. Both the low membrane permeability and the apparent absence of pores may be explained by the high osmotic pressure (600 mOsm) of the concentrated sea water in which *Valonia* lives. As already mentioned, at high osmotic pressures the osmotic permeability diminishes, probably due to restriction of the pores in the cell membrane (Diamond, 1966; Rich and coworkers, 1968).

C. Energy of Activation of Water Transport

The rate of water permeability varies with temperature. By means of the Arrhenius equation

$$\ln k = \frac{\Delta E}{R} \cdot \frac{1}{T} + \ln K \qquad (26)$$

the activation energy, ΔE, for water transfer may be calculated from the slope of the plot of ln k against $1/T$ (K is an arbitrary constant); values for various cells are shown in table 7. While some values are similar to the activation energy for the self-diffusion of water, $4 \cdot 6$ kcal/mol, others are much higher.

Table 7. Activation energy for water transfer across cell membranes

Cell	Activation energy kcal/mol	Reference
Egg of *Arbacia punctulata*	13–17	Lucké and McCutcheon (1932)
Human erythrocyte (by haemolysis method)	$3 \cdot 9$	Jacobs, Glassman and Parpart (1935)
Sheep erythrocyte	$7 \cdot 6$	Widdas (1951)
Lobster nerve	$2 \cdot 5$	} Nevis (1958)
Squid nerve	$5 \cdot 5$	
Ehrlich ascites tumour cell	$9 \cdot 6$	Hempling (1960)
Toad bladder	$9 \cdot 8 \pm 0 \cdot 4$	Hays and Leaf (1962)
Nitella (alga)	$8 \cdot 5 \pm 1 \cdot 5$	Dainty and Ginzburg (1964)
Artificial phospholipid membrane	$12 \cdot 7 \pm 0 \cdot 3$	Price and Thompson (1969)

Hempling (1960) suggested that the high energy of activation could be explained by assuming that the water in the membrane pore has a quasi-crystalline structure and that for transfer, an exchange must occur between free water outside and quasicrystalline water in the membrane. Hays and Leaf (1962) found that when toad bladder is treated with vasopressin the apparent pore diameter rises from $8 \cdot 4$ Å to 36–41 Å and at the same time the activation energy for water transfer is reduced from $9 \cdot 8$ to $4 \cdot 1$ kcal/mol. This suggests that in the wider pore the relative amount of quasicrystalline water (presumably associated with the pore wall) is less than in the narrow pore. The concept of quasicrystalline water in membrane pores thus receives considerable support. However, Everett and coworkers (1970) have recently shown that Derjaguin's (1965) quasicrystalline 'anomalous water' is a solution, probably of silicic acid.

Price and Thompson (1969) have recently obtained an activation energy of $12 \cdot 7$–$13 \cdot 1$ kcal/mol. for the transfer of water through artificial phospholipid membranes.

X. FURTHER SUPPORT AND CRITICISM OF THE PORE THEORY

A. Membrane Pore Size Measured by Reflexion Coefficients

The reflexion coefficient (σ) is a measure of the relative permeability of a membrane to solutes. It ranges from $\sigma = 1$ for complete impermeability, i.e.

the solute is wholly reflected from the membrane, to $\sigma = 0$ for complete permeability, i.e. no reflexion of solute.

There are several methods for estimating σ. In one method, the concentrations of solutions of nonpermeating solute and of permeating solute separated by a membrane are compared, when no net volume flow (of water and solute combined) takes place across the membrane. Then σ is given by

$$\sigma = \frac{-\Delta\Pi_i}{\Delta\Pi_s} \tag{27}$$

where Π_i and Π_s are the osmotic pressures of the nonpermeating and permeating solutes (the negative sign is due to the fact that the differences are necessarily in opposite directions). In practice $\Delta\Pi_s$ is varied while $\Delta\Pi_i$ remains constant and the resulting net flow across the membrane is measured; the value of $\Delta\Pi_s$ when the net flow is zero is obtained by interpolation.

Another ingenious method of measuring σ for nonelectrolytes is based on measurement of streaming potentials (Wright and Diamond, 1969a). The osmotic flow produced by a nonelectrolyte gives rise to a proportionate streaming potential. σ is given by the ratio of the flow or streaming potential produced by a permeating nonelectrolyte to that produced by a standard nonelectrolyte known to be impermeable, for example sucrose. The method is rapid and by its means large numbers of reflection coefficients have been studied (Wright and Diamond, 1969b).

According to the detailed analysis of nonelectrolyte permeability by Wright and Diamond (1969b), there were two principal factors governing the reflection coefficients of nonelectrolytes:

i) lipid solubility as shown by the oil–water partition coefficient; this governed the permeation of the vast majority; and,

ii) molecular size; this applied only to a few very small molecules with low lipid solubility. These effects were supposed to be related to two possible routes of nonelectrolyte movement through the membrane; the 'lipid route' by dissolving in the hydrocarbon layer of the phospholipids and the 'polar route' by passing through local collections of water molecules or pores lying between the hydrocarbon chains. Wright and Diamond emphasized, however, that no molecule could safely be regarded as passing exclusively by the 'polar route'. With this reservation the reflection coefficients of small nonlipid-soluble nonelectrolytes can be used to estimate the size of the pores by determining the relative ease of passage of molecules of different size by the 'polar route'.

Goldstein and Solomon (1960) used the following equation to connect the reflection coefficient of small nonlipid-soluble nonelectrolytes with the available pore area

$$\sigma = 1 - \frac{A_s}{A_w} \tag{28}$$

(where A_s is the pore area available for solute flow and A_w is the area for water flow), i.e. when $A_s = A_w$, $\sigma = 0$, and when $A_s = 0$, $\sigma = 1$. This equation is only approximate as will be discussed in Section X B2.

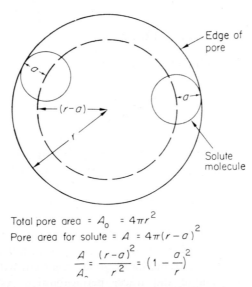

Total pore area $= A_0 = 4\pi r^2$

Pore area for solute $= A = 4\pi(r-a)^2$

$$\frac{A}{A_0} = \frac{(r-a)^2}{r^2} = \left(1 - \frac{a}{r}\right)^2$$

Figure 7. Model of passage of solute molecule (radius a) through a pore (radius r). The area available for passage of the centre of a solute molecule is outlined by the hatched line and is smaller than the area of the pore in the ratio $(1-a/r)^2$:1.

Clearly the available pore area depends on molecular size, the larger the molecular radius (a) the smaller the available pore area (figure 7). A simple equation relating these two quantities is

$$\frac{A}{A_0} = \left(1 - \frac{a}{r}\right)^2 \tag{29}$$

This equation must, however, be corrected for two further effects:

i) laminar flow through the pore means that flow is more rapid in the middle of the pore;

ii) friction between the fluid and the lip of the pore must also be allowed for.

J

Renkin (1954) produced a corrected equation taking account of these effects, thus

$$\frac{A_s}{A_o} = \left[2\left(1-\frac{a_s}{r_p}\right)^2 - \left(1-\frac{a_s}{r_p}\right)^4\right]\left[1-2\cdot104\left(\frac{a_s}{r_p}\right)+2\cdot09\left(\frac{a_s}{r_p}\right)^3\right.$$

$$\left. -0\cdot95\left(\frac{a_s}{r_p}\right)^5\right] = f\left(\frac{a_s}{r_p}\right) \qquad (30)$$

where r_p is the true radius of the pore.

On combining (28) and (30) we obtain

$$\frac{A_s}{A_w} = \frac{f(a_s/r_p)}{f(a_w/r_p)} = 1-\sigma \qquad (31)$$

If σ is measured for a number of solutes whose molecular radius, a_s, is known, then taking 1·5 Å as the average effective radius of the water molecule (see Paganelli and Solomon, 1957), the theoretical relation between σ and a_s can be worked out for various values of r_p. By comparing these theoretical relations with that actually given by a biological membrane, the most appropriate value of r_p for the membrane can be chosen (figure 8). Values of pore radius thus obtained are shown in table 8. They vary from 4 to 8 Å and are thus in remarkably good agreement with those obtained by comparison of osmotic and tracer permeabilities. As mentioned above, however, this calculation takes no account of the passage by the 'lipid route' of a part of the solute used, so that too much weight cannot be placed on the agreement of these methods on the value of the pore size.

Table 8. Pore radii of cell membranes estimated from reflexion coefficients

Cell	Pore radius Å	Reference
Human erythrocyte	4·2	Goldstein and Solomon (1960)
Squid axon	4·3	Villegas and Barnola (1961)
Intestinal mucosa	4·0	Lindemann and Solomon (1962)
Kidney (in tissue slice)	5·6	Whittembury and coworkers (1961)
Ependymal epithelium	8·2	Heisey and coworkers (1963)
Toad skin (outer membrane)	4·5	Whittembury (1962)
(inner membrane)	7·0	
Dog erythrocyte	6·2	Rich and coworkers (1967)

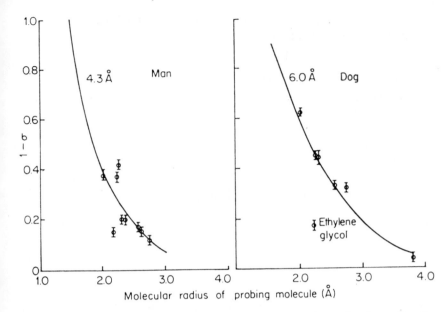

Figure 8. Points show the relation between $(1-\sigma)$ and the molecular radius for a variety of solutes with respect to the membranes of erythrocytes of (a) man, and (b) dog. Continuous lines show calculated relationship according to equation (31) taking $r = 4.3$ Å for man and $r = 6.0$ Å for dog. (From Solomon, 1968).

B. Criticisms of the Pore Theory of Water Transfer

Evidence for the presence of membrane pores has been based on three lines of evidence.

i) Difference between osmotic and tracer water permeability coefficients.

ii) Reflexion coefficients of non-electrolytes.

iii) Activation energy for water transfer.

Objections have been made to all of these lines of evidence and these will be dealt with in turn.

1. *Unstirred Layers*

It has been pointed out by Dainty (1963) and Dainty and House (1966a) that during both osmotic and tracer experiments on biological membranes it is usually assumed that the external solution is well stirred. This is, however, certainly not true in a layer of solution next to the membrane surface in which stirring is incomplete. In this unstirred layer there is a gradient of concentration which reduces the actual concentration gradient within the membrane itself. The permeability coefficient of the membrane thus appears lower than the true value, according to the following equation

$$\frac{1}{P_{\text{apparent}}} = \frac{1}{P_{\text{true}}} + \frac{\delta}{D} \tag{32}$$

where δ is the thickness of the unstirred layer and D is the diffusion coefficient of water in it. Dainty (1963) further showed that, owing to the net flow of water, the unstirred layer is not significant in osmotic experiments, but is important in tracer water experiments; he proposed that the difference described in Section IX B between osmotic and tracer water permeabilities is due to unstirred layers. This view has been confirmed for artificial phospholipid membranes by Cass and Finkelstein (1967), and Everitt, Redwood and Haydon (1969), and also in the marine alga, *Valonia*, by Gutknecht (1967, 1968). However, Shaafi and coworkers (1967) measured the thickness of the unstirred layer in human red cells as 5·5 μ and concluded that this was insufficient to account for the difference between osmotic and tracer permeabilities in this cell. Dainty and House (1966b) have also failed to account for the difference in frog skin, even by an unstirred layer of up to 200 μ. Thus on the basis of this evidence, while pores appear to be excluded in artificial membranes and in *Valonia*, they appear to be present in red cells (see also Solomon, 1968) and frog skin.

2. *Allowance for Solute Permeability in Determining Reflexion Coefficients*

Dainty and Ginzburg (1963) showed that equation (28) is incorrect and should actually read

$$\sigma = 1 - \frac{\omega \overline{V}_s}{L_p} - \frac{A_s}{A_w}$$

where ω is solute mobility coefficient, $\overline{V}_s$ is the partial molar volume of solute and L_p the hydraulic conductivity of the membrane. The extra term expresses the volume flow of *solute* which takes place along with the flow of solvent. However, it has been shown that this term is relatively small compared with σ (Hoshiko and Lindley, 1964; Dainty and Ginzburg, 1964; Rich and coworkers, 1967) so that the calculated reflexion coefficients given in Section X A and the pore radii derived from them are unlikely to be affected. Perhaps a more serious difficulty is the partial passage of the probe molecules by the 'lipid route' (Wright and Diamond, 1969b).

3. *Interpretation of Activation Energy of Water Transfer on the basis of Solubility of Water in Membrane Lipid*

Price and Thompson (1969) have calculated a theoretical activation energy for water transfer on the basis that the hydrocarbon chain element of the phospholipid bilayer has the same properties as bulk hydrocarbons. It was thus concluded that the experimental activation energy was consistent with the hypothesis that water crosses the phospholipid membrane by diffusion after dissolving in it. However, this explanation requires that the water solubility in the membrane be higher than that deduced from olive oil-water partition coefficients. Price and Thompson also noted that their experimental activation energy was consistent with the presence of narrow membrane pores as well as with water solubility in the membrane.

The situation with regard to all three criticisms of the pore theory is thus confused. It seems likely that there will prove to be little difference in reality between a model proposing narrow water filled (probably temporary) apertures between the hydrocarbon chains of the phospholipids and a model proposing high water solubility in a hydrocarbon membrane less than 70 Å thick. As pointed out by Wright and Diamond (1969b), the existence of streaming potentials and electroosmosis in biological membranes implies the presence of channels in which water–ion interaction can occur. Interactions between fluxes of water and of urea, thiourea and acetamide have also been detected (Andersen and Ussing, 1957; Hays and Leaf, 1962).

XI. MECHANISM OF ISOTONIC FLUID TRANSPORT

Minor degrees of coupling between water and nonelectrolyte fluxes have already been mentioned (Andersen and Ussing, 1957; Hays and Leaf, 1962) and these have been associated with interaction in pores in the cell membrane. Across certain epithelia, however, much stronger and more complete interaction occurs between water and NaCl fluxes whereby movement of one NaCl molecule causes the transfer of hundreds of water molecules. Although in intestine, gall-bladder and kidney tubule, water flow can occur against an osmotic gradient, the fluid transported is always isotonic with the bathing solution from which it comes (Diamond, 1964). By a detailed irreversible thermodynamic analysis, Diamond (1962) was able to exclude filtration, normal osmosis, electroosmosis, pinocytosis, and active water transport as possible mechanisms of isotonic water transfer. In later investigations on rabbit gall-bladder, Diamond (1964) showed that the mechanism of water transfer is one in which sodium chloride is transported actively into some confined space within the transporting epithelium; in the space the hypertonic solution so created is enabled to absorb water from the epithelial cells (and ultimately from the luminal fluid) before being eventually discharged on the serosal side. This process was called 'local osmosis'.

Tormey and Diamond (1967) went on to identify the sites of 'local osmosis'. These proved to be the lateral intercellular spaces between the epithelial cells of the gall-bladder; the spaces are shut off from the lumen by tight junctions (zonula occludens) but are open to the serosal side (figure 9). When the gall-bladder secretes actively, they are dilated. Diamond and Bossert (1967) calculated, on the basis that a standing gradient of concentration must exist in the lateral intercellular spaces, that sodium chloride concentrations in the spaces might lie between 300 and 800 mOsm. However, approximate estimates by means of streaming potentials (Machen and Diamond, 1969) have indicated that the average concentration in the spaces is probably around 10 mOsm; the maximum concentrations at the closed end of the spaces would of course be higher than this. In spite of this discrepancy there seems to be no doubt that 'local osmosis' into the intercellular spaces is the mechanism of isotonic water transport in the gall-bladder, and probably also in the intestine and kidney tubule.

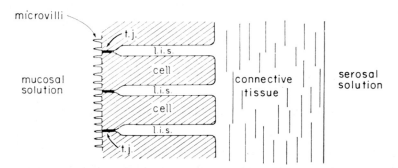

Figure 9. Diagram of gall bladder epithelium (not to scale). Adjacent epithelial cells are separated by long, narrow, lateral intercellular spaces (l. i. s.) open at the serosal end but sealed by tight junctions (t. j.) at mucosal end. Salt and water pass from the mucosal solution to the cells probably passively. Salt is then actively expelled into the lateral intercellular spaces; water follows passively down the osmotic gradient created by the salt since there is time for osmotic equilibrium to be approached (if not achieved) before the fluid leaves the intercellular space. (From Machen and Diamond, 1969).

XII. CONCLUDING REMARKS

Osmosis is by far the most important mechanism of water transfer across cellular membranes although other mechanisms such as pinocytosis, electro-osmosis and contractile vacuoles play subsidiary roles in various cells. By means of osmotic equations, it is possible to predict changes in cell volume produced by changes of osmotic pressure. In making these predictions it is, however, important to take notice of the nonsolvent volume of the cell, which is not osmotically responsive, and also of the osmotic coefficient of the intracellular solutes, especially the proteins. Due to the latter two factors, the volume change of the cell is considerably smaller than that predicted by inverse proportion to the osmotic pressure. The cellular organelles such as the nucleus, mitochondria and microsomes also respond to osmotic changes although especially in mitochondria nonosmotic changes due to metabolic factors also take place.

Studies of the rate of water transfer across the cell membrane both by means of osmotic and tracer water flow lead to the conclusion that some water lies in the hydrocarbon layer of the phospholipid biological membrane. This water may either be in solution in the hydrocarbon layer or lie in narrow and probably temporary water-filled pores. In either case this water component is very important in facilitating water transfer across the membrane and is probably the reason why the biological membrane has a considerably higher water permeability than would be expected from a pure hydrocarbon layer of similar thickness. The water permeability properties of cellular membranes are closely imitated by artificial bimolecular phospholipid membranes. The latter membranes also have similar electrical properties and ionic permeability. The study of artificial membranes is thus contributing greatly to our understanding of biological membranes.

In multicellular epithelial membranes, water transfer is often accompanied by salt transfer in such a way that the total fluid transferred is isotonic with the bathing solutions. Such isotonic transfer appears to be due to excretion of salt into intercellular channels between the epithelial cells. These are so narrow and tortuous that osmotic equilibration takes place between the cells and the fluid in the channels before the latter flows out into the bathing solution on the serosal side of the epithelium. An isotonic effluent is thus produced.

Acknowledgements

This work was supported by a grant from the Medical Research Council.

REFERENCES

Adair, G. S. (1929) *Proc. Roy. Soc. A.*, **126**, 16

Adair, G. S. (1967) Unpublished data quoted by Dick D. A. T. In *Physical Bases of Circulatory Transport.* (Eds. E. B. Reeve and A. C. Guyton, Saunders, Philadelphia, p. 220

Amoore, J. E. and W. Bartley (1958) *Biochem. J.*, **69**, 223

Andersen, B. and H. H. Ussing (1957) *Acta Physiol. Scand.*, **39**, 228

Appelboom, J. W. T., W. A. Brodsky, W. S. Tuttle and I. Diamond (1958) *J. Gen. Physiol.*, **41**, 1153

Ashton, F. T. (1957) *Biol. Bull. Mar. Biol. Lab., Woods Hole*, **113**, 319

Barer, R. and D. A. T. Dick (1957) *Expl. Cell. Res., Suppl.*, **4**, 103

Barry, P. H. and A. B. Hope (1969) *Biophys. J.*, **9**, 729

Barton, T. C. and D. A. J. Brown (1964) *J. Gen. Physiol.*, **47**, 839

Battin, W. T. (1959) *Expl. Cell. Res.*, **17**, 59
Bentzel, C. J. and A. K. Solomon (1967) *J. Gen. Physiol.*, **50**, 1547
Birks, R. I. and D. F Davey (1969) *J. Physiol. (London)*, **202**, 171
Blinks, J. R. (1965) *J. Physiol (London)*, **177**, 42
Blondin, G. A. and D. E. Green (1967) *Proc. Nat. Acad. Sci. U.S.A.*), **58**, 612
Brandt, P. W. (1958) *Expl. Cell. Res.*, **15**, 300
Brandt, P. W. and G. D. Pappas (1960) *J. Biophys. Biochem. Cytol.*, **8**, 675
Bray, G. A. (1960) *Amer. J. Physiol.*, **199**, 915
Bunch, W. H. and G. Kallsen (1969) *Science*, **164**, 1178
Burns, V. W. (1969) *Biochem. Biophys. Res. Commun.*, **37**, 1008
Cass, A. and A. Finkelstein (1967) *J. Gen. Physiol.*, **50**, 1765
Chapman-Andresen, C. (1962) *C.R. Trav. Lab. Carlsberg*, **33**, 73
Chapman-Andresen, C. (1967a) *C.R. Trav. Lab. Carlsberg*, **36**, 161
Chapman-Andresen, C. (1967b) Protoplasma, **63**, 103
Chapman-Andresen, C. and D. A. T. Dick (1961) *C.R. Trav. Lab. Carlsberg*, **32**, 265
Chapman-Andresen, C. and D. A. T. Dick (1962) *C.R. Trav. Lab. Carlsberg*, **32**, 445
Chapman-Andresen, C. and J. R. Nilsson (1967) *C.R. Trav. Lab. Carlsberg*, **36**, 189
Chappell, B. C. and A. R. Crofts (1965) *Biochem. J.*, **95**, 393
Cook, J. S. (1967) *J. Gen. Physiol.*, **50**, 1311
Cope, F. W. (1969) *Biophys. J.*, **9**, 303
Crick, F. H. C. and A. R. W. Hughes (1950) *Expl. Cell. Res.*, **1**, 37
Dainty, J. (1963) *Adv. Botanical Res.*, **1**, 279
Dainty, J. (1965) *Symp. Soc. Exp. Biol.*, **19**, 75
Dainty, J. and B. Z. Ginzburg (1963) *J. Theor. Biol.*, **5**, 256
Dainty, J. and B. Z. Ginzburg (1964) *Biochim. Biophys. Acta*, **79**, 102
Dainty, J. and C. R. House (1966a) *J. Physiol. (London)*, **182**, 66
Dainty, J. and C. R. House (1966) *J. Physiol. (London)*, **185**, 172
Derjaguin, B. V. (1965) *Symp. Soc. Exp. Biol.*, **19**, 55
Diamond, J. (1962) *J. Physiol. (London)*, **161**, 503
Diamond, J. M. (1964) *J. Gen. Physiol.*, **48**, 15
Diamond, J. M. (1966) *J. Physiol. (London)*, **183**, 58
Diamond, J. M. and W. H. Bossert (1967) *J. Gen. Physiol.*, **50**, 2061
Diamond, J. and S. C. Harrison (1957) *J. Physiol. (London)*, **183**, 37
Dick, D. A. T. (1964) *J. Theor. Biol.*, **7**, 504
Dick, D. A. T. (1966) *Cell Water*. London: Butterworths.
Dick, D. A. T. (1969) *J. Gen. Physiol.*, **53**, 836
Dick, D. A. T. and L. M. Lowenstein (1958) *Proc. Roy. Soc. B.*, **148**, 241
Dick, E. G., D. A. T. Dick and S. Bradbury (1970) *J. Cell Sci.* **6**, 451
Dintenfass, L. (1968) *Nature, Lond.*, **219**, 956
Dydynska, M. and D. R. Wilkie (1963) *J. Physiol. Lond.*, **169**, 312
Edelman, I. S. (1961) *Ann. Rev. Physiol.*, **23**, 37
Everett, D. H., J. M. Haynes and P. J. McElroy (1970) *Nature (Lond.)*, **226**, 1033
Everitt, C. T., W. R. Redwood and D. A. Haydon (1969) *J. Theor. Biol.*, **22**, 20
Fensom, D. S. and J. Dainty (1963) *Canad. J. Bot.*, **41**, 685
Gainer, M. (1968) *Bioscience*, **18**, 702
Gamble, J. L. and J. S. Tarr (1966) *Fed. Proc.*, **25**, 631
Gary-Bobo, C. M. (1968) *J. Gen. Physiol.*, **50**, 2547
Gary-Bobo, C. M. and A. K. Solomon (1968) *J. Gen. Physiol.*, **52**, 825
Goldstein, D. A. and A. K. Solomon (1960) *J. Gen. Physiol.*, **44**, 1
Gosselin, R. E. (1967) *Fed. Proc.*, **26**, 987
Gross, E. L. and L. Packer (1967) *Arch. Biochem.*, **121**, 779
Guest, G. M. (1948) *Blood*, **3**, 541
Guest, G. M. and M. Wing (1942) *J. Clin. Invest.*, **21**, 257
Gutknecht, J. (1967) *Science*, **158**, 787
Gutknecht, J. (1968) *Biochim. Biophys. Acta*, **163**, 20
Harding, C. V. and C. Feldherr (1959) *J. Gen. Physiol.*, **42**, 1155

Harris, D. L. (1943) *Biol. Bull. Mar. Biol. Lab., Woods Hole*, **85**, 179
Hays, R. M. and A. Leaf (1962) *J. Gen. Physiol.*, **45**, 933
Hazlewood, C. F., B. L. Nichols and N. F. Chamberlain (1969) *Nature, Lond.*, **222**, 747
Heisey, S. R., D. Held and J. R. Pappenheimer (1963) *Am. J. Physiol.*, **203**, 775
Hempling, H. G. (1958) *J. Gen. Physiol.*, **41**, 565
Hempling, H. G. (1960) *J. Gen. Physiol.*, **44**, 365
Hendry, E. B. (1954) *Edinburgh Med. J.*, **61**, 7
Hill, D. K. (1950) *J. Physiol., Lond.*, **111**, 304
Hinke, J. A. M. (1970) *J. Gen. Physiol.* In press.
Hiramoto, Y. (1969) *Exp. Cell Res.*, **56**, 201
Hoshiko, T. and B. D. Lindley (1964) *Biochim. Biophys. Acta*, **79**, 301
Hughes, A. F. W. (1952) *Quart. J. Micr. Sci.*, **93**, 207
Hunter, A. S. and F. R. Hunter (1961) *Expl. Cell Res.*, **22**, 609
Huxley, H. E., S. Page and D. R. Wilkie (1963) *J. Physiol. (London)*, **169**, 325
Itoh, S. and I. L. Schwartz (1957) *Am. J. Physiol.*, **188**, 490
Jacobs, M. H., H. N. Glassman and A. K. Parpart (1935) *J. Cell. Comp. Physiol.*, **7**, 197
Johnson, J. A. and T. A. Wilson (1967) *J. Theor. Biol.*, **17**, 304
Karnovsky, M. J. (1968) *J. Gen. Physiol.*, **52**, 645
Kitching, J. A. (1938) *Biol. Rev.*, **13**, 403
Kushmerick, M. J. and R. J. Podolsky (1969) *Science*, **166**, 1297
Lang, M. A. and H. Gainer (1969) *J. Gen. Physiol.*, **53**, 323
Le Fevre, P. G. (1964) *J. Gen. Physiol.*, **47**, 585
Lehninger, A. L. (1964) *The Mitochondrion*. W. A. Benjamin; New York
Lindemann, B. and A. K. Solomon (1962) *J. Gen. Physiol.*, **45**, 801
Ling, G. N., M. M. Ochsenfeld and G. Karreman (1967) *J. Gen. Physiol.*, **50**, 1807
Løvtrup, S. (1963) *J. Theor. Biol.*, **5**, 341
Loewenstein, W. R. and Y. Kanno (1963) *J. Cell Biol.*, **16**, 421
Lucké, B. and M. McCutcheon (1932) *Physiol. Rev.*, **12**, 68
McConaghey, P. D. and M. Maizels (1961) *J. Physiol. (London)*, **155**, 28
McDowell, M. E., A. V. Wolf and A. Steer (1955) *Am. J. Physiol.*, **180**, 545
MacGregor, H. C. (1962) *Expl. Cell Res.*, **26**, 520
Machen, T. E. and J. M. Diamond (1969) *J. Membrane Biol.*, **1**, 194
McLaughlin, S. G. and J. A. M. Hinke (1966) *Canad. J. Physiol. Pharmacol.*, **44**, 837
Maffly, R. H. and A. Leaf (1959) *J. Gen. Physiol.*, **42**, 1257
Mast, S. O. and C. Fowler (1935) *J. Cell Comp. Physiol.*, **6**, 151
Mazur, P. (1963) *J. Gen. Physiol.*, **47**, 347
Mellors, R. C., A. Kupfer and A. Hollender (1953) *Cancer*, **6**, 372
Meryman, H. T. (1956) *Science*, **124**, 515
Mikulecky, D. C. (1967) *Biophys. J.*, **7**, 527
Miller, D. M. (1964) *J. Physiol. (London)*, **170**, 219
Mitchison, J. M. and M. M. Swann (1953) *Q. J. Microsc. Sci.*, **94**, 381
Moore, C. and B. C. Pressman (1964) *Biochem. Biophys. Res. Commun.*, **15**, 562
Nachmias, V. T. (1968) *Exp. Cell Res.*, **51**, 347
Naora, H., M. Naora, M. Izawa, V. G. Allfrey and A. E. Mirsky (1962) *Proc. Nat. Acad. Sci., U.S.*, **48**, 853
Nevis, A. H. (1958) *J. Gen. Physiol.*, **41** 927
Nobel, P. S. (1969) *J. Theor. Biol.*, **32**, 375
Ogston, A. G. (1956) *Proc. Int. Wool Textile Res. Conf.*, 1955, Vol. B, C.S.I.R.O. Melbourne p. 92
Olmstead, E. G. (1960) *J. Gen. Physiol.*, **43**, 707
Olmstead, E. G. and J. M. Granum (1964) *Proc. Soc. Exp. Biol. Med.*, **116**, 327
Opie, E. L. (1966) *Proc. Nat. Acad. Sci. U.S.*, **56**, 426
Ørskov, S. L. (1946) *Acta Physiol. Scand.*, **12**, 202
Packer, L., J. M. Wrigglesworth, P. A. G. Fortes and B. C. Pressman (1968) *J. Cell Biol.*, **39**, 382
Paganelli, C. V. and A. K. Solomon (1957) *J. Gen. Physiol.*, **41**, 259

Palade, G. E. (1960) *Anat. Rec.*, **136**, 254
Parpart, A. K. (1940) *Cold Spring Harbor Symp. Quant. Biol.*, **8**, 25
Perutz, M. F. (1948) *Nature, Lond.*, **161**, 204
Ponder, E. (1944) *J. Gen. Physiol.*, **27**, 273
Ponder, E. (1948) *Hemolysis and Related Phenomena.* Grune and Stratton, New York
Ponder, E. (1950) *J. Gen. Physiol.*, **33**, 177
Ponder, E. and D. Barreto (1957) *Blood*, **12**, 1016
Prescott, D. M. and D. Mazia (1954) *Expl. Cell Res.*, **6**, 117
Prescott, D. M. and E. Zeuthen (1953) *Acta Physiol. Scand.*, **28**,
Price, H. D. and T. E. Thompson (1969) *J. Mol. Biol.*, **41**, 443
Renkin, E. M. (1954) *J. Gen. Physiol.*, **38**, 225
Reuben, J. R., L. Girardier and H. Grundfest (1964) *J. Gen. Physiol.*, **47**, 1141
Reuben, J. P., E. Lopez, P. W. Brandt and H. Grundfest (1963) *Science*, **142**, 246
Rich, G. T., R. I. Shaafi, T. C. Barton and A. K. Solomon (1967) *J. Gen. Physiol.*, **50**, 2391
Rich, G. T., R. I. Shaafi, A. Romualdez and A. K. Solomon (1968) *J. Gen. Physiol.*, **52**, 941
Riddick, D. H. (1968) *Amer. J. Physiol.*, **215**, 736
Robbins, E. and A. Mauro (1960) *J. Gen. Physiol.*, **43**, 523
Robinson, J. R. (1950) *Proc. R. Soc. B.*, **137**, 378
Robinson, J. R. (1952) *Proc. R. Soc. B.*, **140**, 135
Rustad, R. C. (1959) *Nature, Lond.*, **183**, 1058
Savitz, D., J. W. Sidel and A. K. Solomon (1964) *J. Gen. Physiol.*, **48**, 79
Schiemer, H. G., G. Gunther and D. Sina (1967) *Frank, Z. Path.*, **76**, 427
Schmidt-Nielsen, B. and C. R. Schrauger (1963) *Science*, **139**, 606
Schumaker, V. N, (1958) *Expl. Cell Res.*, **15**, 314
Shaafi, R. I., G. T. Rich, V. W. Sidel, W. Bossert and A. K. Solomon (1967) *J. Gen. Physiol.*, **50**, 1377
Shaw, J. (1958) *J. Exp. Biol.*, **35**, 920
Sjolin, S. (1954) *Acta Paediat. Stockh.*, **43**, Suppl. 98
Solomon, A. K. (1968) *J. Gen. Physiol.*, **51**, Suppl. 335S
Staudinger, H. (1935) *Ber. Dt. Chem. Ges.*, **68**, 2320
Stoner, C. D. and H. D. Sirak (1969) *J. Cell Biol.*, **43**, 521
Tedeschi, H. (1961) *Biochim. Biophys. Acta*, **46**, 159
Tedeschi, H. and D. L. Harris (1955) *Arch. Biochem. Biophys.*, **58**, 52
Tedeschi, H. and H. J. Hegarty (1965) *Biochem. Biophys. Res. Commun.*, **19**, 558
Tedeschi, H., J. M. James and W. Anthony (1963) *J. Cell Biol.*, **18**, 503
Tormey, J. M. and J. M. Diamond (1967) *J. Gen. Physiol.*, **50**, 2031
Tosteson, D. C. and J. F. Hoffman (1960) *J. Gen. Physiol.*, **44**, 169
Tybjaerg-Hansen, A. (1961) *Nature, Lond.*, **190**, 504
Vargas, F. F. (1968) *J. Gen. Physiol.*, **51**, 123
Villegas, R. and F. V. Barnola (1961) *J. Gen. Physiol.*, **44**, 963
Villegas, R., J. C. Barton and A. K. Solomon (1958) *J. Gen. Physiol.*, **42**, 355
Villegas, R. and G. M. Villegas (1960) *J. Gen. Physiol.*, **43**, No. 5, Pt. 2, Suppl. 1, 73
Wang, J. H., C. B. Anfinsen and F. M. Polestra (1954) *J. Am. Chem. Soc.*, **76**, 4763
White, H. L. and D. Rolf (1962) *Am. J. Physiol.*, **202**, 1195
Whittembury, G. (1962) *J. Gen. Physiol.*, **46**, 117
Whittembury, G., N. Sugino and A. K. Solomon (1961) *J. Gen. Physiol.*, **44**, 689
Widdas, W. F. (1951) *J. Physiol. (London)*, **113**, 399
Wilson, W. L. and L. V. Heilbrunn (1960) *Q. J. Microsc. Sci.*, **101**, 95
Wright, E. M. and J. M. Diamond (1969a) *Proc. R. Scc. B.*, **172**, 203
Wright, E. M. and J. M. Diamond (1969b) *Proc. R. Soc. B.*, **172**, 227

III
Transport at Cell Membranes
and its Regulation

CHAPTER 8

Functional aspects of active cation transport

A. G. Lowe

Department of Biological Chemistry
The University, Manchester 13, England

I.   INTRODUCTION

The aim of this article is to describe some of the important cellular processes for which the active transport of ions can be regarded as being directly or indirectly essential. It must always be dangerous to describe any physiological process as being the function or purpose of ion transport, since purpose is an essentially human characteristic. However, it will become apparent that the physiological processes and related matters outlined below would not be possible without active ion transport, and to this extent it seems correct to describe them as being at least part of the function of active ion transport in living organisms.

It would be impossible in the space available to attempt a comprehensive review of every case in which active transport is important, or of the considerable variety of explanations which have been offered concerning the mechanism by which processes such as aqueous secretions are effected. Such a work has been provided by Ussing, Kruhøffer, Thaysen and Thorn (1960). Instead a brief account of some of the better known phenomena of active transport will be given and the different processes related to one another. Clearly, any active transport from one compartment to another would be fruitless if passive movements between the compartments were not restricted, and it follows that any active transport-dependent process in cells must also be a function of cell membrane permeability. This can be seen most clearly in the case of the active exchange of Na^+ and K^+, the result of which varies greatly among different cells and tissues, and it is therefore necessary to take membrane permeabilities into account when considering the function of active transport in any particular case.

II. THE ACTIVE TRANSPORT OF SODIUM AND POTASSIUM

A. General

The existence of an energy-dependent mechanism for the transport of Na^+ and K^+ has been established beyond doubt in many animal tissues. The work of Caldwell and coworkers (1960a, b) on the squid giant axon showed that the energy source for this cellular K^+ uptake and Na^+ extrusion was dependent on ATP rather than respiration or glycolysis directly. Schatzmann (1953) and subsequently many other workers established that the cardiac glycoside, ouabain, and similar substances were specific inhibitors of the Na–K transport mechanism. Ouabain has subsequently been invaluable as a tool for diagnosing a role for Na–K transport in particular cellular processes, and it enabled Skou (1957, 1960) to identify a membrane-bound enzyme prepared from crab nerve as the transport-ATPase or 'Na–K pump'. This ATPase requires Mg^{2+}, Na^+ and K^+ for maximal activity and is inhibited by ouabain. Furthermore, the concentrations of Na^+ and K^+ required to activate the isolated enzyme are similar to those occurring in nerve axoplasm and the cytoplasm of the many other cells from which the transport-ATPase has been isolated. This information is useful in that if we assume a unique biological mechanism for Na–K transport, any process involving this transport must be associated with a Mg^{2+}-dependent $(Na^+–K^+)$-activated and ouabain-sensitive ATPase.

Since much of our knowledge of Na–K transport is derived from work using the giant axon of the squid, it seems appropriate to examine first the function of the process in nerve. The studies of Hodgkin and Huxley (1952), who used the voltage clamp technique on the squid giant axon, established that, at least under physiological conditions, the ionic currents responsible for propagating the nervous impulse are carried by Na^+ and K^+ ions. The electrical potential across the axon membrane has a sign such that the axoplasm is negative relative to the exterior in the resting state. This resting potential, can be considered to arise from the unequal distribution of the permeable K^+ ion across the axon membrane, since the magnitude of the potential (E) is approximately described by:

$$E = -\frac{RT}{zF} \ln \frac{[K^+]_{in}}{[K^+]_{out}}$$

where R is the gas constant, T the absolute temperature, F the Faraday constant, z the charge on the K^+ ion, and $[K^+]$ in and $[K^+]$ out are the K^+ concentrations inside and outside the axon. On stimulation, the axolemma apparently changes its permeability allowing the previously impermeable Na^+ ions to enter the axoplasm down the electrical and chemical gradient, thereby rendering the axoplasm positive relative to the outside. Increased Na-permeability is only transient, however, so that after about 1 msec K^+ again becomes the dominant permeable cation. Since the membrane potential is then positive inside, where K^+ is at high concentration, there follows a net outward movement of K^+, which restores the membrane potential to its initial negative value. It is important to note that ouabain has no direct effect on the conduction of the nervous impulse, indicating that the Na–K pump is not involved, and this is further confirmed by the fact that giant axons can conduct impulses when the axoplasm is replaced by appropriate ionic solutions lacking ATP, and usually containing K^+. Interestingly, Tasaki and Singer (1966) have shown that squid giant axons can conduct action potentials of nearly normal characteristics when the external solution contains 300 mM hydrazinium chloride and 200 mM calcium chloride (lacking Na^+), while the axoplasm is replaced by 90 mM NaF and 5mM Na phosphate, pH 7·3 (lacking K) with glycerol to maintain osmolarity. This makes it possible that it is the properties of charged groups in the membrane itself, rather than movements of Na and K, that are fundamental to the conduction of the nervous impulse.

Despite this, it is clear that under physiological conditions each nerve impulse is accompanied by a net loss of Na of the order of 5 pmole/cm² and a similar gain of K^+ by the giant axon. To allow continuing nervous function these changes, together with other changes in axoplasmic Na^+ and K^+ arising from passive leakage in the resting nerve, must be restored by metabolism. The Na–K pump can be regarded as instrumental in this process in nerve and in other tissues such as striated muscle, where action potentials also occur.

B. The Maintenance of Cell Volume and Ionic Composition

An important characteristic of living cells is that they contain a high concentration of impermeable proteins and other impermeable anions. These species are necessarily associated with osmotic forces which may be in excess of those arising from impermeable solutes in the outside medium, so that in the absence of a balancing force, swelling and eventual bursting would be expected in a water-permeable cell. It can be demonstrated that if two saline solutions are separated by a membrane which is impermeable to an anionic protein present in one compartment only, but permeable to salt, and the two compartments are kept at constant volume, an equilibrium is established such that the excess osmotic pressure in the protein compartment is balanced by an electrical potential across the membrane. At this, the Gibbs–Donnan equilibrium, the concentration of permeant cations is higher in the protein compartment, the converse being true for permeant anions. The electrical potential across the membrane can be described by the Nernst equation:

$$E = - \frac{RT}{z\mathrm{F}} \ln \frac{[\text{cation}]_{\text{in}}}{[\text{cation}]_{\text{out}}} = \frac{RT}{z\mathrm{F}} \ln \frac{[\text{anion}]_{\text{in}}}{[\text{anion}]_{\text{out}}}$$

In cells such as the erythrocyte, many of the above conditions of the model system apply. The cells contain a high concentration of haemoglobin, and the chloride concentration of the plasma exceeds that of the cells. It is usually assumed that chloride is at the Gibbs–Donnan equilibrium distribution since the erythrocyte membrane is highly permeable to chloride. However, the erythrocyte differs from the model in that its volume is not fixed and therefore swelling of the cell would be expected as water entered down the osmotic gradient. It is evident that some means of 'bailing' out osmotically active material must exist in the cell. That this is the case is suggested by the fact that if erythrocytes are kept at $0°$, where metabolism is minimal, cell volume slowly increases, while cell K^+ decreases and cell Na^+ increases. These changes are slow because of the low membrane permeability to Na and K ions, which must enter the cell together with water and chloride to permit swelling, but when the permeability of erythrocytes is increased by surface-active agents such as digitonin (Davson and Danielli, 1938) rapid swelling followed by lysis occurs.

Under normal conditions it appears that the active transport of Na^+ and K^+ across the erythrocyte membrane counteracts the tendency for a net gain of cellular cations. The studies of Whittam and Ager (1965) and Garrahan and Glynn (1967) have shown that the erythrocyte's Na–K pump extrudes 3 Na$^+$ and takes in 2 K$^+$ for each molecule of ATP hydrolysed. If this Na:K ratio is correct, the pump must be electrogenic, as indicated by the work of Kernan (1962) and Adrian and Slayman (1966) on frog muscle. An electrogenic Na-pump would require that chloride accompanied any Na$^+$ extruded in excess of absorbed K$^+$, so that operation of the Na–K pump would effect a net loss of cellular salt together with osmotically associated water. Even if the pump were not electrogenic, its operation would lead to loss of cellular salt, provided the rate of passive inward leakage of Na$^+$ were less than the outward leakage of K$^+$. The control of erythrocyte volume has been extensively studied by Tosteson and Hoffman (1960).

Volume control by the Na–K pump is certainly important in many, if not all animal cells. For instance, Leaf (1956) showed that slices of pig kidney cortex, rat liver, and rat cerebral cortex swelled when metabolism, and therefore Na–K transport, was inhibited. Swelling was accompanied by increased cellular Na$^+$ and chloride, and by loss of K$^+$, in keeping with the changes described in the erythrocyte.

C. Importance of Cell K Concentration in Metabolism

One consequence of the operation of the Na–K pump is normally a high concentration of cell K$^+$ relative to cell Na$^+$ and to blood K$^+$. Data for effects of Na$^+$ and K$^+$ on enzymic activity are not very plentiful, but table 1 lists some of the enzymes for which effects of alkali metals have been measured. All the listed enzymes are substantially stimulated by K$^+$, and several are inhibited by Na$^+$, although antagonism between Na$^+$ and K$^+$ was usually not tested. It is notable that many of the enzymes in table 1 are catalysts of synthetic reactions involving ATP, and that two are glycolytic enzymes, so that the importance of cytoplasmic K$^+$ in both glucose degradation and general anabolism is manifest. As is predictable from this, glycolysis itself, at least in a cell-free extract from *Lactobacillus arabinosus* (Clark and MacLeod, 1954) is stimulated by K$^+$. Effects of Na$^+$ and K$^+$ have not been investigated for many enzymes and therefore it is quite likely that further studies will reveal stimulatory effects of K$^+$ on many more enzymes, and on cell metabolism in general.

The mechanism by which K$^+$ stimulates enzymes is not clear. However, in the case of enzymes metabolizing ATP it is possible that K$^+$ plays some part in the binding of ATP to the active centre and in catalysing transfer of the terminal ATP phosphate group to the substrate phosphate acceptor (Lowenstein, 1960). This implies that the size of the K$^+$ ion relative to the Na$^+$ ion is critical in such transfer reactions, and a distinction of this type may

Table 1. Enzymes activated by potassium

Enzyme	Source	Effect of Na^+	References
Pyruvate kinase	Muscle	−	Kachmar and Boyer (1953)
Fructokinase	Liver	+	Parks and coworkers (1957)
Adenylate kinase	Erythrocyte membrane	+	Askari and Fratantoni (1964)
Deoxyguanylate kinase	*E. coli*	0(+)	Bessman and Van Bibber (1959)
Formyl glycineamidine phosphoribosyl synthetase	Liver	−	Melnik and Buchanan (1957)
Acetyl CoA synthetase	Heart	0	Von Korff (1953)
S-Adenosyl methionine synthetase	Liver, yeast	−	Mudd and Cantoni (1958)
Pantothenate synthetase	*E. coli*	0	Maas (1952)
Carbamyl phosphate synthetase	Liver	−	Marshall and coworkers (1961)
γ-Glutamyl cysteine synthetase	Wheat	0(−)	Webster and Varner (1954)
Glutathione synthetase	Liver	0	Snoke and coworkers (1953)
D-Alanyl-D-alanine synthetase	*Streptococcus*	0	Neuhaus (1962)
Tyrosine activating enzyme	Liver	0	Schweet and Allen (1958)
Phosphodiesterase	*E. coli*	0	Spahr and Schlessinger (1963)
5′-Adenylic acid deaminase	Erythrocyte	+	Askari (1963)
Formimino tetrahydrofolic acid cyclodeaminase	Liver	0	Tabor and Wyngarden (1959)
Tryptophanase	*E. coli*	−	Happold and Struyvenberg (1954)
L-Serine dehydrase	Liver	0	Nishimura and Greenberg (1961)
Aldolase	Yeast	0	Richards and Rutter (1961)
Malic enzyme	*Lactobacillus*	0	Nossal (1951)
Propionyl CoA carboxylase	Yeast	0	Edwards and Keech (1968)
Phosphotransacetylase	*Clostridium*	−	Stadtman (1952)
AICAR transformylase	Liver	0	Flaks and coworkers (1957)
Aldehyde dehydrogenase	Yeast	−	Black (1951)
Inosine-5′-phosphate dehydrogenase	*Aerobacter*	0	Magasanik and coworkers (1957)
Glycerol dehydrogenase	*Aerobacter*	0	Lin and Magasanik (1960)
β-Galactosidase	*E. coli*	+	Cohn and Monod (1951)
Myosin-ATPase (with EDTA)	Muscle	0	Kielley and coworkers (1956)

Under the heading 'Effect of Na^+', − indicates that Na^+ antagonised the activation by K^+, + indicates that Na^+ stimulated (somewhat less than K^+), 0 indicates that Na^+ had no activating effect and in most cases that possible antagonism with K^+ was not measured.

also be an essential factor in the mechanism of the transport ATPase itself (Lowe, 1968).

D. Cation Gradients and the Transport of Amino Acids and Sugars

There are a number of instances in which the accumulation of amino acids or sugars by cells has been shown to be dependent on the gradient of Na^+ concentration between a cell and its environment. Examples are the accumulation of glycine by duck erythrocytes (Christensen and coworkers, 1952) and the uptake of sugars across the intestinal wall of the toad (Csáky and Thale, 1960). To the extent that cell to plasma Na^+ gradients are dependent on the Na–K pump, such accumulation processes can be regarded as functions of active Na–K transport. This notion is supported by the findings that ouabain inhibits accumulation of, for instance, 3-O-methylglucose across the toad intestine (Csáky and coworkers, 1961) and that anaerobiosis prevents accumulation of 6-deoxyglucose (Crane and coworkers, 1961), while poisoning of respiration by cyanide prevents glycine accumulation in ascites tumour cells (Eddy and Mulcachy, 1965). These results are compatible with direct participation of the Na–K pump, but this possibility is made improbable by the fact that even when accumulation is inhibited, rapid exchange diffusion of sugars or amino acids continues.

Both Na-dependent accumulation and facilitated diffusion of, for instance, 6-deoxyglucose in the intestine are inhibited by phlorizin (Crane and coworkers, 1961) and since facilitated diffusion of glucose in the human erythrocyte is also phlorizin-sensitive, the mechanisms of these processes may be related. Cotransport of glucose with sodium could be explained by the existence of a 'carrier' or 'porter' protein in the cell membrane of intestinal epithelia. Such a carrier would have the property of binding glucose and Na^+ cooperatively, and would effect transport either by providing a specific and static diffusion pathway across the membrane, or by itself diffusing across the lipid phase of the membrane. The kinetics of transport of 6-deoxyglucose across hamster intestine (Crane and coworkers, 1965) are compatible with such a system. Similarly, the kinetics of glycine uptake into Ehrlich ascites tumour cells (Eddy, 1968) indicate cooperative transport of glycine with Na^+ and here there is evidence of one-to-one transport of glycine and Na^+. Vidaver (1964a, b) has also obtained good evidence for cotransport of glycine with Na^+ in pigeon erythrocytes.

III. THE SECRETION AND ABSORPTION OF SOLUTES AND WATER

A. Amphibian Skin

Perhaps the most striking physiological phenomena associated with active ion transport are the secretion and absorption of body fluids in higher animals. The concept of the linkage between active Na-transport and such processes was developed by Ussing (1948, 1954) from studies of amphibian skin, which is capable of absorbing NaCl from dilute solution outside the animal. This uphill transport can be related in mechanism to the regulation of cell volume, and may in fact be an evolutionary development of the ability of cells to control their ion and water contents.

An understanding of the mechanism of absorption of NaCl has been facilitated by studies of the permeability of the frog skin to Na^+ and K^+. Ussing and his coworkers (Ussing and Zerahn, 1951; Koefoed-Johnsen and Ussing, 1958) measured an electrical potential across the frog skin of about 100 mV when both sides were bathed in frog Ringer. Increasing the K^+ concentration inside the skin decreased this potential in accordance with the Nernst equation, indicating that this surface is K-permeable, whereas similar variations in Na^+ concentration did not modify the potential significantly, indicating that Na^+ is impermeable. However, when concentrations of Na^+ and K^+ were varied at the outer surface of the skin, Na^+ was found to be permeable and K^+ impermeable. These experiments make it clear that the frog skin is asymmetric with respect to Na^+ and K^+ permeability, and further studies have indicated (Ussing, 1963, 1964) that this asymmetry is a property of the skin epithelia. Assuming that ions and water can cross the skin only by passing through the epithelia, and that chloride is permeable to both inner and outer epithelial membranes, it can readily be seen that operation of the Na–K pump at the inner epithelial surface will lead to transport of NaCl from the medium outside to that inside the skin. Chloride is expected to accompany the actively transported Na^+ in the same way as during the control of cell volume, but the transcellular transport of NaCl arises from the different Na-permeabilities of the inward and outward-facing cell surfaces and presumably from the situation of the Na–K pump chiefly at the inward surface.

Confirmation that the Na–K pump is the driving force for this salt transport comes from the effects of inhibitors. Francis and Gatty (1938) and Huf and coworkers, (1957) showed that cyanide, inhibitors of glycolysis (e.g. iodoacetate) and uncouplers of oxidative phosphorylation, prevented salt uptake. About 20 Na^+ ions are transported per mole oxygen consumed and associated with active transport (Zerahn, 1956), which indicates a value of about 3 Na^+ ions per mole of ATP synthesized via oxidative phosphorylation in the mitochondria. This is consistent with the stoicheiometry of about 3 Na^+ ions transported for each ATP hydrolysed in the erythrocyte and in the squid giant axon. Further evidence for the direct involvement of the Na–K pump is the fact that ouabain inhibits salt uptake when applied to the inside surface but not the outside of the skin (Koefoed-Johnsen, 1957), while application of K-free Ringer to the inside of the skin also substantially reduces active sodium transport (Huf and Wills, 1951; Ussing, 1954).

In the past there has been some discussion as to whether water absorbed in conjunction with salt is taken up by a separate active transport, or simply by osmotic force developed through salt secretion. The second alternative seems more likely since there is little or no direct evidence for active water transport. This raises the question of how a significant osmotic gradient can be developed by secretion of relatively small quantities of salt into a large volume of Ringer or plasma, where dilution would minimize such a gradient. However, microscopic examination of sections of skin epithelia (Ussing, 1963, 1964) has shown that solute-containing channels exist between epithelial cells, and that these are continuous with the internal bathing solution only via narrow openings, barred by a membrane impermeable to colloids but permeable to water and small ions. If these channels are largely unstirred, then secretion of salt into them could be sufficient to generate the required osmotic pressure there, and hence to provide the driving force for exit of water from the cells. Dilution by cell water would also lead to expulsion of the secretion out of the channels into the bathing solution.

Transport of water across the epithelia must also be dependent on the permeabilities of both outer and inner surfaces of the cells to water. Application of the neurohypophyseal hormone, vasopressin (Fuhrman and Ussing, 1951) to the inside of the frog skin stimulated both active Na-transport and the passive permeability of the skin to water, urea and Na^+ itself. Na-transport is also stimulated by application of atropine to the outside of the skin of *Rana esculanta*, presumably because of an effect on the permeability of this surface (Kirschner, 1955). It therefore appears that the permeability of the skin epithelia to both Na and water is a limiting factor in salt and water transport.

B. Ion Transport in the Kidney

Among the most important roles of active ion transport in higher animals is the regulation of body fluid volume and composition, and the recovery of salts and water from glomerular filtrate in the kidney is an important mechanism for this in many animals. Studies of mammalian kidneys have shown that the handling of salts and water is very different in the proximal tubules, the loop of Henle, the distal tubules and the collecting ducts. In the proximal tubules, which receive filtered plasma from the glomerulus, reabsorption of about 85 per cent Na, 85 per cent water, 70 per cent chloride, 95 per cent bicarbonate and a varying proportion of K^+ occurs (Thaysen, 1960), the absorbed solution being isotonic with the filtrate. This phenomenon can readily be interpreted as operation of a Na–K pump at the peritubular surface, in conjunction with low Na-permeability at this surface, where K^+, water and chloride would be permeable, and permeability to both ions and water at the luminal surface, as illustrated in figure 1. If the Na-permeability of the luminal surface, rather than the epithelial water permeability is rate-limiting in transport, then the absorbed solution would be expected to be isotonic with plasma and to contain chloride and bicarbonate in the proportions found in glomerular filtrate. Contrary to this, bicarbonate is absorbed preferentially to chloride, apparently indicating a higher permeability for the large bicarbonate ion. However, administration of Diamox, an inhibitor of carbonic anhydrase (Berliner and coworkers, 1951) results in reduced filtration rate as well as increased bicarbonate excretion. Reduced filtration seems to indicate that bicarbonate in the cell (generated from CO_2 originating from plasma or metabolism through carbonic anhydrase) is necessary to permit a high rate of chloride absorption. The need for bicarbonate could arise because chloride enters the plasma via a specific exchange mechanism with bicarbonate. Conclusive evidence for such an exchange has not been presented, but it will be seen below that such a mechanism would be helpful in explaining many processes of secretion and absorption.

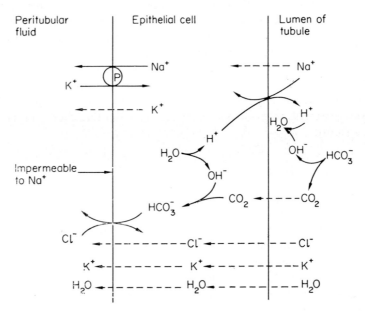

Figure 1. A scheme for the transport of solutes and water across epithelial cells of the proximal kidney tubule. ----→ indicates passive diffusion. Ⓟ indicates transport via the Na–K pump. ⤫ indicates exchange diffusion.

A corollary of the generation of bicarbonate from CO_2 by carbonic anhydrase is the production of protons. Acidification of proximal tubule fluid is observed, and it has been therefore suggested by Thaysen (1960) that protons in the tubular epithelia exhange with Na^+ ions in the tubular fluid, in a process which apparently continues in the distal tubules.

Removal of K^+ from proximal tubular fluid probably occurs by passive diffusion down the gradient generated by absorption of Na^+, chloride, bicarbonate and water. However, Oken and Solomon (1960) found that in *Necturus* kidney the concentration of K^+ in the tubule was about 1·5 times that of the plasma, although this gradient does not exist in proximal tubules (Litchfield and Bott, 1962). Concentration of K^+ in *Necturus* tubules can be accounted for by the electrical potential of about 20 mV between the lumen (negative) and plasma (Giebisch, 1961; Whittembury and coworkers, 1961).

Micropuncture studies have shown that whereas proximal fluid is isosmotic with plasma, early distal fluid is hypotonic (Wirz, 1956). However, analysis of fluid from the bend of Henle's loop (Lassiter and coworkers, 1961) demonstrated that here fluid was 2·8 times hypertonic to plasma, implying that water must be absorbed from the descending arm of Henle's loop. Water absorption from the descending arm is also implied by the fact that fluid/plasma ratios at the bend of Henle's loop are 11 for insulin, 17 for urea and 2·2 for Na^+ (Lassiter and coworkers, 1961). Since tubular fluid becomes hypotonic in the distal tubules, transport of salt without water into the plasma from fluid in the ascending loop of Henle and early distal tubule is to be expected. This would provide a basis for the hypertonic character of the medulla, (Schmidt-Nielsen and O'Dell, 1961) and also the osmotic driving force for absorption of water from the descending loop of Henle into the medulla.

Transport of salt without water in the ascending loop of Henle and early distal tubule requires that the luminal membranes of the tubular epithelia in these regions be impermeable to water. Other features of distal ion transport, which are summarized in figure 2 (a) and (b) are acidification of distal fluid, excretion of K^+, and preferential uptake of chloride rather than bicarbonate. Distal acidification can be interpreted as exchange of luminal Na^+ for cellular protons, as described above for the proximal tubule (Pitts and Alexander, 1945), and it is possible that the occurrence of this exchange mechanism is associated with the low water permeability of the tubular membranes. Orloff and Burg (1960) have also proposed that K^+ is excreted by specific exchange for Na^+ through the same assumed catalytic system. This would explain the observed increase in K^+ excretion and alkalinity of urine during inhibition of carbonic anhydrase (Berliner and coworkers, 1951, 1967), which would normally generate protons competing with K^+ for the exchange system. Similarly, the stopped-flow experiments of Walker and

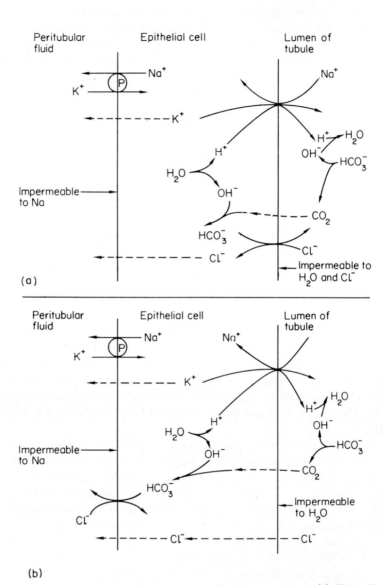

(a)

(b)

Figure 2. Two schemes for the transport of solutes across epithelial cells of the distal kidney tubule. $----\rightarrow$ indicates passive diffusion. $\bowtie$ indicates exchange diffusion. $\underline{\textcircled{P}}$ indicates transport via the Na–K pump.

coworkers (1961) showed that low tubular Na^+ inhibited K^+ secretion into the tubule and are consistent with K^+/Na^+ exchange.

The basis of chloride transport across the distal epithelia is not well understood. However, Rector and Clapp (1962) showed by micropuncture studies that distal fluid chloride was reduced to levels as low as 0·1 mM during sulphate perfusion of salt-depleted rats. The associated gradient of chloride concentration between tubular lumen and plasma could not be accounted for by the transtubular potential of about 60 mV, and is therefore indicative of some type of catalysed chloride transport. Possibly connected with chloride transport are the argentophilic, presumably chloride-rich, cells in the distal epithelia (Okkels, 1929; Ljunberg, 1947). In view of the uptake of both chloride and bicarbonate in the distal tubule and the likelihood that bicarbonate, and possibly also chloride, has a low passive permeability to the epithelia, it seems worthwhile to postulate that these two anions are transported by two types of distal epithelia. Cells of type 1 (figure 2a) might possess a chloride–bicarbonate exchange system at the luminal membrane, which in conjunction with the associated systems and permeabilities shown in figure 2a, would lead to epithelial chloride accumulation and net transport of NaCl from lumen to peritubular fluid. Cells of type 2 (figure 2b) might possess a similar exchange catalyst, this time at the peritubular epithelial membrane, making possible transport of Na^+ with (predominantly) bicarbonate from tubular to peritubular fluid. Bicarbonate would arise by hydration of CO_2, derived from metabolism or by diffusion into the cell from luminal fluid. In both of these cell types, some passive movements of chloride across the luminal membrane might also be involved in solute transport, but in view of the fact that water permeability is low, membrane-catalysed transport of both cations and anions at the luminal membrane is probably more important.

Operation of the distal reabsorption mechanism results in hypotonic distal fluid, whereas in mammals a hypertonic urine may be excreted, implying removal of luminal water without salts. This process apparently occurs in the collecting tubules, which are situated in the medulla. As discussed above, the medulla is rendered hypertonic by secretion of salts by the early distal tubules and ascending loops of Henle, and this hypertonicity presumably provides the osmotic force for water absorption. Urinary hypertonicity is dependent on the presence of vasopressin, which is thought to be necessary to make the collecting ducts permeable to water. Hypotonic urine may also be excreted, and may arise from unchanged distal fluid which has passed through collecting ducts which were impermeable to water.

Another feature of salt-handling in the distal tubules and collecting duct is the secretion of ammonia. Ammonia is probably generated in the epithelial cells chiefly by deamination of glutamine by glutaminase (Van Slyke and coworkers, 1943) and is thought to pass into the urine as the uncharged molecule in conjunction with exchange of cellular protons for luminal Na^+, thus giving urinary NH_4^+ (Sartorius and coworkers, 1949). This effects uptake of Na^+ from the collecting duct without change in osmotic activity, and also leads to urinary acidification. The stopped-flow studies of Jaenike and Berliner (1960) also indicate that exchange of Na^+ for cellular K^+ also takes place in the collecting duct.

As described above, the primary force accounting both directly for recovery of salts, and indirectly for water absorption from glomerular filtrate, is the active transport of Na^+. Direct evidence for participation of the Na–K pump has been found by Schatzmann and coworkers (1958), who showed that ouabain inhibited water absorption by isolated *Necturus* tubules, and by Orloff and Burg (1960) who found that strophanthidin increased Na excretion and flow rate. Additional evidence has been provided by studies of respiration and salt handling by slices of mammalian kidney cortex, although net transport of salts from tubular fluid to medium cannot be demonstrated in such preparations. However, Willis (1968a,b) using hamster and squirrel kidney, has shown that ouabain partially inhibits both oxygen consumption and Na^+ transport in media containing K^+, while Whittam and Willis (1963), Whittembury (1965) and Willis (1966) have demonstrated that ouabain also inhibits about 50 per cent of K^+ uptake by slices of cortex, and that Ringer K^+ is necessary for about half the observed Na transport. Thus participation of the Na–K pump is indicated for at least 50 per cent of Na transport. Willis (1968b) suggests that the remaining Na^+ and K^+ transport is also mediated by the Na–K pump, and that the only partial inhibition by ouabain can be explained by the highly folded nature of the secretory membranes of the tubular cells. The resultant solute channels (cf. frog skin and gall-bladder epithelia) would constitute long diffusion pathways for externally applied solutes and could lead to large differences between the concentration of ouabain in the applied Ringer and in the solution bathing membrane surfaces effecting Na–K transport.

Measurements of the ratio of ouabain-sensitive Na-transport to associated oxygen uptake give values of 7–16 in slices of kidney cortex (Willis, 1968a) and of 18–28 in the whole tissue (Kramer and Deetjen, 1960; Lassen and coworkers, 1961). These are similar to the ratios found for frog skin Na-transport and are therefore compatible with participation of the Na–K pump as discussed above.

C. Osmoregulation in Aquatic Animals

Aquatic animals such as fish and crustaceans provide striking examples of the need for active ion transport to maintain plasma ionic composition constant despite the proximity of the different aqueous environment. Salt exchanges with the environment in both fish and crustacea appear to be restricted mainly to the gills, which necessarily present a fairly large surface area for gaseous exchange with the medium. The efficiency of the osmo-regulatory system is most marked in crustaceans such as the shore crab, *Carcinus maenas*, and the crayfish, *Astacus pallipses*, which are liable to experience substantial changes in the osmotic pressure of the environment. *Astacus*, in particular, can live in tap water even though blood chloride is about 220 mequiv./l. Survival in a dilute medium implies absorption of salts without water to balance salt loss down the plasma–medium concentration gradient. By analogy with the distal kidney tubule, absorptive cells with low permeability to water, a Na–K pump at the cell–plasma membrane, and possibly Na^+–proton and chloride–bicarbonate exchange mechanisms can be predicted. The studies of Maetz and Romeu (1964), Romeu and Maetz (1964) on the goldfish, *Carassius auratus*, and Shaw (1960) on *Astacus pallipses* are consistent with involvement of such systems. These workers found that Na^+ and Cl^- were taken up independently, and that the presence of bicarbonate and ammonium ions in the seawater inhibited uptake of chloride and Na^+ respectively. Maetz and Romeu (1964) also observed that inhibition of car-bonic anhydrase reduces chloride uptake. These authors suggest the existence of a NH_4^+–Na^+ at the external gill membrane, but the results are also com-patible with proton–Na^+ exchange with movement of un-ionized ammonia as postulated in the mammalian kidney.

D. Secretion and Absorption in Tissues other than the Kidney

1. *Salivary glands*

Secretion and absorption of solutes and water can usually be interpreted in terms of the mechanisms suggested for the kidney tubules. The sublingual salivary gland (e.g. of the cat) secretes a solution isotonic with plasma containing approximately 159 mequiv./l Na, 9 mequiv./l K and 16 mequiv./l chloride (Lundberg, 1957). This secretion could reasonably be effected by an Na–K pump situated at the epithelial–salivary duct membrane, which would be water- and K-permeable and Na-impermeable, while the epithelial plasma membrane would be permeable to Na^+, K^+, chloride and water. The level of K^+ in the secretion would represent the steady-state concentration arising from K-leakage into the duct and recovery of cell K by the Na–K pump.

In contrast to sublingual saliva, the parotid secretion is hypotonic to plasma, hypotonicity being greatest at low rates of secretion. To account for this, Thaysen (1960) has suggested that parotid secretion occurs in two steps: first secretion of isotonic saliva similar to the sublingual secretion, and second, reabsorption of salts, possibly in the striated lobular ducts of the parotid gland. Reabsorption could occur by a mechanism similar to that operating in the distal tubules, and the balance between the rates of the secretory and absorptive processes would determine the Na^+ concentration of the final secretion. Interestingly, increasing osmolarity is accompanied by increasing bicarbonate in the saliva, while chloride is actually decreased as secretion increases up to 25 mg/g parotid/min (Thaysen and coworkers, 1954). This may be indicative of a chloride–bicarbonate exchange system.

Another feature of salivary secretion is the transient high salivary K^+ concentration found immediately after stimulation of secretion in submaxillary, parotid and sublingual saliva (Thaysen and coworkers, 1954; Burgen, 1956; Lundberg, 1958). K^+ is lost from the gland cells at this time, whereas K^+ is regained and Na^+ lost at the end of stimulation. This K-transient is probably due to the fact that, in resting glands, epithelia have a low permeability to ions including K^+, so that operation of the Na–K pump depletes the ducts of K^+. Stimulation presumably increases ionic permeability particularly of K^+ at the duct–cell membrane so that K^+ enters the duct down the concentration gradient. Full activity of the Na–K pump at this border would be delayed until the K^+ concentration had built up sufficiently and any overshoot would give rise to the K-transient. Cation permeability at the epithelial–plasma membrane is clearly also necessary for secretion and hence K-exchange at this junction would be expected to increase during stimulation, as is observed in submaxillary gland (Burgen, 1956).

Notably, the rat submaxillary gland is one of the richest known sources of the $(Na^+–K^+)$-activated ATPase (Schwartz and Moore, 1968). This is to be expected in view of the rapid rates of salivary secretion and of the essential role that the Na–K pump is thought to play in secretion.

2. *Sweat*

Human sweat is normally hypotonic to plasma, although osmolarity increases with secretory rate, Na^+ varying from about 5–60 mequiv./l (Robinson and Robinson, 1954). This situation is similar to that with parotid saliva, but sweat differs from this saliva in that chloride concentration parallels tonicity and bicarbonate is always low. Probably the mechanism of secretion is similar to that for parotid saliva and involves an isotonic primary secretion with subsequent reabsorption of salts without water. However, any bicarbonate–chloride exchange appears to be small or absent, although such an exchange would seem appropriate at the plasma–epithelial membrane as a means of eliminating cell bicarbonate and maintaining cell chloride.

3. *Pancreatic secretion*

Human pancreas secretes a solution isotonic with plasma, containing Na^+ as chief cation and with bicarbonate increasing with the rate of secretion inversely to chloride concentration (Bro-Rasmussen and coworkers, 1956). Inhibition of pancreatic secretion by atropine or vagal section (Mellamby, 1925; Thomas and Crider, 1946) does not affect the volume or bicarbonate content of the secretion. This has led Thaysen (1960) to suggest that two types of secretion are involved, the first being of small volume, similar in composition to sublingual saliva, and containing enzymes, and the second being of larger volume and containing bicarbonate as chief anion. The second secretion might involve Na–K pumping and permeabilities similar to those suggested for sublingual saliva, but would differ in having a predominant chloride–bicarbonate exchange system at the duct–epithelial border. Inhibition of carbonic anhydrase (Birnbaum and Hollander, 1953) diminishes both the volume and bicarbonate content of pancreatic secretion, indicating at least that bicarbonate enters the secretion as the ion rather than as CO_2. Ridderstrap and Bonting (1968) have reported evidence for participation of the transport ATPase in pancreatic secretions.

4. *Bile*

The ionic composition of human hepatic bile is approximately 133 mequiv./l Na, 8·7 mequiv./l K, 117 mequiv./l chloride and 3 mequiv./l bicarbonate in addition to bile acids (Hammarsen, 1897). The secretion is isotonic with plasma and its composition is independent of secretory rate, so that a mechanism of secretion similar to that of sublingual saliva seems appropriate. Dog hepatic bile contains chloride and bicarbonate in roughly equal proportion (Ravdin and coworkers, 1932) indicating that in dog a chloride–bicarbonate exchange may occur as in pancreatic secretion.

Hepatic bile is reduced in volume in the gall-bladder, the concentration of residual bile salts increasing about 10-fold (Ravdin and coworkers, 1931, 1932). The gall-bladder lends itself to study of the reabsorbed solution since the intact bladder can be suspended in a moist atmosphere and droplets collecting on the outside of the bladder during perfusion of the inside can be analysed. With this method, Diamond (1962) has shown that the secreted solution is normally isotonic with plasma, and that this tonicity is always equal to that of the luminal solution, even when part of the osmotic pressure is made up by sugars. This indicates that the primary process is salt transport with water accompanying in response to the osmotic gradient. The mechanism of reabsorption is probably similar to that in the proximal kidney tubule. In keeping with a role for the Na–K pump, the reabsorption is inhibited when bladders are exposed to external ouabain or solutions lacking K^+, while acidification of the bile (Diamond, 1962) may indicate that Na^+–proton exchange is operative at the luminal epithelial border. Diamond (1964, 1965), and Diamond and Tormey (1966b) suggested a model of 'local osmosis' to account for the movement of water across the gall-bladder after the discovery of intra-epithelial canals observed in the gall-bladder by electron microscopy (Diamond and Tormey, 1966a). The workings of this system have already been discussed above in connection with frog skin and kidney cortex.

5. Cerebrospinal fluid and aqueous humour

Cerebrospinal fluid has a tonicity similar to plasma, but Na^+ and chloride are in excess of the plasma concentrations, while there is a deficit in K^+ and bicarbonate. The choroidal epithelia lining the fluid cavity are similar in appearance to the brush border epithelia of the intestine, having deeply folded surfaces apparently suited to a secretory function. A possible mechanism of secretion would involve the Na–K pump and membrane permeabilities allowing passive ion movements similar to those suggested for sublingual saliva. To account for the significant bicarbonate content of the fluid a specific chloride–bicarbonate exchange system can be postulated at the fluid–epithelial border. This exchange could account for the observed accumulation of iodide, bromide and thiocyanate from cerebrospinal fluid by the choroidal epithelia (Davson and Pollay, 1963; Pollay and Davson, 1963) since these ions are usually handled like chloride by biological systems. Also, Davson and Luck (1957) observed that acetazolamide inhibits cerebro-spinal fluid flow, indicating that hydration of CO_2 is a rate-limiting process in fluid secretion, and raising the possibility that a chloride–bicarbonate exchange system is present at the plasma–epithelial membrane. The function of this system would perhaps be to maintain cell chloride, despite a chloride-impermeable plasma membrane.

The aqueous humour is usually similar in tonicity and ionic composition to cerebrospinal fluid, although in some species anterior aqueous bicarbonate exceeds the plasma concentration (Maren, 1965). A secretory mechanism similar to that of cerebrospinal fluid is likely. The studies of Vates and co-workers (1964) and Bonting and Becker (1964) have respectively provided evidence for a role for the transport ATPase in the formation of cerebrospinal fluid and aqueous humour.

6. Secretion and absorption in the intestine

The secretions of the duodenum, jejunum, ileum and colon are apparently isotonic with plasma with ionic compositions as in table 2. These are consistent with secretory mechanisms similar to sublingual saliva with a varying contribution from chloride–bicarbonate exchange, and passive movement of water by local osmosis.

Table 2. Electrolyte concentrations in intestinal secretions of dog

Intestinal region	pH	Na^+	K^+	mEq/l Ca^{2+}	Cl^-	HCO_3^-
Jejunum	6·8	142	6·45	3·4	148	19·2
Ileum	7·6	151	4·7	5·5	78	83
Colon	7·99	147	7·2	4·2	77	90

From de Beer *et al.* (1935)

Recovery of salts and water also occurs throughout the intestinal tract. In the ileum salt reabsorption is accompanied by osmotic water (Curran and Solomon, 1957) in keeping with the relatively high permeabilities of this region. Active Na-transport in the rabbit ileum is inhibited by serosal ouabain (Schultz and Zalurky, 1964) and the mechanism of absorption is likely to be similar to that in the proximal kidney tubules. In the colon, luminal Na^+ is reduced to very low concentration while K^+ may rise to about 30 mEq/l (Darrow, 1957). There is a potential difference of about 30 mV (Ussing and Anderson, 1955) between plasma and lumen of the toad large intestine (plasma positive) and this may account, at least in part for the high luminal K^+.

7. Secretion of HCl in the stomach

It is well established that secretion of HCl into the stomach of dog, cat, frog and man occurs through parietal cells in the gastric mucosae as a solution isotonic with plasma. The chief cations are normally protons with some K^+, and chloride is normally the sole anion. It is clear that the process must involve active transport, but the mechanism is not well understood. Application of Diamox, or similar inhibitors of carbonic anhydrase, to dog stomach, reduces both acid secretion and the volume secreted, whereas the concentration and absolute quantity of secreted Na^+ increases (Powell and coworkers, 1962; Berkowitz and Janowitz, 1967). Similarly, in uninhibited stomachs, there is an inverse relationship between the rate of gastric secretion and Na^+ concentration, whereas acidity increases with secretory rate (Takata, 1933; Gray and Bucher, 1941; Linde and Obrink, 1950). Diamox inhibition is taken to indicate that the cellular origin of protons is the hydration of CO_2 to bicarbonate by carbonic anhydrase, the rate of the uncatalysed reaction being insufficient to support maximal secretion. The origin of CO_2 in parietal cells is presumably partly metabolic, the remainder entering from the blood, thus accounting for the greater alkalinity of venous, compared with arterial gastric blood.

Increased Na^+ secretion with Diamox may be indicative of a mechanism of proton secretion involving exchange of mucosal Na^+ for protons in the parietal cell (Teorell, 1947), in which case the primary secretory process could be secretion of Na^+ by the Na–K pump. However, a serious complication is the secretion of Na^+-containing solutions by gastric epithelia other than parietal cells, since this could account for the increase in Na^+ concentration in the overall gastric secretion during reduced output by parietal cells. On the other hand, this factor could not account for the actual increase in absolute quantity of Na secretion in Diamox inhibition, and further, Cooperstein (1959) found that strophanthidin also inhibited acid secretion, implying a direct role for the Na–K pump. It has also been found that the onset of acid secretion is preceded by a K-transient, similar to that observed in glands such as the sublingual (Hollander, 1961). This is also compatible with acid secretion involving the Na–K pump, and Na^+–proton exchange.

Two alternative mechanisms for acid secretion can be formulated by analogy with the mitochondrion. First, it is possible that an ATPase is located in the luminal parietal membrane, with properties similar to the valinomycin-dependent K-activated system in the mitochondrion (Cockrell and coworkers, 1966). This is capable of taking up K in exchange for protons using ATP as energy source. The possible gastric ATPase would be independent of valinomycin, which apparently makes the mitochondrion K-permeable, and would therefore not be necessary if the parietal membrane is normally K-permeable.

A second possibility is primary involvement of an oxidative electron-transport system, similar to that in mitochondria, situated in the parietal cell luminal membrane. Davenport and Chavre (1956) observed that the ratio of H^+ secreted to O_2 consumed was about 4, but much higher ratios of up to 12 H^+/O_2 are probably applicable to the fraction of O_2-consumption actually involved in acid secretion in parietal cells (Davies, 1957). High H^+/O_2 ratios were thought to exclude a redox mechanism for acid secretion, but can now be considered quite compatible with a redox system operating in the fashion proposed by Mitchell (1966). If Mitchell's (1966) chemiosmotic hypothesis for oxidative phosphorylation is correct, this sytem would generate a proton gradient across the membrane, which would necessarily lead to net acid secretion, provided that protons were accompanied by an anion (chloride). If a mechanism of this type is operative, there should be a direct relationship between acid secretion and associated oxygen consumption. Using frog mucosal sacs, Bannister (1965, 1966) has observed that the ratio of acid secreted to O_2 consumed is constant even during metabolic inhibition by oligomycin and arsenate. In the case of arsenate, this result would not be expected if ATP were the driving force for acid secretion, since ATP synthesis is uncoupled from oxidation by arsenate.

However, this result is at variance with the observed inhibition of acid secretion by strophanthidin, so that the question must remain open.

Parietal cells are also thought to transport chloride actively since, in the resting stomach, a potential of about 30 mV is observed between the lumen (negative) and the serosa in isolated frog mucosa (Hogben, 1951) and across the dog mucosa (Rehm and coworkers, 1951). Some light has been thrown on the mechanism of chloride transport by studies of frog gastric mucosa, where acid secretion does not exceed the rate of uncatalysed hydration of CO_2, and hence is not inhibited by normal concentrations of carbonic anhydrase inhibitors. However, at the exceptionally high concentration of 10 mM, methazolamide does inhibit both chloride transport and the transmucosal potential (Hogben, 1965, 1967). This may be taken as an indication of methazolamide inhibition of a chloride–bicarbonate exchange system, by arguing that a bicarbonate substrate and methazolamide inhibition are common to carbonic anhydrase and chloride–bicarbonate exchange. Situation of such an anion exchanger at the plasma–parietal cell surface would serve to replenish cell chloride during secretion, and to eliminate cell bicarbonate. It is possible that a second chloride transporter is present at the mucosal membrane, since secretion is inhibited by sulphate in the stomach lumen, and this could well exert its effect by blocking a chloride transport system.

The flow of water into the acid secreting stomach is probably osmotically obligated in the same way as in other secretions. Restricted channels or canaliculi are present in parietal cells, and are continuous with the luminal solution (Bradford and Davies, 1950), so that local excesses of osmotic pressure could arise during acid secretion. The most commonly used stimulator of acid secretion is histamine. This may operate by increasing the permeability of the parietal cell to water, and by activating some part of the active transport of protons or chloride.

IV. THE ACTIVE TRANSPORT OF Ca

A. General

Active transport of Ca^{2+} has been demonstrated in vesicles of the sarcoplasmic reticulum from striated muscle (Hasselbach and Makinose, 1961, 1963), in erythrocytes (Schatzmann, 1966), in the squid giant axon (Baker and coworkers, 1969; Blaustein and Hodgkin, 1969) and in mitochondria (Vasington and Murphy, 1962). The mechanisms of Ca^{2+} transport have not yet been fully elucidated, but ATP is normally the energy source and in the sarcoplasmic reticulum at least, transport is probably effected by a membrane bound ATPase, whose mechanism may resemble that of the Na–K

K

pump, and which transports two Ca^{2+} per ATP hydrolysed (Hasselbach, 1964; Weber and coworkers, 1966).

In all the above cases, the direction of Ca-transport is such as to reduce the concentration of cytoplasmic Ca^{2+}. The functional significance of this in cells other than muscle is probably protection of cytoplasmic enzymes from inhibition by Ca^{2+}. Enzymes inhibited by low Ca^{2+} concentrations include the transport ATPase (Epstein and Whittam, 1966), the glycolytic enzymes, enolase and pyruvate kinase, and others such as glutaminase and pantothenate synthetase (Dixon and Webb, 1965). In addition, lysosomal phospholipase A_2 is stimulated by Ca^{2+} (Smith and Winkler, 1968), so that a general disruptive effect of Ca^{2+} on normal cell metabolism is indicated.

B. Mitochondria

Substantial accumulation of Ca^{2+} by mitochondria can be readily demonstrated and can be supported either by substrate oxidation when uptake is inhibited by DNP but not by oligomycin, or by ATP, when uptake is unaffected by DNP but inhibited by oligomycin. In both cases, Ca^{2+} uptake is accompanied by loss of mitochondrial protons (i.e. acidification of the medium) unless a permeable anion such as acetate is present, but the precise mechanism of transport is uncertain. However, in the presence of valinomycin, which is known to render erythrocytes and synthetic lipid bilayers permeable to K^+ (Henderson and coworkers, 1969), mitochondria accumulate K^+ in a fashion similar to that for Ca^{2+} (Cockrell and coworkers, 1966). This makes it probable that generation of a proton gradient is the primary process in the mitochondrion and that permeable cations are accumulated in response to the proton gradient. K^+ accumulation is unlikely to have physiological significance, except in so far as the normally K-impermeable mitochondrial membrane protects mitochondria from uncoupling of oxidative phosphorylation by cytoplasmic K^+.

C. Ca in Striated Muscle

An important role for Ca^{2+} has been detailed in both excitation and contraction of striated muscle, and Ca^{2+} is thought to be similarly involved in cardiac and smooth muscle. Various studies have shown that the free Ca^{2+} concentration in striated muscle sarcoplasm is less than $10^{-6}M$ in resting muscle (Weber and Herz, 1963; Portzehl and coworkers, 1964), although the total muscle Ca^{2+} is much higher. However, at the onset of contraction, whether stimulated electrically, by K-depolarization or by caffeine, there is an up to 20-fold increase in the rate of efflux of ^{45}Ca from isolated *Maia* giant muscle fibres (Caldwell, 1964), implying a substantial increase in concentration of cytoplasmic Ca^{2+} during contraction. This in turn must indicate entry of Ca^{2+} into the cytoplasm, either from bound sites in the muscle or from the external medium. A. V. Hill's (1948) early objection to Ca^{2+} as initiator of contraction was that diffusion of Ca^{2+} from the sarcolemma to myofibrils near the centre of muscle fibres would be too slow to account for the brief delay between electrical stimulation and contraction. However, the experiments employing very fine microelectrodes on frog muscle fibres carried out by Huxley and Taylor (1958) and Huxley and Straub (1958) have shown that adjacent half-sarcomeres contract after stimulation by electrodes placed over the Z-line, but not over other regions of the sarcomere. This result has been interpreted as indicating that the T-system (a transverse system of membrane-encased tubules interconnecting myofibrils in a muscle fibre at the Z-line, and with lumens open to the external medium (Page, 1964)), conducts electrical impulses in a way similar to nerve axons. Since the T-system is closely associated spacially with the sarcoplasmic reticulum, it has been proposed (e.g. Ebashi and Endo, 1968) that the T-system activates the reticulum at the terminal cysternae, causing release of Ca^{2+} into the sarcoplasm, and therefore causing the free Ca^{2+} in the sarcoplasm to rise above the threshold necessary for contraction. As demonstrated by Hasselbach and Makinose (1963) and Weber and coworkers, (1964), isolated membrane vesicles originating from the sarcoplasmic reticulum are capable of ATP-dependent Ca-uptake, and can reduce free medium Ca^{2+} to less than $10^{-6}M$. This is below the threshold for contraction and therefore explains the relaxing factor activity of such vesicles.

Once in the sarcoplasm, Ca^{2+} is believed to activate the contractile mechanism, probably via stimulation of actomyosin ATPase (Hasselbach, 1964). Ca-stimulation of actomyosin ATPase is lost after removal of tropomyosin B from the system (Schaub and Ermini, 1969) and troponin in conjunction with tropomyosin, is thought to be the agent effecting the stimulation. This is because direct evidence of an interaction between tropomyosin and troponin, and actomyosin has been obtained by Ebashi and Kodama (1966), who found that the former two proteins conferred EGTA-sensitivity (i.e. Ca-dependence) on the superprecipitation reaction of purified actomyosin. Appropriately, Ebashi and coworkers (1967) found binding of Ca^{2+} to troponin in a 4 : 1 molar ratio and it is thought that the presence of Ca-troponin in some way modifies the interaction between actin and myosin.

The above is only a brief account of some of the available evidence, but it is sufficient to make clear that the concentration of sarcoplasmic Ca^{2+}, and therefore the control of this by active transport, is an essential part of muscular contraction. A much fuller account of this can be found in the review of Endo and Ebashi (1968).

V. CONCLUDING REMARKS

The preceding sections have discussed various cellular processes in which active ion transport plays an important or essential part, but it is also interesting to speculate as to the primary or original role of active transport in primitive organisms during the early evolution of life. A reasonable supposition is that one of the fundamental requirements of the primitive organism was maintenance of an at least partly controlled region in which such substances as proteins and nucleic acids and their precursors were concentrated. Given a membrane or other structure of limited permeability to enclose this region, the problem of colloid osmotic pressure and consequent dilution of the region must arise. Such dilution could be counteracted by active transport of cations and this then may have been the original function of active transport.

Given the existence of a rather unspecific cation transport system in the primitive cell, it is possible to envisage the gradual evolution of the more specific mechanisms for cation transport found in present-day organisms, and of the variety of functions fulfilled by active transport.

REFERENCES

Adrian, R. H. and C. L. Slayman (1966) *J. Physiol.*, **184**, 970
Askari, A. (1963) *Science*, **141**, 44
Askari, A. and J. C. Fratantoni (1964) *Proc. Soc. Exptl. Biol. Med.*, **116**, 751
Baker, P. F., M. P. Blaustein, A. L. Hodgkin and R. A. Steinhardt (1969) *J. Physiol.*, **200**, 431
Bannister, W. H. (1965) *J. Physiol.*, **177**, 140
Bannister, W. H. (1966) *J. Physiol.*, **186**, 89
De Beer, E. J., C. G. Johnston and D. W. Wilson (1935) *J. Biol. Chem.*, **108**, 113
Berkowitz, J. M. and H. D. Janowitz (1967) *Am. J. Physiol.*, **212**, 72
Berliner, R. W., T. J. Kennedy and J. Orloff (1951) *Amer. J. Med.*, **11**, 274
Bessman, M. J. and van Bibber (1959) *Biochem. Biophys. Res. Comm.*, **1**, 101
Birnbaum, D. and F. Hollander (1953) *Am. J. Physiol.*, **174**, 191
Black, S. (1951) *Arch. Biochem. Biophys.*, **34**, 86
Blaustein, M. P. and A. L. Hodgkin (1969) *J. Physiol.*, **200**, 497
Bonting, S. L. and B. Becker (1964) *Invest. Ophthalmol.*, **3**, 523
Bradford, N. M. and R. E. Davies (1950) *Biochem. J.*, **46**, 414
Bro-Rasmussen, F., S. A. Killmann and J. H. Thaysen (1956) *Acta Physiol. Scand.*, **37**, 97
Burgen, A. S. V. (1956) *J. Physiol.*, **132**, 20
Caldwell, P. C. (1960) *J. Physiol.*, **152**, 545
Caldwell, P. C. (1964) *Proc. Roy. Soc. B.*, **160**, 512
Caldwell, P. C., A. L. Hodgkin, R. D. Keynes and T. I. Shaw (1960a) *J. Physiol.*, **152**, 561
Caldwell, P. C., A. L. Hodgkin, R. D. Keynes and T. I. Shaw (1960b) *J. Physiol.*, **152**, 591
Christensen, H. N., T. E. Riggs and N. E. Ray (1952) *Ann. N.Y. Acad. Sci.*, **63**, 983
Clark, J. A. and R. A. Macleod (1954) *J. Biol. Chem.*, **211**, 531
Cockrell, R., E. J. Harris and B. Pressman (1966) *Biochemistry*, **5**, 2326
Cohn, M. and J. Monod (1951) *Biochim. Biophys. Acta.*, **7**, 153
Cooperstein, I. L. (1959) *J. Gen. Physiol.*, **42**, 1233
Crane, R. K., G. Forstner and A. Eicholz (1965) *Biochim. Biophys. Acta.*, **109**, 467

Crane, R. K., D. Miller and I. Bihler (1961) in A. Kleinzeller and A. Kotyk (Eds.), *Membrane Transport and Metabolism*, Academic Press, New York, p. 432
Csáky, T. Z. and M. Thale (1960) *J. Physiol.*, **151**, 59
Csáky, T. Z., H. G. Hartzog and G. W. Fernald (1961) *Am. J. Physiol.*, **200**, 459
Curran, P. F. and A. K. Solomon (1957) *J. Gen. Physiol.*, **41**, 143
Darrow, D. C. (1957) in Q. R. Murphy (Ed.), *Metabolic Aspects of Transport Across Cell Membranes*, University of Wisconsin Press, p. 23
Davenport, H. W. and V. J. Chavre (1956) *Am. J. Physiol.*, **187**, 227
Davies, R. E. (1957) in Q. R. Murphy (Ed.), *Metabolic Aspects of Transport Across Cell Membranes*, University of Wisconsin Press, Madison, p. 277
Davson, H. and J. F. Danielli (1938) *Biochem. J.*, **30**, 316
Davson, H. and C. P. Luck (1957) *J. Physiol.*, **137**, 279
Davson, H. and M. Pollay (1963) *J. Physiol.*, **167**, 247
Diamond, J. M. (1962) *J. Physiol.*, **161**, 442, 474, 503
Diamond, J. M. (1964) *J. Gen. Physiol.*, **48**, 1, 15
Diamond, J. M. (1965) *S.E.B. Symp.*, **19**, 329
Diamond, J. M. and J. McD. Tormey (1966a) *Nature*, **210**, 817
Diamond, J. M. and J. McD. Tormey (1966b) *Fed. Proc.*, **25**, 1458
Dixon, M. and E. C. Webb (1965) in *The Enzymes*, Longmans, Green & Co. Ltd., London, p. 422
Ebashi, S., F. Ebashi and A. Kodama (1967) *J. Biochem.*, **62**, 137
Ebashi, S. and M. Endo (1968) *Prog. Biophys. Mol. Biol.*, **18**, 123
Ebashi, S. and A. Kodama (1966) *J. Biochem.*, **60**, 733
Eddy, A. A. (1968) *Biochem. J.*, **108**, 195
Eddy, A. A. and M. Mulcahy (1965) *Biochem. J.*, **96**, 76P
Edwards, J. B. and D. B. Keech (1968) *Biochim. Biophys. Acta.*, **159**, 167
Epstein, F. H. and R. Whittam (1966) *Biochem. J.*, **99**, 232
Flaks, J. G., M. J. Erwin and J. M. Buchanan (1957) *J. Biol. Chem.*, **229**, 603
Francis, W. L. and O. Gatty (1938) *J. Gen. Physiol.*, **36**, 607
Fuhrman, F. A. and H. H. Ussing (1951) *J. Cell. Comp. Physiol.*, **38**, 109
Garrahan, P. J. and I. M. Glynn (1967) *J. Physiol.*, **192**, 217
Giebisch, G. (1961) *J. Gen Physiol.*, **44**, 659
Gray, J. S. and G. R. Bucher (1941) *Am. J. Physiol.*, **133**, 542
Hammarsen, O. (1897) *Nova Acta Soc. Scient. Uppsaliensis*, **16**, Ser III, 1
Happold, F. C. and A. Struyvenberg (1954) *Biochem. J.*, **58**, 379
Hasselbach, W. (1964) *Prog. Biophys. Biophys. Chem.*, **14**, 169
Hasselbach, W. and M. Makinose (1961) *Biochem. Z.*, **333**, 518
Hasselbach, W. and M. Makinose (1963) *Biochem. Z.*, **339**, 94
Henderson, P. J. F., J. D. McGivan and J. B. Chappell (1969) *Biochem. J.*, **111**, 521
Hill, A. V. (1948) *Proc. Roy. Soc. B.*, **135**, 446
Hodgkin, A. L. and A. F. Huxley (1952) *J. Physiol.*, **116**, 449
Hogben, C. A. M. (1951) *Proc. Nat. Acad. Sci.*, **37**, 393
Hogben, C. A. M. (1965) *Fed. Proc.*, **24**, 135
Hogben, C. A. M. (1967) *Mol. Pharmacol.*, **3**, 318
Hollander, F. (1961) *Gastroenterology*, **40**, 477
Huf, E. G., N. S. Doss and J. P. Wills (1957) *J. Gen. Physiol.*, **41**, 397
Huf, E. G. and J. Wills (1951) *Am. J. Physiol.*, **167**, 255
Huxley, A. F. and R. W. Straub (1958) *J. Physiol.*, **143**, 40P
Huxley, A. F. and R. E. Taylor (1958) *J. Physiol.*, **144**, 426
Jaenike, J. R. and R. W. Berliner (1960) *J. Clin. Invest.*, **39**, 481
Kachmar, J. F. and P. D. Boyer (1953) *J. Biol. Chem.*, **200**, 669
Kernan, R. P. (1962) *Nature*, **193**, 986
Kielley, W. W., H. M. Kalckar and L. B. Bradley (1956) *J. Biol. Chem.*, **219**, 95
Kirschner, L. B. (1953) *J. Cell. Comp. Physiol.*, **45**, 89
Koefoed-Johnsen, V. (1957) *Acta Physiol. Scand.*, **42**, Suppl. 145, 87
Koefoed-Johnsen, V. and H. H. Ussing (1958) *Acta Physiol. Scand.*, **42**, 298

von Korff, R. W. (1953) *J. Biol. Chem.*, **203**, 265
Kramer, K. and Deetjen (1960) *Arch. Ges. Physiol.*, **271**, 782
Lassen, N. A., D. Munck and J. H. Thaysen (1961) *Acta. Physiol. Scand.*, **51**, 371
Lassiter, W. E., C. W. Gottschalk and M. Mylle (1961) *Am. J. Physiol.*, **200**, 1139
Leaf, A. (1956) *Biochem. J.*, **62**, 241
Lin, E. E. C. and B. Magasanik (1960) *J. Biol. Chem.*, **235**, 1820
Linde, S. and K. J. Obrink (1950) *Acta Physiol. Scand.*, **21**, 54
Litchfield, J. B. and P. A. Bott (1962) *Am. J. Physiol.*, **203**, 667
Ljungberg, E. (1947) *Acta Med. Scand.*, Suppl. 186
Lowe, A. G. (1968) *Nature*, **219**, 934
Lowenstein, J. M. (1960) *Biochem. J.*, **75**, 269
Lundberg, A. (1957) *Acta Physiol. Scand.*, **40**, 101
Lundberg, A. (1958) *Physiol. Rev.*, **38**, 21
Maas, W. K. (1952) *J. Biol. Chem.*, **198**, 23
Maetz, J. and F. G. Romeu (1964) *J. Gen. Physiol.*, **47**, 1209
Magasanik, B., H. S. Moyed and L. B. Gehring (1957) *J. Biol. Chem.*, **226**, 339
Maren, T. H. (1967) *Physiol. Rev.*, **47**,
Marshall, M., R. L. Metzenburg and P. P. Cohen (1961) *J. Biol. Chem.*, **236**, 2229
Mellanby, J. (1925) *J. Physiol.*, **60**, 85
Melnik, I. and J. M. Buchanan (1957) *J. Biol. Chem.*, **225**, 157
Mitchell, P. (1966) *Biol. Rev.*, **41**, 445
Mudd, S. H. and G. L. Cantoni (1958) *J. Biol. Chem.*, **231**, 481
Neuhaus, F. C. (1962) *J. Biol. Chem.*, **237**, 778
Nishimura, J. S. and D. M. Greenberg (1961) *J. Biol. Chem.*, **236**, 2684
Nossal, P. M. (1951) *Biochem. J.*, **49**, 407
Okkels, H. (1929) *Bull. Histol. Appl. Physiol.*, **6**, 12
Oken, D. E. and A. K. Solomon (1960) *J. Clin. Invest.*, **39**, 1015
Orloff, J. and M. Burg (1960) *Am. J. Physiol.*, **199**, 49
Page, S. (1964) *J. Physiol.*, **175**, 10P
Parks, R. E., E. Ben-Gershom and H. A. Lardy (1957) *J. Biol. Chem.*, **227**, 231
Pitts, R. F. and R. S. Alexander (1945) *Amer. J. Physiol.*, **144**, 239
Pollay, M. and H. Davson (1963) *Brain*, **86**, 137
Portzehl, H., P. C. Caldwell and C. Ruegg (1964) *Biochim. Biophys. Acta.*, **79**, 581
Powell, D. W., R. C. Robbins J. D. Boyett and B. I. Hirschowitz (1962) *Am. J. Physiol.*, **202**, 293
Ravdin, I. S., C. G. Johnston, J. H. Austin and C. Riegel (1931-2) *Am. J. Physiol.*, **99**, 638
Ravdin, J. S., C. G. Johnston, C. Riegel and S. L. Wright (1932) *Am. J. Physiol.*, **100**, 317
Richards, O. C. and W. J. Rutter (1961) *J. Biol. Chem.*, **236**, 3177
Ridderstrap, A. S. and S. L. Bonting (1968) *Federation Proc.*, **27**, 834
Rector, F. C. and J. R. Clapp (1962) *J. Clin. Invest.*, **41**, 101
Rehm, W. S., L. E. Hokin, T. P. D. E. Skaffenried, F. J. Bajandas and F. E. Coy (1951) *Amer. J. Physiol.*, **164**, 187
Robinson, S. and A. H. Robinson (1954) *Physiol. Rev.*, **34**, 202
Romeu, F. S. and J. Maetz (1964) *J. Gen. Physiol.*, **47**, 1195
Sartorius, O. W., J. C. Roemmelt and R. F. Pitts (1949) *J. Clin. Invest.*, **28**, 423
Schatzmann, H. J. (1953) *Helv. Physiol. Pharmacol. Acta.*, **11**, 346
Schatzmann, H. J. (1966) *Experientia*, **22**, 364
Schatzmann, H. J., E. E. Windhager and A. K. Solomon (1958) *Am. J. Physiol.*, **195**, 570
Schaub, M. and M. Ermini (1969) *Biochem. J.*, **111**, 777
Schmidt-Nielsen, B. and R. O'Dell (1961) *Am. J. Physiol.*, **200**, 1119
Schwartz, A. and C. A. Moore (1968) *Am. J. Physiol.*, **214**, 1163
Shaw, J. (1960) *J. Exp. Biol.*, **37**, 534, 548, 557
Schultz, S. G. and R. Zalusky (1964) *J. Gen. Physiol.*, **47**, 567
Schweet, R. S. and E. H. Allen (1958) *J. Biol. Chem.*, **233**, 1104
Skou, J. C. (1957) *Biochem. Biophys. Acta.*, **23**, 394
Skou, J. C. (1960) *Biochim. Biophys. Acta.*, **42**, 6

Van Slyke, D. D., R. A. Phillips. P. B. Hamilton, R. M. Archibald, P. H. Fucher and A. Hiller (1943) *J. Biol. Chem.*, **150**, 481
Smith, A. D. and H. Winkler (1968) *Biochem. J.*, **108**, 867
Snoke, J. E., S. Yanari and K. Block (1953) *J. Biol. Chem.*, **201**, 573
Spahr, P. F. and D. Schlessinger (1963) *J. Biol. Chem.*, **238**, PC 2251
Stadtman, E. R. (1952) *J. Biol. Chem.*, **196**, 527
Tabor, H. and L. Wyngarden (1959) *J. Biol. Chem.*, **234**, 1830
Takata, C. (1933) *Jap. J. Med. Sci.*, **3**, 33
Tasaki, I. and I. Singer (1966) *Ann. New York Acad. Sci.*, **137**, 792
Teorell, T. (1947) *Gastroenterology*, **9**, 425
Thaysen, J. H. (1960) in *Handbuch Exptl. Pharmakol*, Vol. 13, Springer-Verlag, Berlin, p. 424
Thaysen, J. H., N. A. Thorn and I. L. Schwartz (1954) *Am. J. Physiol.*, 178, 155, 160
Thomas, J. E. and J. O. Crider (1944) *Am. J. Physiol.*, **140**, 574
Tosteson, D. C. and J. F. Hoffman (1960) *J. Gen. Physiol.*, **44**, 169
Ussing, H. H. (1948) *Cold Spring Harbor Symposia on Quantitative Biology*, **13**, 193
Ussing, H. H. (1954) *Symp. Soc. Exp. Biol.*, **8**, 407
Ussing, H. H. (1963-4) *Harvey Lectures*, **59**, 1
Ussing, H. H. and B. Andersen (1955) *Proc. 3rd Int. Congress Biochem.*, Brussels, 1955, p. 434
Ussing, H. H., P. Kruhøffer, J. H. Thaysen and N. A. Thorn (1960) in *Handbuch der Experimentellen Pharmakologie*, Springer-Verlag, Berlin, Vol. 13
Ussing, H. H. and K. Zerahn (1951) *Acta Physiol. Scand.*, **23**, 110
Vasington, F. D. and J. V. Murphy (1962) *J. Biol. Chem.*, **237**, 2670
Vates, T. S., S. L. Bonting and W. W. Oppelt (1964) *Am. J. Physiol.*, **206**, 1165
Vidaver, G. (1964a) *Biochemistry*, **3**, 662
Walker, W. G., C. R. Cooke, J. W. Payne, R. F. Baker and D. J. Andrew (1961) *Am. J. Physiol.*, **200**, 1133
Weber, A. and R. Herz (1963) *J. Biol. Chem.*, **238**, 599
Weber, A., R. Herz and I. Reiss (1964) *Proc. Roy. Soc. B.*, **160**, 489
Weber, A., R. Herz and I. Reiss (1966) *Biochem. Z.*, **345**, 329
Webster, G. C. and J. E. Varner (1954) *Arch. Biochem. Biophys.*, **52**, 22
Whittam, R. and M. E. Ager (1965) *Biochem. J.*, **97**, 214
Whittam, R. and J. S. Willis (1963) *J. Physiol.*, **168**, 158
Whittembury, G. (1965) *J. Gen. Physiol.*, **48**, 699
Whittembury, G., N. Sugino and A. K. Solomon (1961) *J. Gen. Physiol.*, **44**, 689
Willis, J. S. (1966) *J. Gen. Physiol.*, **49**, 1221
Willis, J. S. (1968a,b) *Biochim. Biophys. Acta.*, **163**, 506, 516
Wirz, H. (1956) *Helv. Physiol. Pharmacol Acta.*, **14**, 352
Zerahn, K. (1956) *Acta Physiol. Scand.*, **36**, 300

CHAPTER 9

Protein Biosynthesis: Mechanism, Requirements and Potassium-Dependency

Sidney Pestka

Roche Institute of Molecular Biology
Nutley, New Jersey, U.S.A.

I. GENERAL CONSIDERATIONS

The synthesis of a protein is a complex sequence of highly coordinated events. In all cells synthesizing proteins, the general scheme of information transfer is probably very similar. An outline of this general scheme of information flow is presented in figure 1. The genetically stable component controlling the genotype of each cell is deoxyribonucleic acid (DNA) (Avery, and coworkers, 1944; Watson and Crick, 1953; Beadle, 1957; Spiegelman, 1957; Kornberg, 1961). In many organisms the DNA is complexed with histones and constitutes a major fraction of the chromosomes. The DNA serves as its own template for replication of identical molecules which are passed on from generation to generation as the hereditary material.

The ultimate expression of the genetic information in an organism, its phenotype, is in large part determined by the sum total of the protein constituents of the cell. The information flow from DNA to proteins occurs via an intermediate ribonucleic acid molecule called messenger RNA (*m*-RNA) (Caspersson, 1941; Brachet, 1941; Volkin and Astrachan, 1956; Nomura, Hall, and Spiegelman, 1960; Jacob and Monod, 1961; Nirenberg

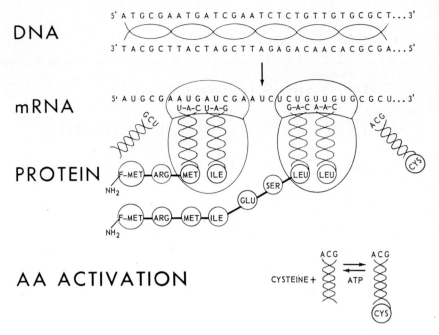

Figure 1. Schematic illustration of transcription of messenger RNA and protein synthesis.

and Matthaei, 1961; Watson, 1963). Although a single homogeneous *m*-RNA transcribed on a cellular DNA template has not been isolated, a vast body of circumstantial evidence supports the existence of this intermediate in protein synthesis. So firm is the faith in its existence, that a significant proportion of investigations in molecular biology presuppose its existence without referring to possible limitations of this assumption. The *m*-RNA molecule is considered to be a copy of the genetic information of a particular portion of the DNA molecule. It is this *m*-RNA molecule which serves as a direct template for protein biosynthesis in all organisms.

All proteins in cells are synthesized on particulate structures called ribosomes (Borsook and coworkers, 1950; Hultin, 1950; Allfrey and coworkers, 1953; Zamecnik and Keller, 1954; Rabinovitz and Olson, 1956; Schweet and coworkers, 1958; McQuillen and coworkers, 1959). The precise structure of ribosomes varies from organism to organism; however, several essential features of their structure appear to be common among the various species. The structure and composition of *Escherichia coli* ribosomes are illustrated in figure 2. Ribosomes from all organisms appear to consist of two separable subunits, one approximately twice as large as the second.

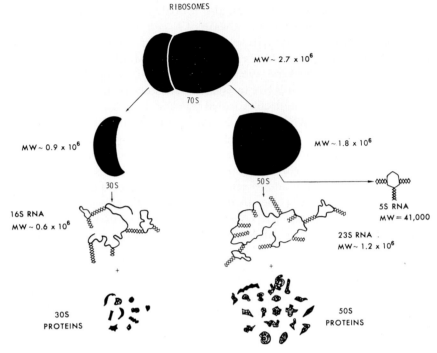

Figure 2. Structure and composition of *E. coli* ribosomes.

Each subunit appears to be composed of a single large strand of RNA and many different individual protein molecules (Kurland, 1960). In addition, the larger subunit contains an additional small ribonucleic acid molecule called 5 S RNA (Rosset and Monier, 1963; Comb and Katz, 1964; Galibert and coworkers, 1965). The subunit structure as well as the presence of 5 S RNA must play an essential role in protein synthesis. For, although organisms have evolved in highly divergent paths, these features of protein synthesis which are present in bacteria have been maintained in the vertebrates. Nevertheless, the ribosomal proteins often vary from species to species and occasionally even within a species. Also bacterial and mammalian 5 S RNAs appear generally similar in structure (Forget and Weissman, 1967; Brownlee and coworkers, 1968), although they differ significantly in primary sequence. Each contains two sequences of oligonucleotides which are complementary to the common pentanucleotide sequence $GprTp\psi pCpG/A$, which appears in *t*-RNA (Zamir and coworkers, 1967).

The *m*-RNA binds to the small subunit of the ribosomes (Takanami and Okamoto, 1963). The binding of *m*-RNA to ribosomes requires Mg^{2+} ions and, in the case of natural templates, also probably an initiation factor (F3), which is isolated from the ribosomal wash of *E. coli* (Brown and Doty, 1968; Iwasaki and coworkers, 1968; Revel and coworkers, 1968). The translation of *m*-RNA molecules occurs on the smaller subunit (Matthaei and coworkers, 1964; Suzuka and coworkers, 1965; Pestka and Nirenberg, 1966a,b).

The translation of the *m*-RNA template codons occurs through the use of adaptor ribonucleic acid molecules called transfer ribonucleic acids (*t*-RNA) (Crick, 1958). These *t*-RNA molecules are relatively small ribonucleic acid moieties approximately 25,000 in molecular weight (Tissières, 1959). Amino acids must first be attached to these *t*-RNA molecules before they can be incorporated into a protein. There is a set of individual *t*-RNA species for each of the twenty natural amino acids. A single amino acid can be esterified to any one of the *t*-RNA species in its set, but it cannot be esterified under usual circumstances to the *t*-RNA species of another amino acid set. The scheme for the acylation of *t*-RNA molecules is illustrated in figure 3. The

$$\text{AMINO ACID} + \text{ATP} \underset{}{\overset{\text{Enzyme}}{\rightleftharpoons}} \text{AA–AMP} + \text{PP}$$

$$\text{AA–AMP} + \text{tRNA} \underset{}{\overset{\text{Enzyme}}{\rightleftharpoons}} \text{AA–tRNA} + \text{AMP}$$

Figure 3. Amino acid activation and acylation of tRNA.

enzyme which performs this function, the aminoacyl-*t*-RNA synthetase, is different for each amino acid. The energy for the synthesis of proteins comes in part from ATP, which is a substrate in the reaction for the synthesis of aminoacyl-*t*-RNAs. The mixed anhydride, the aminoacyl adenylate, is an intermediate in these reactions (Davie and coworkers, 1956; DeMoss and coworkers, 1956; Berg, 1956; Hoagland and coworkers, 1956; Schweet and Allen, 1958). The synthetase carries out both steps of the reaction sequence, namely, the formation of the aminoacyl adenylate and immediately thereafter the formation of aminoacyl-*t*-RNA (Holley, 1957; Ogata and coworkers, 1957; Weiss and coworkers, 1958; Schweet and coworkers, 1958; Berg and Ofengand, 1958; Hoagland and coworkers, 1958).

The *t*-RNA molecules from all organisms appear to have certain features in common. They are approximately 70 to 90 nucleotides in length. They all end in a terminal pG or pC at the 5′-terminus of the molecule. Their 3′-end consists of the sequence CpCpA. Most *t*-RNA molecules that have been sequenced (Madison and coworkers, 1966; Zachau and coworkers, 1966; RajBhandary and coworkers, 1967) can be arranged in a cloverleaf pattern originally suggested by Holley and his coworkers (1965). An alanine *t*-RNA molecule is illustrated in figure 4 (Holley and coworkers, 1965). The loop

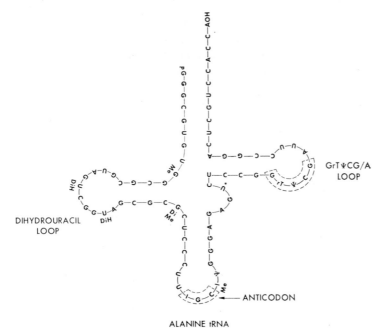

ALANINE tRNA

Figure 4. Structure and sequence of nucleotides from yeast alanine tRNA.

closest to the 3′-terminus of the molecule contains a common sequence, that is a sequence of nucleotides which is common to all *t*-RNAs. This common sequence is GprTpψpCpG/A (Zamir and coworkers, 1967). The opposite loop very often contains one or more dihydrouracil residues. The centre loop contains the anticodon, that sequence which corresponds to the codons for that molecule and for that amino acid. It is the interaction of a trinucleotide template codon with the anticodon of *t*-RNA which determines the translation of *m*-RNA (Nirenberg and Leder, 1964; Crick, 1966; Khorana and coworkers, 1966; Nirenberg and coworkers, 1966). This interaction occurs on the small subunit of the ribosomes.

The translation of an *m*-RNA message occurs in an orderly sequence of events (figure 1). Each trinucleotide sequence represents a code for an amino acid. Transfer RNAs interact with the codon through the anticodon bringing an amino acid into position on the ribosome where the peptide bond is formed. The next non-overlapping codon in sequence is then translated in a similar manner by the binding of another *t*-RNA molecule corresponding to that codon on the ribosome. In this way the *m*-RNA codons are translated in sequence until the entire protein primary sequence is constructed. The *m*-RNA molecules are translated from the 5′-end to the 3′-end direction (Salas and coworkers, 1965). Proteins are correspondingly synthesized from the amino terminal to the carboxy terminal end (Bishop and coworkers, 1960; Yoshida and Tobita, 1960; Dintzis, 1961; Goldstein and Brown, 1961; Canfield and Anfinsen, 1963). The colinearity of the DNA, *m*-RNA, and protein is strictly maintained (Yanofsky and coworkers, 1964).

Although the general scheme outlined is applicable to almost all organisms studied, it is strikingly different in detail from one organism to another. Components of the systems from bacteria are not interchangeable with those from mammalian cells. Nevertheless, some components such as *t*-RNA are usually able to function in cell-free extracts of heterogeneous systems. Furthermore, the specific regulatory and control mechanisms which are applicable to the processes of genetic information transfer, no doubt, vary significantly from organism to organism. Although a great deal is now known about the specific processes of *m*-RNA translation in a variety of organisms, I shall chiefly discuss protein biosynthesis in *E. coli* as the multitude of steps in the messenger RNA translation have been elucidated.

II. SPECIAL CONSIDERATIONS

A. Initiation

The finding of formylmethionyl-*t*-RNA in *E. coli* by Marcker and Sanger (1964) stimulated the investigation of the initiation events of protein biosynthesis which progressed rapidly in many laboratories. Since proteins are constructed from the amino terminal end to the carboxy terminal end, the finding of formylmethionyl-*t*-RNA (f-Met-*t*-RNA) suggested that this molecule functions as an initiator of protein synthesis and that the formylmethionine must be the initial amino terminal amino acid. Many individual laboratories have contributed to the understanding of the mechanism of initiation (Adams and Capecchi, 1966; Clark and Marcker, 1966; Leder and Bursztyn, 1966; Sundararajan and Thach, 1966; Webster and coworkers 1966; Salas and coworkers, 1967). The scheme which appears to best fit this, body of data on initiation is illustrated in figure 5. The *m*-RNA molecule first binds to the small subunit (30 S subunit) of the ribosomes (Henshaw and coworkers, 1965; Joklik and Becker, 1965a, b; McConkey and Hopkins, 1965; Mangiarotti and Schlessinger, 1966; Eisenstadt and Brawerman, 1967; Ghosh and Khorana, 1967; Nomura and coworkers, 1967). The binding occurs at the beginning of the cistron for the protein to be constructed. This may not necessarily correspond to the beginning of the *m*-RNA. In response to an initiator codon, AUG or GUG, f-Met-*t*-RNA is bound to the 30 S-*m*-RNA complex to form a ternary complex. This process requires GTP and three initiation factors, which are proteins that can be isolated from a 1M ammonium chloride wash of *E. coli* ribosomes. Two factors designated F1 and F2 appear to be required for the binding of the f-Met-*t*-RNA to 30 S

INITIATION SEQUENCE

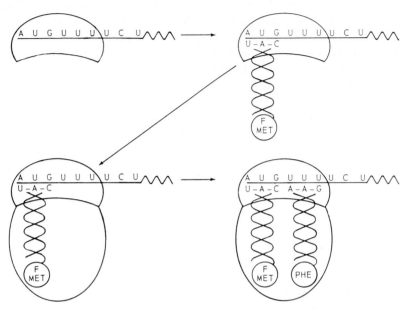

Figure 5. Schematic illustration of the initiation events for protein biosynthesis in *E. coli.*

ribosomes in response to an initiator codon. A third protein factor (F3) seems to be required for the binding of the *m*-RNA to the 30 S subunit (Brown and Doty, 1968; Iwasaki and coworkers, 1968; Revel and coworkers, 1968). The GTP is required for the binding of the f-Met-*t*-RNA and does not appear to be hydrolysed in the initial stages, although the question of hydrolysis at this stage is controversial. A low concentration of magnesium (0·004 M) is necessary. In the presence of high magnesium concentrations (0·02 M) f-Met-*t*-RNA attaches to ribosomes in response to codons in the absence of initiation factors. No monovalent cation is specifically required for these events.

The precise order of the following event has not been established. As depicted in figure 5, however, the 50 S subunit is then bound to the ternary complex of *m*-RNA-30-S ribosome-f-Met-*t*-RNA. The attachment of 50 S subunits can be sensitively measured by an assay which depends on the conversion of aminoacyl-*t*-RNA from a ribonuclease-sensitive to ribonuclease-insensitive state (Pestka, 1968b). Thus the requirements for 50 S attachment can be easily examined. Magnesium ion is required for this attachment, but not potassium. Potassium may stimulate this binding slightly (Pestka, unpublished results). This assay for studying 50 S attachment can also be used to study binding of aminoacyl-*t*-RNA to 70 S ribosomes and codon recognition (Pestka, 1968b). After attachment of the 50 S subunit to the complex, the ribosome is in the 70 S form. It is possible that the level of subunits and 70 S ribosomes are controlled by another factor which forms a complex with the 30 S subunit and prevents its association with the 50 S particle (Subramanian and coworkers, 1968).

Following the binding of the 50 S subunits to the complex, the aminoacyl-*t*-RNA molecule corresponding to the codon adjacent to the initiation codon (internal codon) is bound to the ribosomal complex. While the binding of the initiator *t*-RNA (f-Met-*t*-RNA) requires two initiator factors (F1 and F2), the binding of the aminoacyl-*t*-RNA to the ribosomes in response to the internal codons appears likewise to require two protein factors (Tu, Ts) as well as GTP. The rate of attachment of aminoacyl-*t*-RNA molecules in response to internal codons is stimulated by transfer factors Tu and Ts in the presence of GTP (Lucas-Lenard and Lipmann, 1966; Ravel, 1967; Ertel and coworkers, 1968). This stimulation is most evident at low magnesium (0·004–0·006 M). Although the binding of aminoacyl-*t*-RNA to ribosomes occurs well at high magnesium (0·02 M), the dependency on GTP and the transfer factors is abolished at high magnesium concentration. Potassium and ammonium ions can slightly stimulate the binding of aminoacyl-*t*-RNA to ribosomes in response to codons but binding occurs very well in their absence. It appears that GTP is hydrolyzed in this binding of aminoacyl-*t*-RNA to ribosomes (Weissbach and coworkers, 1970).

The initiation events depicted here apply specifically to *E. coli*. Although a *t*-RNA species which accepts methionine and which can be formylated to form an f-Met-*t*-RNA in mammalian cells exists (Caskey and coworkers, 1967), f-Met-*t*-RNA has been found only in mammalian cell mitochondria (Smith and Marcker, 1968). At present there is no evidence that initiation of proteins in the cytoplasm of mammalian cells (excluding the mitochondria) follows a similar pattern as that in *E. coli*. Mammalian cells may not require a special mechanism for initiation. It is possible that mammalian cells contain only monocistronic messages and therefore require no specific initiation mechanism within a message. Bacterial messages can be either mono- or polycistronic.

As a whole, the events so far described have no particular requirement for monovalent cations. However, aminoacyl-*t*-RNA binding to the 30 S subunits and to 70 S ribosomes (Pestka and Nirenberg, 1966a,b; Pulkrabek and Rychlik, 1968) can be stimulated by potassium ions. It is possible that this stimulation of aminoacyl-*t*-RNA binding to ribosomes in response to codons represents the entering of aminoacyl-*t*-RNA into a potassium-dependent site.

B. Peptide Bond Formation

Although binding of aminoacyl-*t*-RNA to 70 S ribosomes in response to a template has been considered as a single step, it probably consists of several events. The *t*-RNA first binds to ribosomes in response to codons on the 30 S subunit (codon recognition step). The aminoacyl-*t*-RNA is then appropriately oriented so that the aminoacyl end is closely associated with the ribosomes (Pestka, 1967; 1968b).

After the binding of both f-Met-*t*-RNA and aminoacyl-*t*-RNA (Phe-*t*-RNA in figure 6I) to ribosomes, the next event in sequence is the formation of the peptide bond. This occurs by transfer of the f-Met group to Phe (figure 6II). The peptide transferase, which carries out this reaction, appears to be an integral part of the 50 S subunit (Rychlík, 1966; Gottesman, 1967; Monro, 1967; Maden and Monro, 1968; Pestka, 1968a). Magnesium is required for peptide bond formation as it is for *t*-RNA binding. In addition, potassium has been shown to be an essential requirement for peptide bond synthesis. Although the potassium requirement in protein synthesis has been long recognized (Lubin and Ennis, 1964; Schlessinger, 1964), this requirement

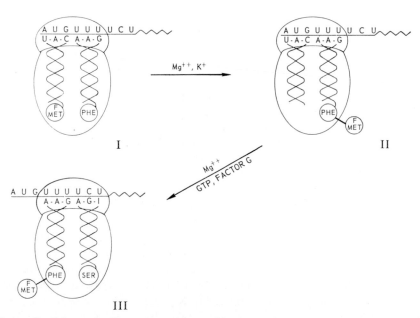

Figure 6. Schematic illustration of peptide bond formation and translocation events of protein biosynthesis in *E. coli*.

has now been localized primarily to the binding of the aminoacyl end of aminoacyl *t*-RNA to ribosomes (Pestka, 1969a,b). Whether peptide bond formation, *per se*, has a potassium requirement is uncertain, although binding of at least one substrate (aminoacyl end of aminoacyl *t*-RNA) of the peptidyl transferase to ribosomes requires potassium.

In the absence of potassium, peptide bond formation is absent. The formation of diphenylalanine and oligophenylalanine as a function of potassium concentration is shown in figure 7. As can be seen, in the absence

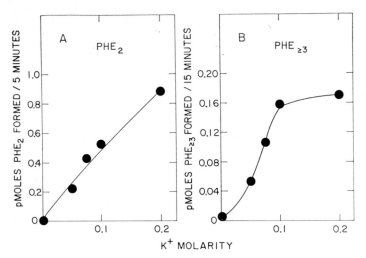

Figure 7. Effect of potassium concentration on peptide bond formation (PHE_2 synthesis) and translocation ($PHE_{\geqslant 3}$ synthesis). (From S. Pestka, (1970) *Arch. Biochem. Biophys.*, **136**, 89–96).

of K^+, there is no detectable peptide bond synthesis. N-acetyl-phenyl-alanine-puromycin formation from N-acetyl-phenylalanyl-*t*-RNA and puro-mycin in the presence of ribosomes can be considered an analogue of natural peptide bond synthesis (Haenni and Chapeville, 1966; Lucas-Lenard and Lipmann, 1967; Weissbach and coworkers, 1968). Its synthesis as a function of potassium concentration (figure 8) indicates that K^+ is necessary for its formation, as also shown by Maden and Monro (1968). Although ammonium ions can substitute for K^+ *in vitro*, K^+ is probably the active ion *in vivo*.

The K^+ requirement for peptide bond synthesis has been chiefly investigated in bacterial cell-free extracts. In mammalian cells, however, there appears to be a similar requirement for K^+ (Levine and coworkers, 1966; Lubin, 1967).

Although K^+ is required for peptide bond synthesis, the precise mode of

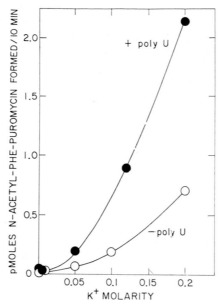

Figure 8. Effect of potassium on peptide bond formation as measured by synthesis of N-acetyl phenylalanyl-puromycin in the presence and absence of poly U. (From S. Pestka (1970) *Arch. Biochem. Biophys.*, **136**, 80–88).

action of K^+ is undefined. Potassium could be required in order for the peptide transferase to function. This enzyme would then be potassium-dependent or potassium-activated. On the other hand, it is conceivable that K^+ could make available a new site of the ribosome for peptidyl and/or aminoacyl-*t*-RNA. Entrance of aminoacyl-*t*-RNA molecules into this site would be prerequisite for peptide bond formation. Furthermore, both possibilities may be valid. There is evidence that K^+ can stimulate aminoacyl-*t*-RNA binding to 30 S subunits (Pestka and Nirenberg, 1966a) as well as to 70 S ribosomes (Pulkrabek and Rychlík, 1968; Pestka, unpublished observations). Alternative explanations may also be possible.

During early logarithmic growth when protein synthesis is occurring at maximal rate, intracellular K^+ is maximal (Schultz and Solomon, 1961; Jones and coworkers, 1965). Ribosomes isolated from cells at early logarithmic growth are also more active than ribosomes from other phases of growth. These alterations may simply reflect effects of K^+ concentrations.

C. Translocation

Translocation is a term which has been introduced (Nishizuka and Lipmann, 1966) to explain the processes of controlled movement during protein biosynthesis (figure 6II to III). The ribosome and *m*-RNA move with respect to each other as the template codons are being translated in sequence from the 5′-end to the 3′-end of *m*-RNA. Closely associated with translocation is the rejection of deacylated *t*-RNA from the ribosome after it has contributed its nascent peptide to the aminoacyl-*t*-RNA.

Since there has been no way of measuring translocation unassociated from peptide bond formation up to the present, estimates of translocation have generally depended on measuring peptide bond synthesis under specified restricted conditions (Erbe and Leder, 1968; Pestka, 1968c). In figure 9 are summarized the reactions which occur in studying translocation with the use of Phe-*t*-RNA and a polyuridylic acid (poly U) template (Pestka, 1968a; 1969c). In the presence of 0·02 M Mg^{2+}, binding of Phe-*t*-RNA to 70 S ribosomes occurs readily (figure 9I) in the absence of any supernatant factors. In fact, no supernatant factors or initiation factors are detectable on these ribosomes washed extensively in 1 M NH_4Cl. In the presence of K^+, diphenylalanyl-*t*-RNA formation occurs (figure 9II), but very little synthesis

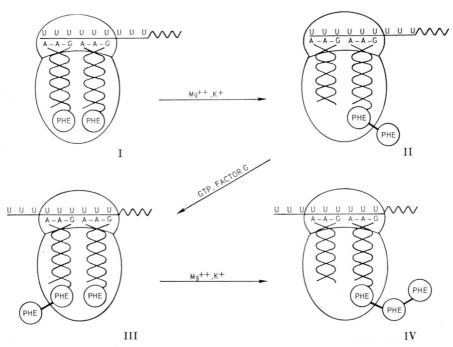

Figure 9. Schematic illustration of the events of protein synthesis as measured by di-, oligo-, and polyphenylalanine synthesis. (From S. Pestka (1969) *J. Biol. Chem.*, **244**, 1533–1539)

of triphenylalanyl-*t*-RNA or any higher homologues is detectable. Thus this system lends itself very well to an examination of the requirements for translocation above those requirements for binding and peptide bond synthesis (figure 9I and II). Although translocation is not measured directly, each translocation event (figure 9II to III) would be followed by the synthesis of an additional peptide bond beyond diphenylalanine (figure 9III to IV). Therefore, determination of the peptides of chain length equal to or greater than triphenylalanine would be in essence measuring translocation. For, after each translocation step, the synthesis of a peptide bond takes place. Such studies have indicated that enzymatic translocation requires transfer factor G as well as GTP, which is hydrolysed in the process (Nishizuka and Lipmann, 1966; Pestka, 1968c, 1969c). Since it has not yet been feasible to examine translocation uncoupled from peptide bond formation, its specific cationic requirements have not been ascertained. Similar studies have been performed with other templates such as polyadenylic acid (poly A) (Pestka, 1969c) and defined oligonucleotides (Erbe and Leder, 1968).

Investigations with poly U and Phe-*t*-RNA have shown that translocation can occur at a slow rate in the absence of factor G and GTP (Pestka, 1969c). This was not demonstrable with the use of a poly A template. Since secondary structure of poly A is substantial compared to poly U (Fresco and Klemperer, 1959; Leng and Felsenfeld, 1966), which contains little secondary structure, the possibility of translocation may be related to secondary structure; and factor G and GTP may, in part, function to alter the secondary structure of the template to permit this movement.

Other studies of translocation have utilized the stimulation of formation of phenylalanyl-puromycin as a criterion (Brot and coworkers, 1968; Tanaka and coworkers, 1968). In the presence of ribosomes, Phe-*t*-RNA and a poly U template, a fraction of Phe-*t*-RNA bound to ribosomes can react with puromycin. This fraction is considered to be in a donor state. The unreactive fraction is considered to be in an acceptor state. In the presence of factor G and GTP, much of the unreactive bound Phe-*t*-RNA can be converted into a reactive form, which has been interpreted as translocation of Phe-*t*-RNA from acceptor to donor states. Similar studies have been performed in bacterial and mammalian cell-free extracts with the use of ribosomes containing labeled peptidyl-*t*-RNA (Smith and coworkers, 1965; Skogerson and Moldave, 1968).

D. Termination

Evidence for specific terminator codons has come from genetic and biochemical experiments (Brenner and Stretton, 1965; Brenner and coworkers 1965; Bretscher and coworkers, 1965; Weigert and Garen, 1965a,b; Ganoza and Nakamoto, 1966; Brenner and coworkers, 1967; Last and coworkers, 1967). At the completion of the protein which remains attached to *t*-RNA, a termination signal initiates the release of the completed protein (figure 10). One of three codons can signify release: UAA, UAG, or UGA. To date there is no evidence that a specific *t*-RNA, which is involved in release, recognizes these codons. There appear to be release factors which respond to these codons (Capecchi, 1967; Caskey and coworkers, 1968; Scolnick and coworkers, 1968). One factor (R1) corresponds to UAA and UAG; the other (R2), to UAA and UGA. In the presence of a terminator codon and release factor the completed protein is released from the ribosome and *t*-RNA. Termination and release can function in the absence of other supernatant factors (Tu, Ts, or G). Specific requirements for release independent of *t*-RNA binding have not yet been clearly delineated. It is possible that release is a form of abortive peptide bond formation where the completed protein is transferred to water rather than to aminoacyl-*t*-RNA, which is absent from the ribosome in the presence of a terminator codon. The inhibition of release by chloramphenicol, sparsomycin, streptomycin and tetracycline (Scolnick and coworkers, 1968) suggests that an intermediate peptidyl-transfer to water or some other moiety possibly occurs. On the other hand, when a suppressor aminoacyl-*t*-RNA is present on the ribosome in response

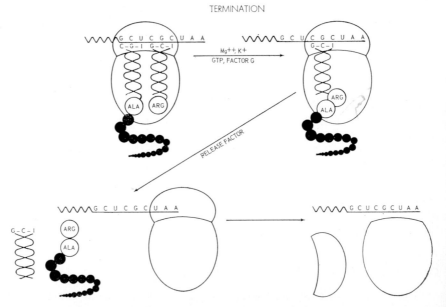

Figure 10. Schematic illustration of the termination events of protein biosynthesis in *E. coli*.

to one of the terminator codons, release can be suppressed and chain elongation continued (Capecci and Gussin, 1965). Thus it appears that for release factors to function no new aminoacyl-*t*-RNA need be bound to the ribosome. Conceivably release factors and suppressor *t*-RNAs may compete with each other for ribosomal binding sites.

Although *in vitro* release can occur in the presence of any one of the terminator codons, which of the codons functions naturally *in vivo* is unknown. With respect to mammalian cells virtually nothing is known about the termination coding mechanism; nor is it known if any specific triplet coding mechanism is operative.

III.   CONCLUDING REMARKS

An outline of the mechanism and requirements for protein synthesis has been detailed. Shown in Tables 1 and 2 is a list of the components and

Table 1. The elements of protein synthesis

1.	Ribosomes, subunits
	a *r*-RNA
	b Ribosomal proteins
	c 5 S RNA
	d Peptide transferase
2.	*t*-RNA, AA-*t*-RNA, F-Met-*t*-RNA
3.	ATP
4.	Synthetases
5.	Transformylase
6.	Deformylase
7.	Initiation factors (F_1, F_2, F_3)
8.	*m*-RNA codons
9.	Mg^{2+}
10.	K^+
11.	Transfer factors (Tu, Ts, G)
12.	GTP
13.	Release factors (R_1, R_2)

Table 2. The events of protein synthesis

1.	Acylation (amino acid activation)
2.	Formylation
3.	Deformylation
4.	Initiation
5.	Translation (recognition)
6.	Peptide synthesis
7.	Translocation
8.	Termination (release)

events, respectively, of protein biosynthesis. All the steps of protein synthesis that have been measured independently of other reactions have a magnesium requirement. The binding of the aminoacyl end of aminoacyl t-RNA to ribosomes shows an absolute dependency on the presence of potassium. Thus the overall process of peptide bond formation also requires potassium, since the binding of at least one of the substrates for peptide bond synthesis requires potassium. This stage of protein synthesis, therefore, can possibly be controlled by the level of intracellular potassium. Thus it is possible that intracellular potassium may sometimes regulate protein biosynthesis.

REFERENCES

Adams, J. M. and M. R. Capecchi (1966) *Proc. Natl. Acad. Sci. U.S.*, **55**, 147
Allfrey, V. G., M. M. Daly and A. E. Mirsky (1953) *J. Gen. Physiol.*, **37**, 157
Avery, O. T., C. M. MacLeod and M. McCarty (1944) *J. Exptl. Med.*, **79**, 137
Beadle, G. W. (1957) In W. D. McElroy and B. Glass (Eds.), *The Chemical Basis of Heredity*, Johns Hopkins Press, Baltimore, p. 3
Berg, P. (1956) *J. Biol. Chem.*, **222**, 1025
Berg, P. and E. J. Ofengand (1958) *Proc. Natl. Acad. Sci. U.S.*, **44**, 78
Bishop, J. O., J. Leahy and R. S. Schweet (1960) *Proc. Natl. Acad. Sci. U.S.*, **46**, 1030
Borsook, H., C. L. Deasy, A. J. Haagen-Smit, G. Keighley and P. H. Lowy (1950) *J. Biol. Chem.*, **187**, 839
Brachet, J, (1942) *Arch. Biol. (Liege)*, **53**, 207
Brenner, S., L. Barnett, E. R. Katz and F. H. C. Crick (1967) *Nature*, **213**, 449
Brenner, S. and A. O. W. Stretton (1965) *J. Mol. Biol.*, **13**, 944
Brenner, S., A. O. W. Stretton and S. Kaplan (1965) *Nature*, **206**, 994
Bretscher, M. S., H. M. Goodman, J. R. Menninger and J. D. Smith (1965) *J. Mol. Biol.*, **14**, 634
Brot, N., R. Ertel and H. Weissbach (1968) *Biochem. Biophys. Res. Commun.*, **31**, 563
Brown, J. C. and P. Doty (1968) *Biochem. Biophys. Res. Commun.*, **30**, 284
Brownlee, G. G., F. Sanger and B. G. Barrell (1968) *J. Mol. Biol.*, **34**, 379
Canfield, R. E. and C. B. Anfinsen (1963) *Biochemistry*, **2**, 1073
Capecchi, M. R. (1967) *Proc. Natl. Acad. Sci. U.S.*, **58**, 1144
Capecchi, M. R. and G. N. Gussin (1965) *Science*, **149**, 417
Caskey, C. T., B. Redfield and H. Weissbach (1967) *Arch. Biochem. Biophys.*, **120**, 119
Caskey, C. T., R. Tompkins, E. Scolnick, T. Caryk and M. Nirenberg (1968) *Science*, **612**, 135
Caspersson, T. von (1941) *Naturwissenschaften*, **29**, 33
Clark, B. F. C. and K. A. Marcker (1966) *J. Mol. Biol.*, **17**, 394
Comb, D. G. and S. Katz (1964) *J. Mol. Biol.*, **8**, 790
Crick, F. H. C. (1958) *Symp. Soc. Exptl. Biol.*, **12**, 138
Crick, F. H. C. (1966) *J. Mol. Biol.*, **19**, 548
Davie, E. W., V. V. Koningsberger and F. Lipmann (1956) *Arch. Biochem. Biophys.*, **65**, 21
DeMoss, J. A., S. M. Genuth and G. D. Novelli (1956) *Proc. Natl. Acad. Sci. U.S.*, **42**, 325
Dintzis, H. M. (1961) *Proc. Natl. Acad. Sci. U.S.*, **47**, 247
Eisenstadt, J. M. and G. Brawerman (1967) *Proc. Natl. Acad. Sci. U.S.*, **58**, 1560
Erbe, R. W. and P. Leder (1968) *Biochem. Biophys. Res. Commun.*, **31**, 798
Ertel, R., N. Brot, B. Redfield, J. Allende and H. Weissbach (1968) *Proc. Natl. Acad. Sci. U.S.*, **59**, 861
Forget, B. G. and S. M. Weissman (1967) *Science*, **158**, 1695

Fresco, J. R. and E. Klemperer (1959) *Ann. N.Y. Acad. Sci.*, **81**, 730
Galibert, F., C. J. Larsen, J. C. Lelong and M. Boiron (1965) *Nature*, **207**, 1039
Ganoza, M. C. and T. Nakamoto (1966) *Proc. Natl. Acad. Sci. U.S.*, **55**, 162
Ghosh, H. P. and H. G. Khorana (1967) *Proc. Natl. Acad. Sci. U.S.*, **58**, 2455
Goldstein, A. and B. J. Brown (1961) *Biochim. Biophys. Acta.*, **53**, 438
Gottesman, M. E. (1967) *J. Biol Chem.*, **242**, 5564
Haenni, A. L. and F. Chapeville (1966) *Biochim. Biophys. Acta.*, **114**, 135
Henshaw, E., M. Revel and H. Hiatt (1965) *J. Mol. Biol.*, **14**, 241
Hoagland, M. B., E. B. Keller and P. C. Zamecnik (1956) *J. Biol. Chem.*, **218**, 345
Hoagland, M. B., M. L. Stephenson, J. F. Scott, L. I. Hecht and P. C. Zamecnik (1958) *J. Biol. Chem.*, **231**, 241
Holley, R. W. (1957) *J. Amer. Chem. Soc.*, **79**, 658
Holley, R. W., J. Apgar, G. A. Everett, J. T. Madison, M. Marquisee, S. H. Merrill, J. R. Penswick and A. Zamir (1965) *Science*, **147**, 1462
Hultin, T. (1950) *Exptl. Cell Res.*, **1**, 376
Iwasaki, K., S. Sabol, A. J. Wahba and S. Ochoa (1968) *Arch. Biochem. Biophys.*, **125**, 542
Jacob, F. and J. Monod (1961) *J. Mol. Biol.*, **3**, 318
Joklik, W. K. and Y. Becker (1965a) *J. Mol. Biol.*, **13**, 496
Joklik, W. K. and Y. Becker (1965b) *J. Mol. Biol.*, **13**, 511
Jones, W. B., A. Rothstein, F. Sherman and J. N. Stannard (1965) *Biochim. Biophys. Acta.*, **104**, 310
Khorana, H. G., H. Büchi, H. Ghosh, N. Gupta, T. M. Jacob, H. Kössel, R. Morgan, S. A. Narang, E. Ohtsuka and R. D. Wells (1966) *Cold Spring Harbor Symp. Quant. Biol.*, **31**, 39
Kornberg, A. (1961) *Enzymatic Synthesis of DNA*, John Wiley, New York, p. 1
Kurland, C. G. (1960) *J. Mol. Biol.*, **2**, 83
Last, J. A., W. M. Stanley, Jr., M. Salas, M. B. Hille, A. J. Wahba and S. Ochoa (1967) *Proc. Natl. Acad. Sci. U.S.*, **57**, 1062
Leder, P. and H. Bursztyn (1966) *Proc. Natl. Acad. Sci. U.S.*, **56**, 1579
Leng, M. and G. Felsenfeld (1966) *J. Mol. Biol.*, **15**, 455
Levine, H., M. R. Trindle and K. Moldave (1966) *Nature*, **211**, 1302
Lubin, M. (1967) *Nature*, **213**, 451
Lubin, M. and H. L. Ennis (1964) *Biochim. Biophys. Acta.*, **80**, 614
Lucas-Lenard, J. and F. Lipmann (1966) *Proc. Natl. Acad. Sci. U.S.*, **55**, 1562
Lucas-Lenard, J. and F. Lipmann (1967) *Proc. Natl. Acad. Sci. U.S.*, **57**, 1050
Maden, B. E. H. and R. E. Monro (1968) *Eur. J. Biochem.*, **6**, 309
Madison, J. T., G. A. Everett and H. Kung (1966) *Science*, **153**, 531
Mangiarotti, G. and D. Schlessinger (1966) *J. Mol. Biol.*, **20**, 123
Marcker, K. A. and F. Sanger (1964) *J. Mol. Biol.*, **8**, 835
Matthaei, H., F. Amelunxen, K. Eckert and G. Heller (1964) *Ber. Bunsengesellschaft*, **68**, 735
McConkey, E. H. and J. W. Hopkins (1965) *J. Mol. Biol.*, **14**, 257
McQuillen, K., R. B. Roberts and R. J. Britten (1959) *Proc. Natl. Acad. Sci. U.S.*, **45**, 1437
Monro, R. E. (1967) *J. Mol. Biol.*, **26**, 147
Nirenberg, M., T. Caskey, R. Marshall, R. Brimacombe, D. Kellogg, B. Doctor, D. Hatfield, J. Levin, F. Rottman, S. Pestka, M. Wilcox and F. Anderson (1966) *Cold Spring Harbor Symp. Quant. Biol.*, **31**, 11
Nirenberg, M. and P. Leder (1964) *Science*, **145**, 1399
Nirenberg, M. W. and J. H. Matthaei (1961) *Proc. Natl. Acad. Sci. U.S.*, **47**, 1588
Nishizuka, Y. and F. Lipmann (1966) *Arch. Biochem. Biophys.*, **116**, 344
Nomura, M., B. D. Hall and S. Spiegelman (1960) *J. Mol. Biol.*, **2**, 306
Nomura, M., C. V. Lowry and C. Guthrie (1967) *Proc. Natl. Acad. Sci. U.S.*, **58**, 1487
Ogata, K., H. Nohara and T. Morita (1957) *Biochim. Biophys. Acta.*, **26**, 656
Pestka, S. (1967) *J. Biol. Chem.*, **242**, 4939
Pestka, S. (1968a) *J. Biol. Chem.*, **243**, 2810
Pestka, S. (1968b) *J. Biol. Chem.*, **243**, 4038

Pestka, S. (1968c) *Proc. Natl. Acad. Sci. U.S.*, **61**, 726
Pestka, S. (1969a) *Biochem. Biophys. Res. Comm.*, **36**, 589
Pestka, S. (1969b) *Cold Spring Harbor Symp. Quant. Biol.*, **34**, 395
Pestka, S. (1969c) *J. Biol. Chem.*, **244**, 1533
Pestka, S. and M. Nirenberg (1966a) *J. Mol. Biol.*, **21**, 145
Pestka, S. and M. Nirenberg (1966b) *Cold Spring Harbor Symp. Quant. Biol.*, **31**, 641
Pulkrabek, P. and I. Rychlík (1968) *Biochim. Biophys. Acta.*, **155**, 219
Rabinovitz, M. and M. E. Olson (1956) *Exptl. Cell Res.*, **10**, 747
RajBhandary, U. L., S. H. Chang, A. Stuart, R. D. Faulkner, R. M. Hoskinson and H. G. Khorana (1967) *Proc. Natl. Acad. Sci. U.S.*, **57**, 751
Ravel, J. M. (1967) *Proc. Natl. Acad. Sci. U.S.*, **57**, 1811
Revel, M., M. Herzberg, A. Becarevic and F. Gros (1968) *J. Mol. Biol.*, **33**, 231
Rosset, R. and R. Monier (1963) *Biochim. Biophys. Acta.*, **68**, 653
Rychlík, I. (1966) *Biochim. Biophys. Acta.*, **114**, 425
Salas, M., M. B. Hille, J. A. Last, A. J. Wahba and S. Ochoa (1967) *Proc. Natl. Acad. Sci. U.S.*, **57**, 387
Salas, M., M. A. Smith, W. M. Stanley, Jr., A. J. Wahba and S. Ochoa (1965) *J. Biol. Chem.*, **240**, 3988
Schlessinger, D. (1964) *Biochim. Biophys. Acta.*, **80**, 473
Schultz, S. G. and A. K. Solomon (1961) *J. Gen. Physiol.*, **45**, 355
Schweet, R. S. and E. H. Allen (1958) *J. Biol. Chem.*, **233**, 1104
Schweet, R. S., F. C. Bovard, E. Allen and E. Glassman (1958) *Proc. Natl. Acad. Sci. U.S.*, **44**, 173
Schweet, R. S., H. Lamfrom and E. H. Allen (1958) *Proc. Natl. Acad. Sci. U.S.*, **44**, 1029
Scolnick, E., R. Tompkins, T. Caskey and M. Nirenberg (1968) *Proc. Natl. Acad. Sci. U.S.*, **61**, 768
Skogerson, L. and K. Moldave (1968) *J. Biol. Chem.*, **243**, 5361
Smith, A. E. and K. A. Marcker (1968) *J. Mol. Biol.*, **38**, 241
Smith, J. D., R. R. Traut, G. M. Blackburn and R. E. Monro (1965) *J. Mol. Biol.*, **13**, 617
Spiegelman, S. (1957) In W. D. McElrey and B. Glass, *The Chemical Basis of Heredity*, Johns Hopkins Press, Baltimore, p. 232
Subramanian, A. R., E. Z. Ron and B. D. Davis (1968) *Proc. Natl. Acad. Sci. U.S.*, **61**, 761
Sundararajan, T. A. and R. E. Thach (1966) *J. Mol. Biol.*, **19**, 74
Suzuka, I., H. Kaji and A. Kaji (1965) *Biochem. Biophys. Res. Commun.*, **21**, 187
Takanami, M. and T. Okamoto (1963) *J. Mol. Biol.*, **7**, 323
Tanaka, N., T. Kinoshita and H. Masukawa (1968) *Biochem. Biophys. Res. Commun.*, **30**, 278
Tissières, A. (1959) *J. Mol. Biol.*, **1**, 365
Volkin, E. and L. Astrachan (1956) *Virology*, **2**, 149
Watson, J. D. (1963) *Science*, **140**, 17
Watson, J. D. and F. H. C. Crick (1953) *Nature*, **171**, 964
Webster, R. E., D. L. Engelhardt and N. D. Zinder (1966) *Proc. Natl. Acad. Sci. U.S.*, **55**, 155
Weigert, M. G. and A. Garen (1965a) *Nature*, **206**, 992
Weigert, M. G. and A. Garen (1965b) *J. Mol. Biol.*, **12**, 448
Weiss, S. B., G. Acs and F. Lipman (1958) *Proc. Natl. Acad. Sci. U.S.*, **44**, 189
Weissbach, H., D. L. Miller and J. Hachmann (1970) *Arch. Biochem. Biophys.* **137**, 262
Weissbach, H., B. Redfield and N. Brot (1968) *Arch. Biochem. Biophys.*, **127**, 705
Yanofsky, C., B. C. Carlton, J. R. Guest, D. R. Helinski and U. Henning (1964) *Proc. Natl. Acad. Sci. U.S.*, **51**, 266
Yoshida, A. and T. Tobita (1960) *Biochim. Biophys. Acta.*, **37**, 513
Zachau, H. G., D. Dütting and H. Feldmann (1966) *Hoppe-Seylers Z. Physiol. Chem.*, **347**, 212
Zamir, A., R. W. Holley and M. Marquisee (1967) *J. Biol. Chem.*, **240**, 1267
Zamecnik, P. C. and E. B. Keller (1954) *J. Biol. Chem.*, **209**, 337

CHAPTER 10

Regulation of ion transport by hormones

E. Edward Bittar

Department of Physiology,
University of Wisconsin,
Madison, Wisconsin, USA

I. INTRODUCTION

This chapter is not intended to be a survey of the pertinent literature but rather an appraisal of the various theories which have been put forward to explain the nature of the primary mechanism underlying hormonal actions on ion transport across cellular membranes. A deliberate attempt will therefore be made to emphasise the conceptual approach and to deal with trends of ideas in illustration of which the natriferic actions of vasopressin, insulin and aldosterone at the cellular level will be described.

The view that hormonal regulation of ion transport is a fundamental feature of homeostasis is by no means new, but it is only within the last few years that fairly convincing evidence in favor of this view has been marshalled. Broadly speaking, hormones may be regarded as substances that regulate the rate of pre-existing chemical reactions by interacting with the cell membrane or with an intracellular enzyme. Certain hormones are also capable of inducing enzyme synthesis by means of exerting a primary action on the genome. It is, however, impossible, to draw a sharp line of demarcation between the two mechanisms, i.e. of plasma membrane interaction and of intracellular enzyme activation or induction of new enzyme formation. The fact that a hormone penetrates into the cell and is then able to affect cell metabolism does not of course dispose of the possibility of an interaction between the hormone and the plasma membrane. A prime example of this is seen in aldosterone which modifies the permeability of the membrane such as in toad bladder, and induces *de novo* protein synthesis in this tissue. Unfortunately too little is known of the behavior of biological membranes toward hormones, so one cannot press this argument too far.

II. HORMONAL INTERACTION WITH THE CELL MEMBRANE

It has long been thought that hormones influence ion movements across cellular membranes by modifying the permeability of the membrane to ions. Implicit in this concept is that hormones interact with a receptor in or on the cell surface, thereby producing a conformational change in the membrane protein. Since the protein and the lipid in the membrane are interdependent, it would be natural to expect a change in the structure of the protein to be accompanied by a change in the structure of the lipid. Thus, the question to be decided is whether the behavior of the membrane toward a hormone is governed by a change in the lipid layer. At present one can only guess at the answer: The work of Bangham, Standish and Weissman (1965) showed that steroidal interaction with the lipid phase of 'bangosomes' leads to ion permeability changes in the membrane. A more fruitful approach is perhaps that adopted by Rodbell (1966) and Blecher (1966) who found that insulin and phospholipase bring about similar changes in the ultrastructure of natural membranes. One now wonders if this involves phosphatidylserine, since this lipid vastly increases the activity of the Na^+-K^+-ATPase (Wheeler and Whittam, 1970).

Practically nothing is known of the nature of the hormonal receptor. There are, however, several good reasons for believing that it is a protein and not a mucopolysaccharide. Firstly, a protein offers a considerably greater range of groupings; and secondly, it is capable of having many conformations in space. At present a protein is believed to consist of at least two distinct, non-overlapping sites or sub-units, i.e. the catalytic, and the regulatory site. Monod, Changeux and Jacob (1961) coined the term 'allosteric' in order to describe a protein in which the conformational change is due to binding of the ligand (the effector) to the regulatory site and in which the conformational change—an *allosteric transition*—alters the properties of the catalytic site of the protein. This led to the suggestion by

Jardetzky (1966) and Hill (1969) that the transport enzyme is an allosteric protein, i.e. the Na^+-K^+-ATPase exists as two sub-units. However, there is no concrete evidence yet for any direct involvement of the transport enzyme in the ionic changes that have been described to take place, when vasopressin, aldosterone or insulin are applied to a target tissue. This is not without significance, for it could mean:

i) that the primary site of action of these hormones is the plasma membrane but not the transport enzyme inside this structure, or

ii) that the results obtained with isolated enzyme preparations do not at all parallel the behavior of the enzyme *in situ*.

An alternative point of view which has nothing to do with the plasma membrane, is that hormones such as aldosterone and insulin regulate the rate of Na exchange occurring across the inner membranes of the cell. This fresh way of looking at the problem will be given fuller consideration later on.

The discovery of the adenyl cyclase system by Sutherland and Rall (1957) is of special biological importance. It indeed constitutes the first piece of evidence for the existence of a receptor in a biological membrane. The finding that several physiological substances, such as adrenaline and vasopressin, affect cell metabolism by interacting with the adenyl cyclase, a reaction which in turn results in the production of cyclic-3',5'-AMP, lends credence to the cytoskeletal theory of Sir Rudolph Peters (1956). In a nutshell, this theory states that a hormone need not penetrate into the cell in order to produce a widespread and integrated response by the cell. That a hormone may achieve this effect by simply modifying ion transport across the plasma membrane has also been suggested by Hechter (1965) who following a similar line of reasoning maintains that hormones produce membrane conformation, thereby disturbing the water lattice. Since the bulk of the evidence in support of this theory comes from studies with vasopressin, the action of this hormone will be dealt with first.

III. THE NATRIFERIC ACTION OF VASOPRESSIN

It has gradually become clear that vasopressin influences cell permeability by regulating the concentration of cyclic-3',5'-AMP inside the cell (Orloff and Handler, 1964). Evidence favoring this concept derives mainly from the experiments of Orloff and Handler (1962) showing that cyclic AMP mimics vasopressin in its action on toad bladder. However, evidence put forward by Cuthbert and Painter (1968) indicates that the resistance of frog skin is not reduced but raised by cyclic AMP. Interpretation of this result is not straightforward, but it could mean that the protein *kinases* in frog skin and toad bladder are not necessarily the same. It is clear enough that the older theory of a thiol-disulfide interchange (Schwartz, Rasmussen, Schoessler, Silver and

Fong, 1960; Rasmussen, Schwartz, Young and Marc-Aurele, 1963) is untenable. The plausible view is that an SS–SH interchange is the first in a chain of events leading to the stimulation by vasopressin of water flow and Na transport. This argument though difficult to test has led to attempts at an estimation of the pK of the grouping in the receptor. These have yielded values of 7·45 to 8, which correspond to the pK of –SH groups, α-NH_2 or imidazole groups (see Gulyassy and Edelman, 1965). But this is not the entire problem, for one has to answer the question whether the receptor lies on the inner or outer side of the cell membrane, in particular because the actions of both cyclic AMP and vasopressin are pH dependent. Cyclic AMP is known to exert an optimal effect at pH 7, whereas vasopressin does so at pH 8·4 (Edelman, Petersen and Gulyassy, 1964). Now if it be assumed that alkalinization of the external medium causes a rise in the internal pH, it would then follow that cyclic AMP cannot act efficiently as the mediator of the action of vasopressin. One way out of the difficulty is to consider the possibility that the internal pH of the target cell does not rise when the surrounding medium pH is increased. Until precise information on the internal pH becomes available it would seem unwise to either accept or challenge the cyclic AMP theory.

A further question arises: Is there only one receptor as advocated by Orloff and Handler (1964) and Rasmussen (1969), or are there two receptors? The likelihood of there being two receptors arises from the studies of Heller, Bentley, Morel, Maetz and others who approached the problem from the point of view of biological evolution. Thus, for example, it was found that treatment of salamander with oxytocin produces a natriferic effect but no increase in water flow (Bentley and Heller, 1964). An analogous situation exists in the fresh water fish *Carassius auratus* (see Maetz, 1963). More recently, the actions of analogues of oxytocin–vasopressin on kidney tubule, skin and urinary bladder of *Rana esculenta* have been examined in detail. The results reported by Morel, Jard, Bourguet and Bastide (1969) are consistent with the view that the natriferic and water flow effects of the neurohypophyseal hormones involve two separate and specific receptors. Nevertheless, it is fair to say that decisive evidence as to which theory is correct will probably not come from studies of structure–activity relationships but from efforts at isolating and defining the nature of the receptor. One helpful way toward achieving this end would be to turn to biological models which are simpler than toad bladder or frog skin, yet sensitive to the hormone in question. This is illustrated by the work of Bittar (1966a) involving the use of single muscle fibers from the spider crab, *Maia squinado*. He found that neither internal nor external application of vasopressin at pH 7 affected the efflux of radiosodium. By contrast, vasopressin at pH 7·4 and 8·4 slowed down the emergence of sodium but only when applied

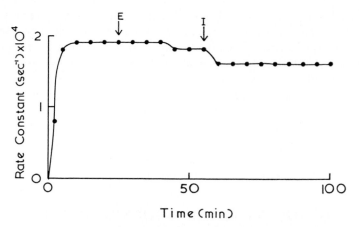

Figure 1. Lack of effect on ^{22}Na efflux from a *Maia* fiber of externally and internally applied 8-Lysine vasopressin (100 mU and 10 U/ml, respectively) when the external pH is 7·0. Ordinate: fraction of ^{22}Na lost per second (From Bittar, 1966a).

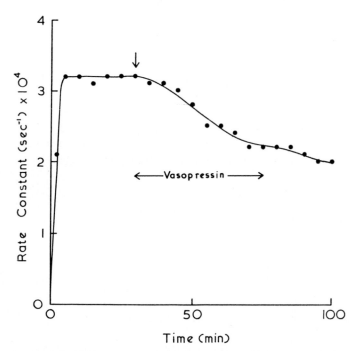

Figure 2. Inhibiting effect on ^{22}Na efflux from a *Maia* fiber of externally applied vasopressin (100 mU/ml) at pH 8·4. Arrow indicates time at which hormone was applied to the fiber (From Bittar, 1966a).

externally. These results with 8-lysine vasopressin are shown in figures 1 and 2. Though not considered a target cell, the *Maia* fiber, or more precisely, its efflux mechanism is unquestionably sensitive to the hormone. This suggests that the fiber can be employed as a model for studying vasopressin's ability to reduce the efflux rate when the external pH is shifted further toward the alkaline side. The fact remains, however, that the behavior of the Na pump of the *Maia* fiber toward vasopressin is the very opposite of that of frog skin or toad bladder, namely an asymmetric cell. The slowing of the efflux by vasopressin is to be expected, since this is also true of aldosterone (see p. 306). Whether crab muscle has an adenyl cyclase system and whether vasopressin *reduces* cyclic AMP production in this tissue is not yet known.

IV.   THE NATRIFERIC ACTION OF INSULIN

The insulin molecule consists of two chains, A and B, which are joined by two disulfide bonds. Only when the hormone maintains a native conformational state are these bonds highly reactive (Rosa, Rossi and Donato, 1968). This observation reinforces the prevailing view that the biological activity of insulin is closely related to a unique structural feature of the molecule, the identity of which remains unknown. A promising approach to the study of this problem is the method developed by Warner (1965). Known as the hexagonal concept of protein conformation because the arrangement of the peptide chain O-atoms is hexagonal, this method has already been applied in the investigation of the conformation of vasopressin and now insulin. As illustrated in figure 3, the two insulin chains or sub-units are linked by disulfide bridges, i.e. A-7 to B-7 and A-20 to B-19.

Studies of the action of insulin on the plasma membrane have been largely motivated by the idea that insulin regulates transport processes as the result

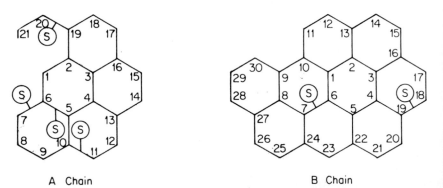

Figure 3. Hexagonal model of insulin showing the –S–S– linkages in the A and B chains or subunits (courtesy of Dr. D. T. Warner).

of its anabolic property. These processes are glucose, amino acid and electrolyte transport. Consequently, it has never been quite certain whether insulin exerts a single or a multiple effect at the level of the plasma membrane. Only recently has the view arisen that the Na pump may well be the primary site at which insulin acts. Historically speaking, the starting point of the evidence leading to this conclusion is the report by Flückiger and Verzár (1954) that insulin decreases the Na content of rat diaphragm, an observation later confirmed by Creese, D'Silva and Northover (1958). Further work by Creese and Northover (1961), Creese (1964; 1968) and independently by Zierler (1957, 1959), Kernan (1962) and Zierler, Rogus and Hazlewood (1966) produced evidence that insulin stimulates Na extrusion by skeletal muscle *in vitro*. Interestingly enough, it was at one time thought that the primary effect of insulin was on K influx (see for example Smillie and Manery, 1960) but this idea is no longer tenable since K influx is largely coupled to Na efflux.

It may be argued that what occurs in an *in vitro* preparation may not necessarily be also true *in vivo*. This is readily dismissed because as first convincingly demonstrated by Kernan, the stimulating action of insulin is dependent on the presence of a high internal Na. Seeking to investigate this important point, Bittar (1967) injected insulin into a crab and some 15 hours later injected the hormone directly into a cannulated fiber. Instead of producing a rise in Na efflux, as in the case of a fiber untreated *in vivo* with the hormone, insulin caused a gradual fall in Na efflux. This is illustrated by figure 4, which also shows that when insulin was applied externally about 20 minutes later, the efflux rose rather sharply. Under these conditions the membrane was found to be hyperpolarized, a finding which agrees with that of Zierler (1959) and Kernan (1962). This idea, that insulin stimulates Na efflux when the Na pump is depressed, is significant. Firstly, it indicates that insulin is a homeostatic hormone the function of which is to keep the concentration of 'free' Na inside the cell relatively constant. The word 'free' is worth noting, first because it is well recognised that a large fraction of the internal Na is sequestered, and second, because the kinetic results obtained by Bittar, Dick and Fry (1968) in preliminary experiments using germ cells, indicate that insulin mobilises the secluded Na, thereby leading to rate constants that show no decline. That the main site of action of insulin is not the plasma membrane is strongly suggested by the fact that insulin is without effect on the Na^+-K^+-ATPase of rat muscle, irrespective of the hormone being applied *in vitro* or *in vivo* (Rogus, Price and Zierler, 1969).

It is important to know whether insulin affects phosphate movements across the plasma membrane and if these are closely linked to the behavior of the Na pump. Sacks and Sinex (1953) found that insulin stimulates the uptake of phosphate by rat diaphragm. Later work by Clauser, Volfin and

L

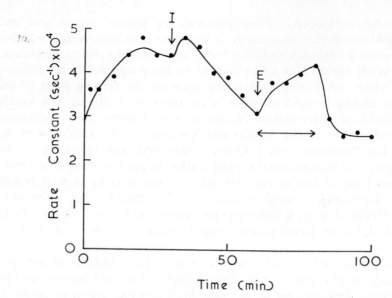

Figure 4. Effects of insulin on Na efflux from a *Maia* fiber which was isolated from a crab some 15 hr after injecting it with a large dose of insulin. Injected insulin (sarcoplasmic concentration about 1 U/ml) caused slowing of the pump, whereas external application of the hormone (1 U/ml) had the opposite effect (From Bittar, 1967).

Eboué-Bonis (1962), Eboué-Bonis, Chambaut, Volfin and Clauser (1963) and Walaas, Walaas and Wick (1969) showed this to be the case. But as reported by Walaas and coworkers (1969) the total ATP content of rat muscle is not increased by insulin. Although this is a vital piece of information it must be remembered that the available methods for the assay of ATP do not distinguish between the free and bound ATP fractions. Conceivably insulin increases the free fraction of ATP by stimulating the rate of adenine nucleotide exchange across the mitochondrial inner membrane. Supposing this to be so, one would then expect the *adenylate translocase* inhibitor, atractylate, to stop insulin from exerting such an effect. Experiments designed to examine this possibility have not yet been reported. Speculation apart, the stimulating effect of insulin on orthophosphate movements in muscle is of a smaller magnitude when the Na pump is inhibited by ouabain (Walaas and co-workers, 1968). Other tissues, for example squid giant axon when stimulated, show a rise in orthophosphate uptake which is also reduced by ouabian (Baker, 1963).

The interesting work of Dormandy on red cells has not yet been mentioned. Dormandy believes that insulin in physiological concentrations causes a shift in the redox potential gradient across the cell-extracellular interface. The experimental evidence brought forward by him indicates that insulin causes an uptake by red cells of protons present in the bathing medium (Dormandy, 1965, 1966). To accept this evidence is obviously to challenge the widely held view that red cells are irresponsive to insulin. It should not be difficult to test Dormandy's argument by measuring the internal pH of crab or barnacle muscle before and after the application of insulin.

To what extent Na–K exchanges are controlled by the internal pH is completely unknown. Early experiments by Keynes (1965) on frog muscle have been interpreted by him to mean that Na and H ions compete for the available transport sites. But clearly any meaningful study of this aspect of ion transport would have to take into consideration not the overall cell pH but rather the regional pH. Should the pH gradients be quite large, one would then be justified in supposing that a component of Na efflux is controlled by this gradient, as for example occurs in oil membranes with cephalin as the carrier (Moore and Schechter, 1969).

V. THE NATRIFERIC ACTION OF ALDOSTERONE

Present knowledge of the mode of action of aldosterone is almost wholly based on studies of the isolated toad bladder of *Bufo marinus* (see Sharp and Leaf, 1966) and rat kidney (Williamson, 1963; Fimognari, Fanestil and Edelman, 1967). It has been demonstrated beyond reasonable doubt that aldosterone in physiological concentrations increases the rate at which Na leaves the epithelial cell, but there is disagreement among investigators about the basic mechanism producing this action. There are two schools of thought. One school led by Edelman states that aldosterone acts primarily by increasing the supply of ATP to the Na pump, while the other led by Leaf and Crabbé maintains that aldosterone stimulates Na transport by increasing the permeability of the apical or mucosal surface of the epithelial membrane to Na. Although one is in no position yet to say which school is right, it is not unlikely that both views pertaining to asymmetric cells are correct. One way of testing the ATP theory is to measure the ATP content of toad bladders before and after treatment with aldosterone. This has already been done by Fimognari, Kasbekar and Edelman (1965) who reported a 30 per cent rise in the concentration of ATP. This result, however, could not be repeated by Sharp and Leaf (1966), which may have been due to methodological differences. Again, it is worth stressing that even if no measurable rise in ATP content is found, it will ultimately be necessary to determine whether or not aldosterone causes an increase in the concentration of the free ATP.

However popular the ATP theory may be, it is vulnerable because it claims that a high ATP/ADP ratio near the inner side of the cell membrane is the basis of the rise in Na efflux occurring in a target cell when treated with aldosterone. The possibility that factors other than just a high ATP/ADP ratio are involved in the mechanism of action of the steroid is suggested by an old experiment by Keynes (1960) who found that ATP when injected into a squid axon does not cause stimulation of Na efflux but its suppression. This effect of injected ATP could be prevented by pretreatment of the axon with magnesium, implying that injected ATP, i.e. any additional free ATP, complexes the free Mg^{2+} or facilitates the binding of Mg^{2+} to the inner membranes. In the light of this observation it is thus plausible to argue that if aldosterone does in fact raise the concentration of free ATP, it is then essential that this be accompanied by a rise in the concentration of free Mg^{2+}.

Unlike insulin, but similar to vasopressin, externally applied aldosterone reduces the efflux of radiosodium from a *Maia* muscle fiber (Bittar, 1966b). To respond in this manner, the *Maia* fiber has first to be exposed *in vivo* to a dose of the steroid. Figure 5 shows that injected aldosterone also causes rapid suppression of Na efflux from a *Maia* fiber. This is taken as evidence that the natriferic action of aldosterone is not dependent upon the presence of a cell surface factor (see however Ballard and Tomkins, 1969).

A far more obvious lesson learnt from these experiments with aldosterone is that by merely applying the induction principle it has become possible to elicit a response from crustacean muscle to aldosterone. This appears to be

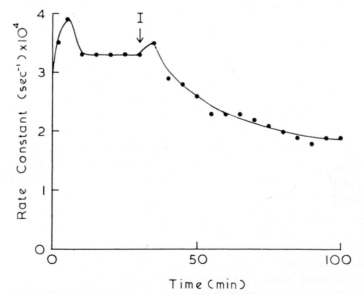

Figure 5. Inhibition of Na efflux from a *Maia* fiber by injected aldosterone (sarcoplasmic concentration about 10^{-5} M). The crab had been injected some 15 hours earlier with a dose of aldosterone (Bittar, 1966b).

true as well of the larvae of *Ambystoma tigrinum* (Alvarado and Kirschner, 1964), so that there are grounds for the belief that this novel experimental approach may be applicable to aquatic animals in general.

VI. IONOPHOROUS ANTIBIOTICS AS CATION CARRIERS

An ionophore, according to Pressman, Harris, Jagger and Johnson (1967) is an agent that can carry ions across membranes in the form of lipid-soluble complexes. Valinomycin, for example, causes K accumulation by mitochondria and hence it is an ionophore. So is nigericin but its effect on cation movements is the opposite of valinomycin. This led Pressman to conclude that there are two classes of ionophores, one behaving like valinomycin, i.e. affecting an inwardly directed K pump of the mitochondrion, and the other like nigericin, affecting proton permeability of the membrane.

As first shown by Harris and Pressman (1967), the action of nigericin is not limited to mitochondria. This monobasic acid has the ability to bring about rapid uptake of protons in exchange for K ions in both canine and human erythrocytes. More important, nigericin has been found by these workers to reduce the Na content of canine red cells and to raise that of human red cells. This is clearly a telling argument against the idea that nigericin acts simply by producing 'holes' in membranes. Instead, observation seems to suggest that nigericin alters the Na content of the cell by stimulating or inhibiting the rate of Na exchange between the free and the sequestered fractions of internal Na. Justification for the view that a portion of the internal Na of red cells is sequestered is provided by the early work of Sheppard (1951) and Solomon (1952). That this may indeed be the mode of action of nigericin is now confirmed by experiments with single toad oocytes, i.e. cells in which an appreciable fraction of the internal Na is known to be secluded by the inner membranes. Nigericin, as indicated by figure 6, exerts a biphasic effect on ^{24}Na efflux. The kinetics indicate that the increase in cell membrane permeability to Na is followed by mobilization of the sequestered fraction, resulting in a partially saturated pump. This action of nigericin has been found by Bittar (1970) to be K-dependent. It thus looks very much as if ionophores such as nigericin, the structure of which is shown in figure 7, and hormones such as insulin, act at the same locus. The question now is, does nigericin liberate mitochondrial Na? There is circumstantial evidence which suggests that it does. For example, mitochondria treated with gramicidin eject Na when nigericin is added to the bathing medium (Graven, Estrada-O. and Lardy, 1966). There is, moreover, autoradiographic evidence from the studies of Dick, Fry and Rogers (1968) that the internal Na of the oocyte is largely extranuclear in location. Nigericin may thus turn out to be a unique tool with which to localize the sequestered fraction of Na in intact cells.

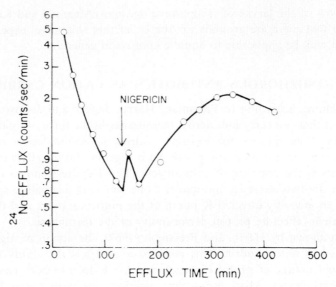

Figure 6. The biphasic stimulating action of nigericin (10 μg/ml) at pH 7 on ²⁴Na efflux from the oocyte of the toad, *Bufo bufo*. Ordinate: rate at which ²⁴Na leaves the oocyte (From Bittar, 1970).

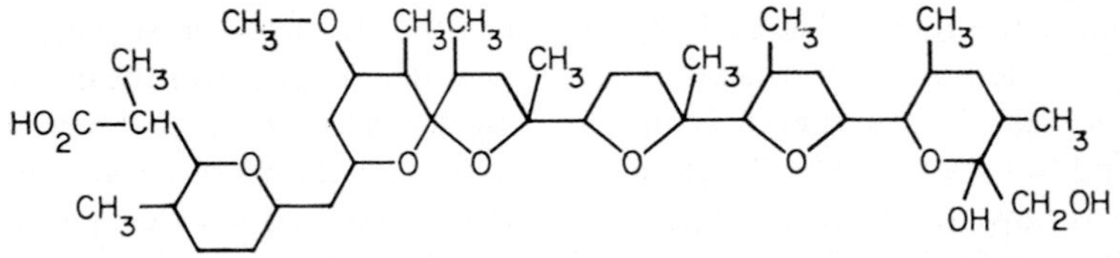

Figure 7. Structure of nigericin (From Steinrauf, Pinkerton and Chamberlin, 1968).

VII. HORMONAL INTERACTION WITH THE GENOME

There is today a considerable literature showing that hormones can stimulate protein synthesis by means of an action on the processes of transcription (e.g. aldosterone) or translation (e.g. insulin). Obviously there are several other ways in which hormones might be able to influence and coordinate protein synthesis., such as

　i) by influencing the transfer of mRNA from the nucleus to the cytoplasm;
　ii) by slowing or enhancing the degradation of mRNA; and
　iii) by changing the ionic environment in the nucleus, the mitochondria, or in the surroundings of the ribosomes. A further possibility is that hormones

may influence the phosphorylation of histones (see, for example, Langan, 1969). So far, none of these concepts have been adequately explored.

A. Aldosterone

The part played by the nucleus in the mechanism of action of aldosterone on Na transport in an asymmetric cell, for example the epithelial cell of toad bladder, is clearly indicated by the observation that the stimulating, natriferic effect of the steroid is blocked by actinomycin. Morphologic evidence showing that tritiated aldosterone is concentrated in the nucleus of this cell has been put forward by Porter, Bogoroch and Edelman (1964), but it is well to recall that this is not proof of its nuclear site of action. Evidence of this sort can be interpreted to mean that the turnover rate of aldosterone in this organelle is very sluggish, and perhaps more sluggish than in mitochondria (Williams and Baba, 1967).

A matter of some interest is that Edelman views the nuclear proteins with which aldosterone interacts as originating from the cytosol. Working with nuclear fractions of rat kidney and other organs, Swanek, Highland and Edelman (1969) have succeeded in isolating what appear to be aldosterone receptors. Supposedly these are the *repressors* which following activation by aldosterone enable the *operator* to initiate transcription of new species of mRNA.

It is important to know whether aldosterone has any influence on the movements of Na across the nuclear envelope. Experiments done with isolated nuclei, for example calf thymus cells or amphibian oocytes may or may not mirror the *in situ* situation. Should this prove to be an obstacle, autoradiography involving the use of ^{22}Na (see for example Appleton, 1966) may provide some indication of the Na content of nuclei before and after the application of aldosterone.

B. Insulin

Sinex, MacMullen and Hastings (1952) were the first to describe the stimulation by insulin of protein synthesis in skeletal muscle. This effect of insulin has been the subject of study by Wool and his colleagues (see Wool and Cavicchi, 1966), who hold the view that it takes place independently of the amino acid uptake caused by the hormone. This is a knotty and controversial question which remains unresolved, particularly after Goldstein and Reddy (1966) had reported that stimulation by insulin of protein synthesis in skeletal muscle is the direct result of the accumulation of amino acids.

From the work of Eboué-Bonis and coworkers (1963) and of Wool (1965) it is known that insulin can stimulate protein synthesis in the absence of RNA synthesis. This had been deduced from the fact that actinomycin fails to abolish the stimulating effect of insulin on the formation of new protein in rat muscle. More recent studies by Martin and Wool (1968) using rat diaphragm, disclose that the insulin-sensitive site in the ribosome is the 60 S subunit. Whether insulin induces *de novo* synthesis of a 'specific protein' as suggested by Wool and whether this protein becomes an integral part of the cell surface is far from clear.

Since the work of Stadie, Haugaard and Vaughan (1953) it has been known that skeletal muscle binds insulin rather avidly. Present evidence coming from experiments with ^{131}I-insulin suggests that insulin interacts with two different receptors located in the surface of the muscle fiber membrane (Garratt, Cameron and Menzinger, 1966). There is also evidence that insulin crosses into the muscle fiber interior and that the hormone is bound by organelles, notably the nuclei and mitochondria (Edelman and Schwartz, 1966).

Now if it is true that insulin reaches the interior of the muscle fiber, then the problem before us is to decide whether stimulation of Na efflux by the hormone is the result of an action on the sarcolemma *per se* (i.e. the surface of the plasma membrane) or on some intracellular process or organelle. An alternative viewpoint would be one which accommodates both possibilities, namely, that unless there exists a 'surface factor', insulin cannot influence Na transport. As already demonstrated by Bittar (1967), injected insulin leads to a transitory rise in Na efflux from the *Maia* fiber. And when insulin is injected into a fiber previously exposed *in vivo* to the hormone, there occurs instead a gradual fall-off in efflux which can be promptly reversed by external application of the hormone. Put in another way, these observations justify the conclusion that insulin stimulates the Na pump only when the pump is depressed and only when the hormone is applied externally. This is in keeping with the idea that the mechanism by which insulin acts involves a surface factor. This factor might well be residues of tryptophan, which as envisaged by Rieser and Maturo (1969) behave at the surface of rat diaphragm as the recognition site of insulin. The possibility that the availability of this surface factor does not depend on RNA synthesis is borne out by the discovery that actinomycin does not prevent insulin from stimulating Na efflux, as illustrated by figure 8. The kinetics of the efflux show quite clearly that external application of insulin stops the rate constants from falling. This result can be accounted for by supposing that insulin freed the fraction of internal Na which injected insulin had secluded. Another interpretation would be that insulin affects the movements of Ca^{2+} across the plasma membrane (see for example Kafka and Pak, 1969), thereby altering the permeability of the membrane to Na.

Other suggestions relating to the primary point of action of insulin have been put forward. For example, Lockwood, Voytovich, Stockdale and

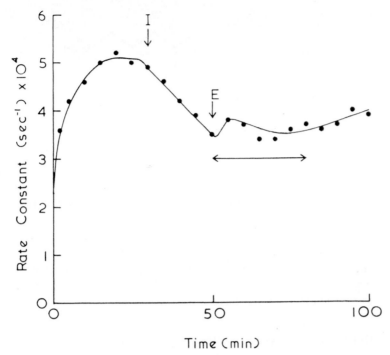

Figure 8. Failure of actinomycin D to completely abolish the stimulating action of externally applied insulin on Na efflux from a *Maia* fiber previously exposed *in vivo* to both insulin and actinomycin over a period of 15 hours.

Topper (1967), working with a mouse mammary gland preparation, found that as little as 5 μg/ml insulin causes a rise in DNA polymerase activity, which becomes maximal two days later. This is all the more interesting because as shown by Goldberg and Atchley (1966) insulin weakens the intrastrand linkages of the DNA double helix and because as shown by Dancheva (1966) insulin causes a rise in the ATP content of nuclei. This line of reasoning is strengthened by the observation that thymus nuclei respire and show a net synthesis of ATP (McEwen, Allfrey and Mirsky, 1963).

C. Ecdysone

Ecdysone the prothorax gland hormone of insects is a C_{27}-7(8)-en-6-oxo steroid (see figure 9), which controls growth and differentiation as the result of its action on the nucleus. Whether or not this effect of ecdysone is related to a primary action on the ionic content of the nucleoplasm or on the process of transcription is still uncertain. Kroeger (1966) produced evidence which

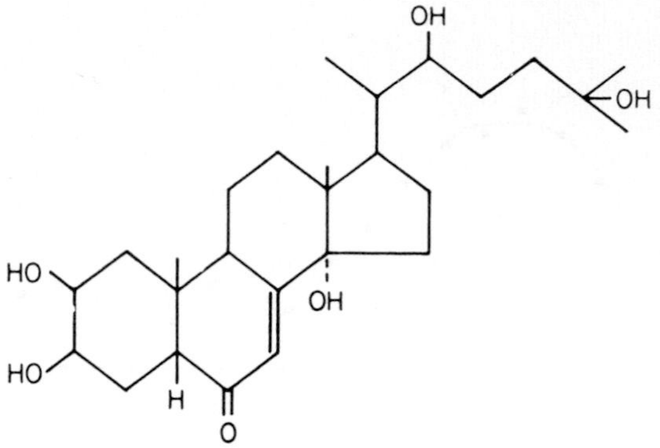

Figure 9. Structure of ecdysone (From Karlson, Hoffmeister, Hummel, Hocks and Spiteller, 1965).

can be interpreted as indicating that ecdysone raises the K concentration of the nucleoplasm. However, attempts by Congote, Sekeris and Karlson (1969) to reproduce this result have so far failed.

VIII. HORMONAL INTERACTION WITH THE MITOCHONDRION

A question of increasing physiological interest is whether the behavior of the Na pump toward insulin (and aldosterone) is directly related to the response of mitochondria to the hormone. Not unnaturally, there has been much speculation about the possibility that the mitochondrion is the primary site of action of insulin. The fact is, however, that unequivocal evidence for the involvement of the mitochondrion is still lacking, presumably because of the failure of experiments with tissue slices or isolated mitochondria to genuinely parallel the *in vivo* situation (see for example Haugaard and Haugaard (1964) for evidence of an *in vivo* effect).

The usual way in which the effect of insulin on mitochondria has been studied is by measurement of the P:O ratio. As early as 1938 Krebs and Eggleston reported that respiration in pigeon breast muscle 'mince' is stimulated by insulin. The view that oxidative phosphorylation is increased by insulin also emerged from the work of Polis, Polis, Kerrigan and Jedeikin (1949), Goranson and Erulkar (1949), and Vester and Stadie (1957). (The old literature dealing with this subject had been fully reviewed by Stadie, 1944, 1954). Vester (1957) tried to explain this by supposing that the hormone facilitates the binding of Mg^{2+} to membranes. It is, however, from the studies of Hall, Sordahl and Stefko (1960) that it became clear that oxidative phosphorylation is stimulated by insulin only when mitochondria exhibit poor respiratory control. Restoration by insulin of the P:O ratio may be regarded on theoretical grounds as a recovery mechanism which is responsible for insulin's action on a depressed pump. If this is the correct view, then the conclusion arrived at by Sacks (1952) that the increase in phosphate turnover produced by insulin is due to the stimulation of oxidative phosphorylation by the hormone does not require any modification.

Bessman (1966) suggested that insulin stimulates ATP formation by attaching hexokinase to mitochondria. The evidence cited by Bessman is based on the thesis work of Bachur showing that high concentrations of hexokinase lead to increased ATP synthesis. Equally interesting is the theory advanced by Conn (1966) that insulin stimulates the activity of creatine kinase, thus

$$\text{CrP} + \text{ADP} \xrightarrow[\text{Mg}^{2+}]{\text{insulin}} \text{Creatine} + \text{ATP}.$$

This seems not unreasonable in view of the finding by Jacobs, Heldt and Klingenberg (1964) that muscle mitochondria contain this enzyme.

One really need not consider every point about how insulin might affect mitochondrial metabolism but three possibilities ought to be examined. First, insulin might exert an effect on mitochondrial DNA (and protein synthesis). Second, insulin might stimulate the *adenylate translocase*, resulting in increased release of ATP. And third, insulin might be involved in directly regulating the Na exchange rate across the mitochondrial inner membrane. It is not enough to show that insulin causes mitochondria to swell (Lehninger and Neubert, 1961; Campbell and Mertz, 1963) and that this swelling is partly reversed by adding ATP. Nor does it seem fruitful to study the movements of Na in isolated mitochondria since they seem to have very little sodium, a feature which suggests that membrane permeability changes must have occurred during the procedure of isolation. In a word, the Na exchange mechanism involving mitochondria, the existence of which has come to light as the result of experiments with nigericin, can only be properly examined by employing single, intact cells. Recognition of this fact is in the author's view the beginning of future progress.

IX. CONCLUDING REMARKS

Evidence has been brought forward to show that vasopressin, aldosterone and insulin can modify Na transport in both symmetric and asymmetric cells. Aldosterone and insulin but not vasopressin act, it would seem, at a point away from the Na pump. In the case of insulin this could well be the mitochondrion at a time when the P:O ratio is low, as well as the internal

Na exchange mechanism when the pump is slowing down. The author tends to the view that experiments with single giant cells are bound to provide some clues as to whether insulin and aldosterone regulate the free ATP fraction by influencing the adenine nucleotide *translocase*. Work with such simpler systems may also settle the question whether regulation of the free Na fraction in the cell by insulin and aldosterone affects not only the degree of activity of the Na pump but also that of the mitochondrion.

REFERENCES

Allfrey, V. G., R. Meudt, J. W. Hopkins and A. E. Mirsky (1961) *Proc. Natl. Acad. Sci. U.S.*, **47**, 907
Alvarado, R. H. and L. B. Kirschner (1964) *Nature*, **202**, 922
Appleton, T. C. (1966) *J. Histochem. Cytochem.*, **14**, 414
Baker, P. F. (1963) *Biochim. Biophys. Acta*, **75**, 287
Ballard, P. L. and G. M. Tomkins (1969) *Nature*, **224**, 344
Bangham, A. D., M. M. Standish and G. Weissman (1965) *J. Mol. Biol.*, **13**, 253
Bentley, P. J. and H. Heller (1964) *J. Physiol.*, **171**, 434
Bessman, S. P. (1966) *Am. J. Med.*, **40**, 740
Bittar, E. E. (1966a) *Biochem. Biophys. Res. Commun.*, **23**, 96
Bittar, E. E. (1966b) *Biochem. Biophys. Res. Commun.*, **23**, 868
Bittar, E. E. (1967) *Nature*, **214**, 726
Bittar, E. E. (1970) MS. in prep.
Bittar, E. E., D. A. T. Dick and D. J. Fry (1968) *Nature*, **217**, 1280
Blecher, M. (1966) *Biochem. Biophys. Res. Commun.*, **23**, 68
Campbell, W. J. and W. Mertz (1963) *Am. J. Physiol.*, **204**, 1028
Clauser, A., P. Volfin and D. Eboué-Bonis (1962) *Gen. Comp. Endocrinol*, **2**, 369
Congote, L. F., C. E. Sekeris and P. Karlson (1969) *Exptl. Cell Res.*, **56**, 338
Conn, R. B. (1966) *Nature*, **211**, 195
Creese, R. (1964) *Nature*, **201**, 505
Creese, R. (1968) *J. Physiol.*, **155**, 343
Creese, R., J. L. D'Silva and J. Northover (1958) *Nature*, **181**, 1278
Creese, R. and J. Northover (1961) *J. Physiol.*, **155**, 343
Cuthbert, A. W. and Painter, E. J. (1968) *J. Physiol.*, **199**, 593
Dancheva, K. I. (1966) *Nature*, **212**, 1361
Dick, D. A. T., D. J. Fry and A. W. Rogers (1968) *J. Physiol.*, **197**, 2P
Dormandy, T. L. (1965) *J. Physiol.*, **180**, 708
Dormandy, T. L. (1966) *J. Physiol.*, **183**, 378
Eboué-Bonis, D., A. M. Chambaut, P. Volfin and H. Clauser (1963) *Nature*, **199**, 1183
Edelman, I. S., R. Bogoroch and G. A. Porter (1963) *Proc. Natl. Acad. Sci. U.S.*, **50**, 1169
Edelman, I. S., M. J. Petersen and P. F. Gulyassy (1964) *J. Clin. Invest.*, **43**, 2185
Edelman, P. M. and I. L. Schwartz (1966) *Am. J. Med.*, **40**, 695
Fimognari, G. M., D. D. Fanestil and I. S. Edelman (1967) *Am. J. Physiol.*, **213**, 954
Fimognari, G., D. K. Kasvekar and I. S. Edelman (1965) *Fed. Proc.*, **24**, 344
Flückiger, von E. and F. Verzár (1954) *Helv. Physiol Pharmac. Acta*, **12**, 50
Garratt, C. J., J. S. Cameron and G. Menzinger (1966) *Biochem. Biophys. Acta*, **115**, 179
Goldberg, M. L. and W. A. Atchley (1966) *Proc. Natl. Acad. Sci. U.S.*, **55**, 989
Goldstein, S. and W. J. Reddy (1966) *Fed. Proc.*, **25** (2), 441
Goranson, E. S. and S. D. Erulkar (1949) *Arch. Biochem.*, **24**, 406
Graven, S. N., S. Estrada-O. and H. A. Lardy (1966) *Proc. Natl. Acad. Sci. U.S.*, **56**, 654
Gulyassy, P. F. and I. S. Edelman (1965) *Proc. Second Int. Cong. Nephrology*, Washington, p. 605

Hall, J. C., L. A. Sordahl and P. L. Stefko (1960) *J. Biol. Chem.*, **235**, 1536
Harris, E. J. and B. C. Pressman (1967) *Nature*, **216**, 918
Haugaard, E. S. and N. Haugaard (1964) *J. Biol. Chem.*, **239**, 705
Hechter, O. (1965) In *Mechanisms of Hormone Action* (Ed. P. Karlson) Academic Press, New York, p. 61
Hill, T. L. (1969) *Proc. Natl. Acad. Sci. U.S.*, **64**, 267
Jacobs, H., H. W. Heldt and M. Klingenberg (1964) *Biochem. Biophys. Res. Commun.*, **16**, 516
Jardetzky, O. (1966) *Nature*, **211**, 969
Kafka, M. S. and C. Y. Pak (1969) *J. Gen. Physiol.*, **54**, 134
Karlson, P., H. Hoffmeister, H. Hummel, P. Hocks and G. Spiteller (1965) *Chem. Ber.*, **98**, 2394
Kernan, R. P. (1962) *J. Physiol.*, **162**, 129
Keynes, R. D. (1960) In *Regulation of the Inorganic Ion Content of Cells*, Ciba Foundation Study Group No 5, Little, Brown & Company, Boston, p. 91
Keynes, R. D. (1965) *J. Physiol.*, **178**, 305
Krebs, H. A. and P. Eggleston (1938) *Biochem. J.*, **32**, 913
Kroeger, H. (1966) *Exptl. Cell Res.*, **41**, 64
Langan, T. A. (1969) *Proc. Natl. Acad. Sci. U.S.*, **64**, 1276
Lehninger, A. and D. Neubert (1961) *Proc. Natl. Acad. Sci. U.S.*, **47**, 1929
Lockwood, D. H., A. E. Voytovich, F. E. Stockdale and Y. J. Topper (1967) *Proc. Natl. Acad. Sci. U.S.*, **58**, 658
Maetz, J. (1963) In *Comparative Aspects of Neurohypophyseal Morphology and Function. Symp. Zool. Soc. (London)* **9**, 107
Martin, T. E. and I. G. Wool, (1968) *Proc. Natl. Acad. Sci. U.S.*, **60**, 569.
McEwen, B. S., V. G. Allfrey and A. E. Mirsky (1963) *J. Biol. Chem.*, **238**, 758
Monod, J., J.-P. Changeux and F. Jacob (1961) *Cold Spring Harbor Symp. Quant. Biol.*, **23**, 389
Moore, J. H. and R. S. Schechter (1969) *Nature*, **222**, 476
Morel, F., S. Jard, J. Bourguet and F. Bastide (1969) In *Protein and Polypeptide Hormones*. Excerpta Medica Foundation, Amsterdam, p. 219
Orloff, J. and J. S. Handler (1962) *J. Clin. Invest.*, **41**, 702
Orloff, J. and J. S. Handler (1964) *Am. J. Med.*, **36**, 686
Peters, R. (1956) *Nature*, **177**, 426
Pressman, B. C., E. J. Harris, W. S. Jagger and J. H. Johnson (1967) *Proc. Natl. Acad. Sci. U.S.*, **58**, 1949
Polis, D. B., E. Polis, M. Kerrigan and A. Jedeikin (1949) *Arch. Biochem.*, **23**, 505
Porter, G. A., R. Bogoroch and I. S. Edelman (1964) *Proc. Natl. Acad. Sci. U.S.*, **52**, 1326
Rasmussen, H. (1969) In *Protein and Polypeptide Hormones*. Excerpta Medical Foundation, Amsterdam, p. 247
Rasmussen, H., I. L. Schwartz, R. Young and J. Marc-Aurele (1963) *J. Gen. Physiol.*, **46**, 1171
Rieser, P. and J. M. Maturo (1969) *Arch. Int. Med.*, **123**, 267
Rodbell, M. (1966) *J. Biol. Chem.*, **241**, 130 and 241
Rogus, E., T. Price and K. L. Zierler (1969) *J. Gen. Physiol.*, **54**, 188
Rosa, U., C. A. Rossi and L. Donato (1968) In *Pharmacology of Hormonal Polypeptides and Proteins* (Eds. N. Back and L. Martini), Plenum Press, New York, p. 336
Sacks, J. (1952) In *Phosphorus Metabolism*, Vol. 2, Johns Hopkins University, p. 653
Sacks, J. and F. M. Sinex (1953) *Am. J. Physiol.*, **175**, 353
Schwartz, I. L., H. Rasmussen, M. A. Schoessler, L. Silver and C. T. D. Fong (1960) *Proc. Natl. Acad. Sci. U.S.*, **46**, 1288
Sharp, G. W. G. and A. Leaf (1966) *Physiol. Rev.*, **46**, 593
Sheppard, C. W. (1951) *Science*, **114**, 85
Sinex, F. M., J. MacMullen and A. B. Hastings (1952) *J. Biol. Chem.*, **198**, 615
Smillie, L. B. and J. F. Manery (1960) *Am. J. Physiol.*, **198**, 67
Solomon, A. K. (1952) *J. Gen. Physiol.*, **36**, 57

Stadie, W. C. (1944) *Yale J. Biol. Med.*, **16**, 539

Stadie, W. C. (1954) *Physiol. Rev.*, **34**, 52

Stadie, W. E., N. Haugaard and M. J. Vaughan (1953) *J. Biol. Chem.*, **200**, 745

Steinrauf, L. K., M. Pinkerton and J. W. Chamberlin (1968) *Biochem. Biophys. Res. Commun.*, **33**, 29

Sutherland, E. W. and T. W. Rall (1957) *J. Biol. Chem.*, **232**, 1077

Swanek, G. E., E. Highland and I. S. Edelman (1969) *Nephron*, **6**, 297

Vester, J. W. (1957) *J. Biol. Chem.*, **227**, 669

Vester, J. W. and W. C. Stadie (1957) *J. Biol. Chem.*, **227**, 669

Walaas, E., O. Walaas and A. Wick (1969) In *Protein and Polypeptide Hormones*. Excerpta Medica Foundation, Amsterdam, p. 164

Warner, D. T. (1965) In *Mechanisms of Hormone Action* (Ed. P. Karlson), Academic Press, New York, p. 83

Wheeler, K. P. and R. Whittam (1970) *J. Physiol.*, **207**, 303

Williams, M. A. and W. I. Baba, (1967) *J. Endocr.* **39**, 543

Williamson, H. E. (1963) *Biochem. Pharmacol.*, **12**, 1449

Wool, I. G. (1965) In *Mechanisms of Hormone Action* (Ed. P. Karlson), Academic Press, New York, p. 98

Wool, I. G. and P. Cavicchi (1966) *Proc. Natl. Acad. Sci. U.S.*, **56**, 991

Zierler, K. L. (1957) *Science*, **126**, 1067

Zierler, K. L. (1959) *Am. J. Physiol.*, **197**, 515

Zierler, K. L., E. Rogus and C. F. Hazlewood (1966) *J. Gen. Physiol.*, **49**, 433

IV
The Cell Surface

CHAPTER 11

Cell membrane surface potential as a transducer

D. Gingell

Department of Biology as Applied to Medicine
Middlesex Hospital Medical School,
London, W.1., England

I. INTRODUCTION

The concept of biological membranes functioning as transducers is not a
novelty. Transduction implies a qualitative change in the nature of the signal

carrying information across the membrane. Certain well documented cellular activities are encompassed by this definition: for example, the action of acetylcholine in bringing about a non-specific ionic permeability increase at the neuromuscular junction. The input in this case is ACh and the output is a small ion flux. Another example concerns the common mechanism of action of a wide variety of hormones which are capable of activating the enzyme adenyl cyclase in certain membranes (Sutherland and coworkers, 1968). This converts ATP to cyclic AMP in the membrane after which cyclic AMP is released into the cell where it initiates the characteristic response of the activating hormone. Here the input is hormonal, transduction centres on adenyl cyclase and the out is probably cyclic AMP. Whereas cyclic AMP appears to cause permeability changes associated with the action of vaso-pression, the majority of adenyl cyclase mediated transductions do not seem to result in permeability changes. A rather different system has been studied by Dean and Matthews (1968) who found that substances which stimulate insulin secretion evoke action potentials in islet cells. They suggest that Na^+ or Ca^{2+} entry during depolarization may accelerate the mechanism respon-sible for insulin release.

I would like to suggest that there is apparently a further class of trans-ductive process which operate in some cells, where the surface membrane electrostatic potential plays an important role, mediating between environ-mental action and cellular response.

In the following section the nature of membrane electrostatic potential is briefly explained. Next, environmental factors which can modify the surface potential are considered in the role of possible input signals. Events within the membrane resulting from changed surface potential, which may generate an output signal to the cytoplasm will then be discussed. The theoretical section ends with interactions between output signal and the internal environ-ment. In the second part some of the concepts developed will be utilised in an attempt to throw light on a variety of cellular responses.

II. THEORETICAL CONSIDERATIONS

A. Genesis of Membrane Surface Potential

The electrostatic surface potential of all cells so far studied is a localized negative magnitude which exists because of the negative ionic charge at the surface. There is good evidence for the view that in most cases this charge is due to the dissociation of carboxyl groups carried by N-acetyl neuraminic acid (see review by Cook, 1968). In contrast to the bioelectric potential, which is a bulk phase potential due to charge separation across the mem-brane resulting from unequal distributions of diffusible ions on either side of the membrane, the surface potential falls off rapidly to zero in a direction normal to the surface. This is because oppositely charged counter-ions from the bulk solution are electrostatically attracted to the negative surface, where they tend to neutralise its charge. The counter-ion 'atmosphere' is in a state of dynamic equilibrium, being subjected to electrostatic forces tending to accumulate ions at the interface and thermal motion tending to randomize their distribution. The system is referred to as an ionic double layer, in which the fixed surface charge density is equal in magnitude but opposite in sign to the mobile counter-ion charge, the whole being electrically neutral. Neutralization is achieved over a distance normal to the surface equal to the thickness of the counter-ion space charge. In physiological saline the potential falls to $1/e$ (e is the base of natural logarithms) of its value at the surface in a distance of around 8 Å and this is often referred to as the 'double layer thickness' or Debye–Hückel length. A simple graphic representation of the relationship between ionic charge and electrostatic potential in the double layer is given in figure 1.

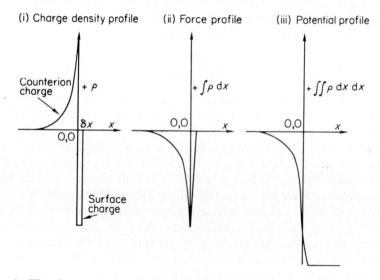

Figure 1. The charge density distribution in a double layer is shown schematically in (i). Overall neutrality requires that $\int_{-\infty}^{0} \rho dx + \int_{0}^{\delta x} \rho dx = 0$. Poission's equation (Appendix, equation 2) states that the second integral of charge with respect to distance is proportional to potential. This is shown in (iii) where integration is performed by summation of the area beneath the curves in (i) and (ii). The first integration, (ii) gives the electrical force exerted on unit charge.

This shows graphically that there is a potential step across one double layer and that potentials throughout the double layer have the same sign as that of the fixed surface charge.

B. Factors Affecting Surface Potential: Input Signals

One of the most intriguing features of the surface potential, viewed as a possible transducer, is the wide variety of environmental changes by which its magnitude is altered. Changes in ionic strength, dielectric constant of the medium, surface charge density, electric fields, relative motion in magnetic fields and close approach to other surfaces would tend to modify the cell surface potential.

1. *Ionic Strength, Dielectric Constant and Surface Charge Density*

The effect of these three parameters on the surface potential is evident from the well known linear approximation equation which holds for potentials $\psi_0 \leqslant 25$ mV.

$$\psi_0 = \frac{4\pi\sigma}{D\kappa} \tag{1}$$

where ψ_0 is the surface potential, σ is the surface charge density, D is the dielectric constant of the medium and $1/\kappa$ is the Debye–Hückel length, which is related to the ionic strength by the expression

$$\kappa^2 = \frac{8\pi e^2 Ac}{10^3 DkT} \tag{2}$$

where A is Avogadro's constant, c is the concentration of symmetrical monovalent electrolyte, k is Boltzmann's constant, T is the absolute temperature and e is the charge on the electron.

It can be seen that surface potential is a monotonic function of surface charge, dielectric constant and ionic concentration (strictly ionic strength for complex electrolytes). Surface potential is directly proportional to surface charge density, inversely proportional to the dielectric constant of the medium outside the cell and inversely proportional to the root of the ionic strength of the medium. Therefore, alteration of surface charge either spontaneously or by adsorption of polyelectrolytes, by changes in ionic strength or dielectric constant of the medium, should affect the value of the cell surface potential (Gingell 1967a,b, 1968).

Curves of ψ_0 against decreasing molarities of monovalent salt in the medium are shown in figure 2 for the case where surface potential is 25 mV in physiological saline. The linear equation (1) and the non-linear equation

$$\sinh \frac{e\psi_0}{2\kappa T} = \sigma \left(\frac{2\kappa TDAc}{10^3 \pi} \right)^{-\frac{1}{2}} \tag{3}$$

which holds good for potentials in excess of 25 mV and molarities in excess of 0·1 molar are given.

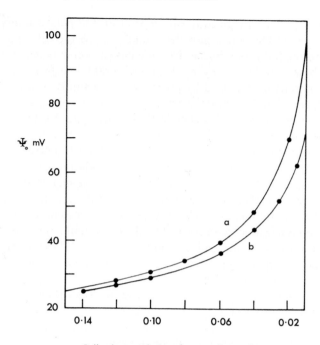

Figure 2. Change in surface potential, ψ_0 at constant surface charge density, with molar concentration of symmetrical monovalent electrolyte in the bulk phase. Curve a is a linear approximation (equation 1) curve b is the exact equation (3).

Equation (3) reduces to (1) for $\psi_0 < 25$ mV. It is seen that an increase in concentration from $0 \cdot 03$M KCl to $0 \cdot 15$M KCl reduces the surface potential from 50 mV to 25 mV. A reduction would also follow the adsorption of net positively charged polyions, provided the centre of gravity of the charge approached sufficiently close to the surface.

There appears to be little information available on the results of changing the dielectric constant, but it is interesting and perhaps significant that dimethylsulphoxide increases membrane permeability and is unusual in that it lowers the dielectric constant of water when present in small amounts, the pure substance having a dielectric constant of $45 \cdot 5$ at $40 \degree$C. It has been suggested (Rammler and Zaffaroni, 1967) that membrane proteins may undergo conformational rearrangement in the presence of dimethylsulphoxide. This may be partly due to lowered dielectric constant.

2. *Electric and Magnetic Fields*

The effects of exposing a cell to an external homogeneous electric field can be considered to occur in two stages. First, an instantaneous capacitive current will flow, whose magnitude will depend on the polarizability of the membrane material. This will occur inside and outside the cell, but current will not flow through the membrane. Its ionic basis is ion movement involved in setting up new counter-ion layers at the membrane interfaces, due to the change in membrane field caused by polarization. Secondly, since biological membranes have high but finite electrical resistances passage of current through the cell membrane will occur in the steady state. Furthermore, the relatively high concentration of ions in the mobile double layer, especially in more dilute solution, will afford a low resistance shunt to current, resulting in surface conduction. Although the steady state condition may be complex, there will be changes in the potential gradient within the membrane, maximal at the ends of the cell in the field direction and zero at the equator. The surface potential of the anodal region of the membrane will become more positive, the cathodal more negative, with respect to the cell interior (nearly isopotential due to the very small intracellular current flowing). Thus the membrane electrostatic fields will change in opposite senses at opposite ends of the cell in the direction of the external field.

Although biological effects of a magnetic field on cells will not be discussed (see Barnothy, 1964; Kholodov, 1966 for references), it is perhaps of interest to consider some possible effects of static and oscillating magnetic fields on cell membranes. Ionic movements will be induced by relative motion of the cell through a magnetic field. (It is assumed that the field will not be disturbed by paramagnetic or diamagnetic atoms in the cell.) At equilibrium there will be a potential difference (p.d.) across the poles of the cell in the direction of the field almost identical to that across the cytoplasm in the field direction. The electrical p.d. is given to a first approximation by

$$E = \frac{d\psi}{dt}$$

E is the electrostatic potential difference across the cell poles and $d\psi/dt$ is the rate of cutting lines of magnetic force by the cell.

This p.d. will exist at equilibrium because there is no applied opposing electric field but merely an opposing force due to the interaction of the magnetic lines of force with moving charges, which is equal and opposite to the force exerted on the charges by the electrostatic field which is created by their displacement. This argument is also applicable to the movement of the cell since induced potential depends on relative velocity of cell and field.

Assuming a homogeneous field of 10^5 oersteds and a relative motion[*] perpendicular to the direction of the field of 100 cm/sec the induced potential is approximately 10^{-7} V for a spherical cell of 5μ radius. This small potential would probably be reduced further by approach of ions of opposite sign to

*Due to movement of field or object.

the local surface potential. However, even a small change in potential may be important in some situations, for example it is known that specially receptive cells of electric fish can detect a potential gradient of $0.03\mu V/cm$.

In an oscillating field the induced EMF will tend to drive mobile counter-ions around the cell, while imposing an equal force of opposite direction on fixed charges, thus tending to rotate the cell. As a consequence of this induced motion, the counter-ions will experience a force driving them centrifugally or centripetally, according to the phase of the field. Counter-ions on each side of a localized region of the membrane will tend to move in the same direction so that when the outer surface potential of a local region rises, that of its inner interface will fall, and vice versa, in an oscillatory manner. It follows that there will be a continuous change in the electro-static field in the membrane in phase (up to a limiting frequency depending on the time constant of the membrane) with the external field intensity. The magnitude of the effect which might be expected has not yet been calculated.

3. *Membrane Apposition: A Model*

It has been shown (Gingell 1967a,b) that close apposition of charged surface membranes in physiological saline would increase the surface potential. The system was treated as a plane-parallel double layer problem where surfaces bearing fixed charges move together in $0.145M$ NaCl. The treatment differs from that of Verwey and Overbeek (1948) who considered the interaction of colloidal particles, where the particle surface charge is reversibly absorbed. In their case apposition of the particles is accompanied by a desorption of surface charge and maintenance of constant surface potential. However, a more appropriate biological analogue would involve interaction between fixed charges largely due to dissociated carboxyl groups of sialic acid at the cell surface. When such surfaces are brought close together the charge density remains constant and in consequence the surface potential must increase. In physiological saline little increase occurs at separations greater than 30 Å but steep increases occur at smaller distances. The actual increment depends on the geometry of the model and interaction potentials are predictably reduced if there is a layer of proteinaceous material outside the lipid of the membrane. In this model (figure 3) equal fixed charges are considered to reside in planes I and II and are treated as if their charge were uniformly distributed. Between the charged planes and the lipid there is a layer which represents a surface coat, whose penetrability to counter-ions in solution is less than unity. The thickness of this layer is t and its degree of impenetrability (i.e. the excluded counter ion space) is α. Potentials at the charged planes are designated ψ_0, at the lipid surface ψ_t and at the potential midway between the surfaces ψ_d. The charged planes are separated by a distance $2d$. Utilizing the equations for ψ_0 and ψ_t derived, (see

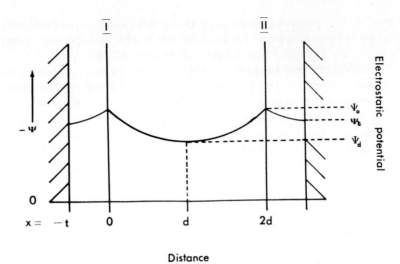

Figure 3. Schematic representation of the electrical potential profile between identical interacting plane parallel double layers under conditions of constant surface charge density. Fixed charges reside in planes I and II, and are separated from a surface, impenetrable to solute ions, by a partially penetrable layer of thickness t. The distance between the charged planes is $2d$.

Appendix 1) values of these functions for variable values of α, t and $2d$ have been computed (Gingell, 1968). The method of computation involves keeping ψ_0 at infinite separation constant, thereby approximating to an electrophoretic zeta potential of 15, 20 or 25 mV, and then calculating the required fixed surface charge density (σ) necessary to generate the chosen potential given pairs of values of α and t. Having thus obtained a value for σ, the potentials ψ_0 and ψ_t and their increments $\delta\psi_0$ and $\delta\psi_t$ are calculated for

Table 1. Highly permeable surface layer, $\alpha = 0.4$

$2d$ (Å)	$\delta\psi_0$	$\delta\psi_t$	$\delta\psi_0$	$\delta\psi_t$	$\delta\psi_0$	$\delta\psi_t$
4	22·90	15·20	19·11	5·40	18·60	2·03
8	12·91	8·57	11·16	3·15	10·91	1·19
12	7·51	4·99	6·62	1·87	6·50	0·71
16	4·45	2·95	3·96	1·12	3·89	0·43
20	2·66	1·76	2·38	0·67	2·34	0·26
24	1·60	1·06	1·44	0·41	1·42	0·16
28	0·96	0·64	0·87	0·25	0·86	0·09
60	0·02	0·01	0·02	0·00	0·02	0·00
	$d'=10$ Å		$d'=20$ Å		$d'=30$ Å	
	$\sigma=0.1047 \times 10^5$		$\sigma=0.1156 \times 10^5$		$\sigma=0.1174 \times 10^5$	
	$\psi_{t(60)}=16.61$		$\psi_{t(60)}=7.07$		$\psi_{t(60)}=2.73$	

Moderately permeable surface layer, $\alpha=0\cdot6$

2d (Å)	$\delta\psi_0$	$\delta\psi_t$	$\delta\psi_0$	$\delta\psi_t$	$\delta\psi_0$	$\delta\psi_t$
4	28·54	21·47	22·85	9·02	21·79	4·03
8	15·30	11·51	12·89	5·09	12·41	2·30
12	8·67	6·53	7·50	2·96	7·26	1·34
16	5·06	3·81	4·44	1·75	4·31	0·80
20	3·00	2·26	2·64	1·05	2·58	0·48
24	1·79	1·35	1·60	0·63	1·55	0·29
28	1·08	0·81	0·96	0·38	0·94	0·17
60	0·02	0·02	0·02	0·01	0·02	0·00
	$d'=10$ Å		$d'=20$ Å		$d'=30$ Å	
	$\sigma=0\cdot9394\times10^4$		$\sigma=0\cdot1048\times10^5$		$\sigma=0\cdot1075\times10^5$	
	$\psi_{t(60)}=18\cdot82$		$\psi_{t(60)}=9\cdot88$		$\psi_{t(60)}=4\cdot63$	

Slightly permeable surface layer, $\alpha=0\cdot8$

2d (Å)	$\delta\psi_0$	$\delta\psi_t$	$\delta\psi_0$	$\delta\psi_t$	$\delta\psi_0$	$\delta\psi_t$
4	40·01	34·48	31·61	18·41	28·52	10·30
8	19·52	16·82	16·34	9·65	15·30	5·53
12	10·58	9·12	9·16	5·41	8·67	3·13
16	6·03	5·20	5·31	3·14	5·06	1·83
20	3·53	3·04	3·14	1·85	3·00	1·08
24	2·10	1·81	1·87	1·11	1·79	0·65
28	1·26	1·08	1·13	0·67	1·08	0·39
60	0·02	0·02	0·02	0·01	0·02	0·01
	$d'=10$ Å		$d'=20$ Å		$d'=30$ Å	
	$\sigma=0\cdot8136\times10^4$		$\sigma=0\cdot9024\times10^4$		$\sigma=0\cdot9397\times10^4$	
	$\psi_{t(60)}=21\cdot56$		$\psi_{t(60)}=14\cdot78$		$\psi_{t(60)}=14\cdot78$	

Computed membrane surface potential changes due to the interaction of ionic double layers associated with approaching cell membranes. Rises in electrostatic potential (mV) at the charged planes, $\delta\psi_0$, and at the inner boundary of the penetrable layer (if present), $\delta\psi_t$ are shown for small separations, 2d, between the charged planes in physiological saline. In all cases, the surface potential in the charged plane at infinite separation is 25 mV. Values of the thickness of the partly penetrable layer, t', its degree of impermeability, α, and corresponding values of fixed surface charge density, σ (e.s.u./cm²), are given. $\psi_{t(60)}$ refers to the potential at the inner side of the permeable surface layer when the charged planes of the membranes are separated by 60 Å=2d, a value which differs little from the potential at infinite separation. A value of $\psi_{t(60)}$ is included for each value of t', being the potential from which the interaction increments $\delta\psi_t$ rise. (From Gingell, 1968).

various distances of separation of the surfaces (figure 4 and table 1). This process is repeated for every pair of values α, t giving a new value for σ each time. Higher surface charge densities are needed to maintain a given potential as α decreases or t increases.

Though lack of experimental evidence makes it difficult to evaluate the quantitative change in surface potential in terms of cellular response

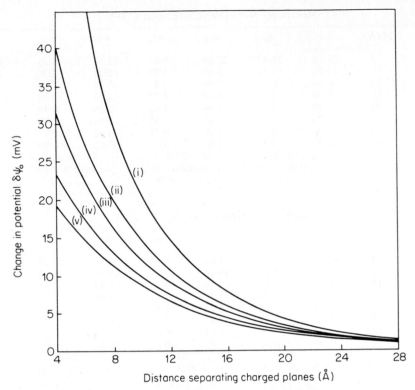

Figure 4. This figure shows computed rises in potential, $\delta \Psi_o$, in the plane of fixed charges borne at the surfaces of approaching membranes in physiological saline. In the first (i) fixed charges reside at an interface which is impenetrable to counterions, while in subsequent cases fixed charges are located at the outer boundary of a partly penetrable layer (e.g. protein) at the surface of the membrane. The thickness of this layer is t and its degree of impenetrability is α. The values of these parameters are as follows: (ii) $\alpha = 0.8$, $t = 10\text{Å}$; (iii) $\alpha = 0.8$, $t = 20\text{Å}$; (iv) $\alpha = 0.4$, $t = 10\text{Å}$; (v) $\alpha = 0.4$, $t = 20\text{Å}$. Values of surface charge density are calculated in each case such that the surface potential (Ψ_o) at infinite separation of the membranes is—25mV. (From Gingell, 1968)

mechanisms it may be that in some instances a change in surface potential of only a few millivolts could lead to a major transformation of membrane properties. One has only to look at excitable membranes for a paradigm, where a potential change of only 4 mV can cause an exponential rise in sodium conductance (Hodgkin and Huxley, 1952).

It is of interest that at moderate degrees of impenetrability of the surface layer, potential changes of a few millivolts occur at the charged plane and at the impenetrable plane even when the thickness of the surface layer is

20–30 Å. Appreciable electrostatic interaction is therefore predicted where the unit membranes are separated by relatively large distances. For example, if each membrane is characterised by $t=30$ Å, $\alpha=0.8$ and $2d=20$ Å, the impenetrable lipid layers would be separated by 80 Å, in which case the corresponding change in ψ_0 is 3 mV. If it is considered that the penetrable material overlies the 70–80 Å unit membrane and corresponds to the electron transparent lanthanum staining layer demonstrated at the surfaces of various cells (Rambourg and Leblond, 1967; Lesseps, 1967), the opportunity for intercellular electrostatic interaction might even occur where the electron microscopic separation of unit membranes considerably exceeds 20 Å.

4. *Interaction in Dilute Media*

The interaction function has been tabulated for the case $\alpha=1.0$ in dilute media (figure 5). The quantitative picture is clearly very different. In 10^{-3}M monovalent symmetrical electrolyte the separation of charged surfaces by $2d=400$ Å will increase ψ_0 by a factor of two, while at 50 Å a sixteen-fold increase is indicated on the simple theory. However, such elevated potentials would be severely truncated by surface pH rise. The local rise in negative

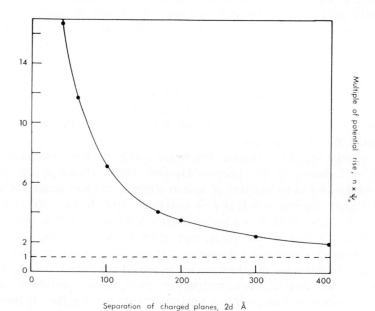

Figure 5. Change in surface potential, under conditions of constant surface charge density, resulting from the approach of identical plane-parallel double layers with impenetrable surfaces, in 0·001 M monovalent salt at 20°C. The curve is uncorrected for changes in surface pH.

potential tends to increase the local H^+ concentration, according to Boltz-mann's distribution, and consequently reduces the degree of dissociation of the ionogenic carboxyl groups responsible for the potential. It can be calcu-lated that a change in ψ_0 from -25 to -100 mV would reduce the local pH from $7\cdot0$ to $5\cdot28$, producing negligible change in the dissociation of groups whose $pK_a \leqslant 4\cdot5$. Rises above -200 mV, however, produce significant association of such groups and reciprocally prevents the attainment of higher potentials by reducing the effective surface charge. Interactions in physiological saline would hardly be affected by pH effects because the calculated interaction potentials are much lower.

5. *Limitations of the Model*

It should be emphasized that the foregoing analysis is only a rough guide to the probable events accompanying the approach of cell surfaces. More-over, even if electrostatic changes of the order predicted do occur, there is no direct evidence that they are instrumental in transmitting the environ-mental event to the cytoplasm.

Further knowledge of the outermost layers of the cell, as well as a more realistic mathematical approach involving discreteness of charge and greater understanding of the factors involved in interaction, may show that the model is quantitatively inaccurate. For example, chemical heterogeneity of the surface and its molecular topography would both be expected to affect surface potential distribution. A more realistic view may well involve a matrix of charges extending through the surface layer in which case mathe-matical difficulties can arise (Wall and Berkovitz, 1957; MacGillivray, 1969). Parsegian (1969) has solved the non-linear Poisson–Boltzmann equation for such interaction.

The structure of the plasma membrane which has been assumed for this analysis is based on the Danielli–Davson model. However, our detailed knowledge of plasma membrane ultrastructure is still very uncertain and the well known arguments will not be reiterated here (Korn, 1966; Chapman and coworkers, 1968). It has rather naively been assumed that the potential in only two regions is important; that at the outermost surface and that at the postulated lipid surface. However, the details of structure are not really a major difficulty for the theory since potential rises must almost certainly occur if fixed charge planes or matrices are brought into apposition; just how the structure would respond is a separate question altogether. If there were a lipid core the field across it might align phospholipid dipoles, for example. Alternatively, the field change through the possibly more freely permeable outermost region may trigger off macromolecular configurational changes whose effect may be transmitted deeper into the membrane. The result might be a change in permeability or nature of the cytoplasmic interface of the membrane, either of which could effectively transmit information about the environmental change to the cytoplasm, as discussed in the following section.

C. Generation of the Output Signal

Some factors which would be expected to alter cell surface membrane potential have been briefly reviewed in the first section. The possible significance of surface potential acting as a transducer, eliciting a change in the properties of the surface membrane in such a way that cytoplasmic response could be initiated (figure 6) will now be examined.

There are a number of ways in which a change in membrane surface potential could affect the membrane and cause input signals from the environment to be transformed into output signals to the cytoplasm. Both passive ionic permeability changes and conformational changes in the membrane resulting in permeability changes would supply an output signal of changed ionic fluxes between membrane and cytoplasm. On the other hand, transfer of information could occur without permeability changes by means of a change at the cytoplasmic interface of the membrane.

1. *Passive Ionic Fluxes*

Passive fluxes through a membrane may depend on the sign and magnitude of the surface potential which acts as a barrier to the diffusion of co-ions but accelerates counter-ions. This occurs because the kinetic energy of ions in solution is of the same order as the electrostatic potential energy barrier of the double layer. Passow (1964, 1965) has pointed out that the pH dependence of ionic permeability in the red blood cell is explicable on this basis: pH, in determining the degree of dissociation of fixed ionogenic groups in the membrane modulates the membrane charge density, and in consequence controls transmembrane ionic fluxes.

Since the membrane electrostatic field is a function of the concentration gradients of co-ions and counter-ions within the membrane, redistribution of counter-ions in the membrane following changes in surface potential would also change the membrane potential profile. (For a discussion of potential profiles, see Goldman, 1943; Polissar, 1954; Frankenhaeuser, 1960; Barr, 1965. Hodgkin and Huxley (1952) proposed that the potential gradient controls K^+ and Na^+ permeabilities of nerve membrane, while the effect of surface potential change on the gradient was invoked by Hodgkin and Chandler (1965) to explain anomalous effects of indifferent ions on Na^+ ion permeability in axons.

If the usual constant field approximation (Goldman, 1943) is made, the membrane potential may be represented as falling linearly between its

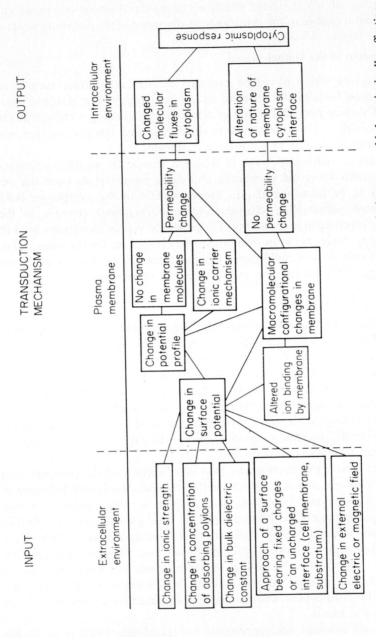

Figure 6. Diagram to show some ways in which extracellular environmental changes could, by principally affecting the membrane surface potential, lead to a cytoplasmic response. (Modified from Wolpert and Gingell, 1968)

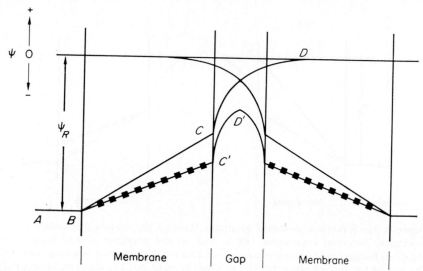

Figure 7. Schematic representation of the potential changes predicted to occur on close approach of two identical membranes, under conditions of constant surface charge density. Each membrane bears fixed charges at its outer surface. The bioelectric potential across the membranes is Ψ_R, with respect to the bulk external phase and *CD* represents the double-layer potential function when the membranes are at infinite separation. As they come together the potential profile changes from *ABCD* to *ABC′D′* and the potential gradient in each membrane changes from *BC* to *BC′* (From Gingell, 1967b)

interfaces (figures 7 and 8). It can be seen that a given membrane field can result from hyperpolarization of the outer surface or depolarization of the inner surface. For example, hyperpolarization induced by the approach of another cell might be equivalent to reduction of the bioelectric potential in terms of the resulting membrane field. Since reduction of the internal negative potential of nerve membrane by a few millivolts is sufficient to trigger cation permeability changes, hyperpolarization of the outer surfaces resulting from membrane apposition might also initiate comparable changes. Similarly an increase in ionic strength or adsorption of polycations to the outer surface might be equivalent to an increase in bioelectric potential. The fact that membrane electrostatic fields produced by modification of the ionic double layer potentials and the bulk phase potential can be functionally equivalent is evident from recent work on synthetic phospholipid bilayers (Mueller and Rudin, 1968). These can be forced into an excitable state, characterized by precise current voltage characteristics, either by setting up a bulk phase potential produced by different salt concentrations on each side of the membrane or by absorbing polycationic material to one side, in exactly the manner anticipated.

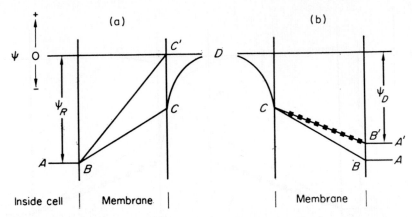

Figure 8. (a) Schematic potential gradients through the plasma membrane due to bioelectric potential (ψ_R) alone (*BC'*), and in the presence of additional fixed negative charges at the outer surface (*BC*). *CD* is the double-layer potential function. Decreasing the surface potential *CC'* by absorption of positive charges, for example would make the gradient across the membrane steeper. (b) Change in potential gradient through the plasma membrane (*BC* to *B'C*) due to reduction of the bioelectric potential from ψ_R to ψ_D. (From Gingell, 1967b)

2. *Configurational Changes*

If phospholipid or protein dipoles are sensitive to the local potential gradient, changes in this could lead to configurational changes resulting in sharp qualitative and quantitative membrane permeability changes. Charged groups of macromolecules as well as dipoles would tend to orientate according to the local potential gradient and electrophoresis of mobile ionic species would occur. The possibility that the molecular steric structure of ion carriers can be affected by the field has been suggested by Mueller and Rudin (1968). Polissar (1954) calculated that if the potential gradient in a membrane is of the order of 10^5 V/cm then the conformation of a macromolecule, with a dipole moment in the range of 170–1400 Debye units (in the range reported for proteins) which has no preferential orientation in the absence of a field, will be particularly sensitive to an increase in the field but insensitive to a decrease. Although Polissar's model is highly schematic it is of interest in suggesting that specificity of response to potential gradient changes might reside in membrane macromolecules and might be a highly characteristic property of individual membranes. A slower response to change in potential gradient may result from the electrophoresis of molecules, either ions or macromolecules, across the membrane; for example, if the outside becomes more negative, negatively-charged molecules will experience a force tending to move them towards the inner surface.

An extremely enlightening study by Bangham and his coworkers has provided a convincing example of a permeability change in a synthetic membrane in response to changes in surface potential, which they interpret tentatively in terms of a conformational rearrangement. (Bangham and coworkers, 1965a,b; Bangham and Papahadjopoulos, 1966; Papahadjopoulos and Bangham, 1966). These workers employed smectic mesophase phospholipid micelles, believed to be structurally similar to the postulated bimolecular lipid structure of cell membranes. By altering the negative surface charge density (measured as zeta potential, which is an experimental approximation to the surface potential) with small amounts of long chain cations and anions, they were able to show that the self exchange flux of $^{42}K^+$ is a function of zeta potential, and hence a function of surface potential. Similar results were found for cationic local anaesthetics. From the data of Bangham and coworkers (1965a), it can be seen that a change in negative zeta potential from 20 to 15 mV causes an 80 per cent rise in potassium $^{42}K^+$ permeability. As the curve rises steeply, similar changes in negative zeta potential at higher potentials would be expected to have an even more marked effect. There is no evidence to suggest that the micellar membranes can distinguish between alkali metal cations: exchange of cations is completely inhibited when the surface charge is positive, in contrast with anion permeability which remains high regardless of the surface charge and seems to be determined by the radius of the hydrated ion. These authors concluded that the rate of cation entry is more rapid than would be expected simply in terms of passive diffusion due to the increased cationic counter-ion concentration near an anionic surface, and compare the elevated permeability of artificial micelles with that following depolarization of a nerve. This is suggestive of a qualitative modification in permeability properties, reflecting perhaps a conformational change in the lipid molecules.

Changes in membrane permeability following surface potential induced molecular configurational changes need not be the source of output signals from the transductive system. For example, it is easy to imagine that a reorientation of macromolecular fixed charges might result in an exchange of Ca^{2+} with the cytoplasm nearby, resulting, for example, in the contraction or relaxation of cytoplasmic proteins whose state, in muscle, is known to be critically determined by Ca^{2+} ion concentration (Hasselbach, 1964). Allosteric configurational changes in relation to membrane physiology has recently received theoretical attention from Changeux and coworkers (1967) and Hill (1967) who show that interaction of a lipoprotein membrane unit (protomer) with another macromolecule or simply a proton can induce a very sharp cooperative transition in protomer structure, and that different conformational states of the protomer may be coexistant. This could provide local presentation of new groups, perhaps enzymatic, at the cytoplasmic

M

interface, so that enzyme products such as cyclic AMP might provide the output signal.

3. *Changes in Ion Binding Ratios*

A third possible effect of changing the surface potential involves changes in the ratios of ions near the surface or bound to it. The concentration c_i of and ion i, valency z at a region where the potential is ψ with respect to a region of bulk concentration c_i, is given by Boltzmann's relation

$$c_i = c_i \exp(-z_j\, e\psi/RT)$$

Consequently if a cell surface becomes more negative with respect to the bulk external solution the ratio of divalent to monovalent cations in the double layer increase due to the $\exp(-z\psi)$ term. It might be anticipated, for example, that if Ca^{2+} is involved in linkages between membrane protomers a decrease in negative potential of the surface would displace Ca^{2+} in favour of Na^+, modify the interprotomeric relationship and alter membrane permeability. Another possibility is that enzyme activity, which is sensitive to divalent cations, would be switched on or off as the local electrostatic potential alters.

D. Membrane Output Signal as Cytoplasmic Effector

Here there are two separate issues. First, in the case of cells which appear to respond to changes in membrane surface potential, what is the mechanism of action of the effector output signal? Secondly, what evidence is there, in general, relating cytoplasmic metabolic responses to ionic concentration changes in other systems?

In the case of excitable cells, there is no doubt that the membrane is sensitive to the potential differential between its surfaces. If an increase in ionic permeability is the primary response to a change in the potential difference the output signal could be viewed as the flux change. Only in neuromuscular transmission does this output signal have an effector action on the cytoplasm, thus involving the whole cell. Although the molecular events linking excitation with contraction of the myofibrils are difficult to ascertain (Podolsky, 1968) it seems likely that ionic fluxes occur across the tight junctions between the T-system tubules and sarcoplasmic reticulum vesicles in response to a wave of electrotonic potential change passing down the T-system membrane, and that these fluxes somehow lead to the liberation of Ca^{2+} from the sarcoplasmic reticulum elements into the myofibrillar space. Here Ca^{2+} release may be considered an intermediate step in the effector action of an ionic flux output resulting in myofibrillar contraction.

The role of Ca^{2+} as a member of an effector concatenation is also clear in the case of induced cortical contraction of amphibian eggs which is discussed in the biological section.

Although the mechanism of action is obscure, the fact that small ions can profoundly affect cellular behaviour is becoming evident. Bygrave (1967) has presented evidence that the rate of many glycolytic steps is limited by cytoplasmic Ca^{2+} concentration and Wyatt (1964) has stressed that cations might have a metabolic homeostatic function. Kroeger and Lezzi (1966) showed that alkali metal cation concentration in the diptera nuclei can activate chromosomal puffing, and a reversible effect on the rates of nucleic acid and protein synthesis in response to cytoplasmic potassium has been demonstrated by Lubin (1967). It is possible that the activation of adenyl cyclase might result from a change in membrane potential. Catecholamine hormones are known to act on liver cell surface membranes and transductively initiate the synthesis of AMP by means of adenyl cyclase. The effector chain results in the activation of a phosphorylase essential for glucose production and the same effector can also initiate the hydrolysis of triglycerides in liver. An interesting hypothesis has been put forward by Rasmussen and Tenenhouse (1968) that the really significant event associated with the conversion of ATP to cyclic AMP may in fact be release of Ca^{2+} from Ca^{2+}—ATP. This cation, rather than cyclic AMP, may trigger off a variety of cellular responses. Specificity of response must reside in the cell if Ca^{2+} evokes different responses in different situations.

III. BIOLOGICAL INTERPRETATIONS

A. Pinocytosis Induction in Free-living-cells

In this section it will be shown that pinocytosis in large free living amoebae is induced under conditions which reduce the negative surface potential, though the situation is less clear in the case of tissue cells. A fall in membrane resistance to ions which accompanies pinocytosis in amoebae almost certainly reflects a change in membrane structure. The ensuing fluxes or possibly a change in the nature of the cytoplasmic interface may represent the output from the transducer, resulting in a cessation of locomotion and channel formation.

The environmental conditions which give rise to pinocytosis in large free-living amoebae have been extensively and carefully investigated by Chapman-Andresen (1962) as well as by Schumaker (1958) and by Brandt and Pappas (1960).

It has been clearly established that whereas molecules bearing formal static charges may induce pinocytosis uncharged molecules do not. Polyelectrolytes in pH regions where they are net cationic are known to adsorb

M*

to the cell surface, where considerable concentration occurs (Schumaker, 1958; Brandt and Pappas, 1960; Gingell, 1967a). Certain smaller cationic molecules, such as the dyes alcian blue and methylene blue also adsorb. On the other hand, small ions which do not adsorb to any significant extent, such as amino acids and inorganic salts, are also inducers.

The similar inducing actions of adsorbing and non-adsorbing charged molecules have presented something of a paradox, which has never received a satisfactory physicochemical explanation. I was struck by the possibility that reduction of membrane surface potential might be the operative factor linking the action of salts and adsorbing polycations: a change in surface potential might be anticipated to alter membrane properties, resulting in pinocytosis. A physical basis indicating how salts and polyions could modify membrane surface potential has already been given in the theoretical section, and I would like to briefly examine the pertinent experimental facts relating to the major classes of inducers.

1. *Inorganic Salts*

If the effect of simple organic salts on membrane surface potential is the key-event in induction, two simple predictions follow. First, the inductive effect should depend on the ionic strength of cations for a negative surface, in which case anions should be relatively unimportant. Secondly, induction should be proportional to the square of the valency of the inducing cation. (Verwey and Overbeek, 1948). The former requirement is observed; for example, molar Na_2SO_4 contains twice the concentration of Na^+ ions present in molar $NaNO_3$ and has twice the Lewis–Randall ionic strength. As anticipated, equal ionic strengths of the two salts have similar pinocytotic inducing abilities. Under similar conditions, it can be seen from the results of Chapman-Andresen (1958) that unbuffered chlorides, nitrates and sulphates of lithium, sodium, potassium and ammonium have similar pinocytotic effects in solutions containing identical concentrations of the cation, regardless of the nature of the anion. This constitutes one of the best pieces of evidence for the surface potential theory. However, although solutions containing similar ionic strengths of monovalent alkali metal and ammonium cations have remarkably similar inductive properties, there is a notable exception. Hydrogen ion is apparently not an inductive cation; amoebae have not been seen to pinocytose as the pH is lowered in the absence of other inducing agents. The reason for this is not immediately clear, but it may be due to hydrogen ion sensitivity of membrane structural integrity, reflecting perhaps the dissocation equilibria of internal ionogenic groups involved in hydrogen bonding. Compared with univalent ions, the action of ions of higher valency is much less easy to interpret. Ca^{2+} and perhaps Mg^{2+} have special effects on membrane permeability which will be mentioned later, while most metal ions of higher valency are characteristically toxic, so that it is difficult to compare the efficacy of such ions with theoretical prediction.

2. *Amino Acids and Proteins*

The second major group of pinocytotic inducers includes the organic ampholytes, amino acids and proteins. Whereas the former have not been shown to adsorb at the cell surface, proteins do adsorb in conditions favourable for pinocytosis. The action of amino acids is therefore essentially salt-like, depending on ionic strength effects, but they are included in this section because their charge is dictated by essentially the same acid–base equilibria governing the charge on proteins. When the environmental pH is less than the pH corresponding to the isoelectric point, pH < pI, an ampholyte molecule is net cationic, whereas when pH > pI, it is net anionic.

Basic amino acids such as lysine and arginine are net cationic in solution below pH 10 due to their high pI values of 9·74 and 10·76, respectively, and as expected are good inducers. Acidic amino acids such as glutamic and aspartic acids, are net cationic below their pI values of 3·22 and 2·77, respectively. Chapman-Andresen has reported induction by these acids at pH 8, but as equimolar NaOH was used to adjust the pH and this pH optimum is characteristic of alkali metal cations, the induction is easily attributable to Na^+. Cells are unable to respond at very low pH, so that the region where pH < pI for dicarboxylic acids cannot be investigated. The neutral amino acids tested by Chapman-Andresen with pI in the range 4–6 give no induction below pH 9. Slight pinocytosis at higher pH is doubtless due to Na^+ from the NaOH used to adjust pH. The field near a symmetrical electrical dipole is small compared with the field of an ion of equal unit charge so an appreciable positive electrostatic field exists around an unequal dipole whose positive charge is greater than its negative charge, a situation realised only when pH < pI.

All proteins and polyelectrolytes which induce pinocytosis do so only on the acid side of their isoelectric points (see Chapman-Andresen, 1962). Their pinocytotic effect is maximal near pH 4 where electrostatic interaction between polyion and surface would probably be maximal because at pH 4 the degree of dissociation of surface carboxyl groups of pK_a equal to 3·0 is still 0·9, while proteins become more strongly cationic as pH falls. As anticipated, the predominantly carboxylic protein pepsin whose isoelectric point is near unity is a non-inducer, while polylysine which is almost entirely cationic induces at pH > 8. Chapman-Andresen's experiments concerning the action of inorganic salt on induction by proteins may in general be interpreted in terms of a reduction in electrostatic binding of protein to the cell surface, since increased ionic strength reduces the potential energy of electrostatic interaction between the anionic cell surface and cationic poly-ions. Quantitative salt elution of fluorescent labelled RNAase with first order kinetics from the surface of *Amoeba proteus* has been demonstrated (Gingell, 1967a). The addition of salt has two theoretically distinct actions; elution of electrostatically bound protein tending to inhibit pinocytosis, and an in-creased ionic strength tending to promote it, both processes being differently affected by pH. This accounts reasonably well for the complex results.

Although the simple conception outlined accounts for many experimental facts of pinocytosis it is notably inadequate with regard to divalent cations. A central role is given to calcium by Nachmias (1968) who has put forward the hypothesis that pinocytosis is induced by removal of bound calcium from the cell surface, for example, as a result of competition by monovalent cations for surface anionic sites. Evidence in support of calcium removal is adduced from the stated pinocytosis inducing ability of 1 mM disodium EDTA (ethylenediaminetetracetic acid salt) at pH 6·3 on *Chaos chaos*, (Brandt, 1958) as well as motile inhibition produced by ATP, which is also a chelating agent for divalent cations.

However it was found (Gingell, unpublished results) that in a medium containing $5·36 \times 10^{-2}$ mM KCl, 1·37 mM NaCl, $4·76 \times 10^{-2}$ mM $NaHCO_3$, 20 mM $CaCl_2$ and of pH 6·2, *Amoeba proteus* exhibited weak pinocytosis, absent in a control lacking calcium. Pinocytosis was still seen at 5 mM Ca^{2+} in a small proportion of cells. Pinocytosis could be induced in 1 mM ATP or EDTA only at alkaline pH. At pH 9·0 the total Na^+ concentration is 3·0 mM, including contribution from the sodium hydroxide used to adjust pH. Although this is insufficient to initiate pinocytosis in divalent cation free media without chelating agents, removal apparently covers the threshold for initiation by chelating Ca^{2+} associated with the surface. The results of Brandt and Freeman (1967) accord with this view. It is clear from these results that cal-cium can induce pinocytosis, yet antagonizes induction by monovalent cations.

These complexities may be relieved if at least two separate actions are attributed to calcium. It is suggested that one aspect of the action of Ca^{2+} is its role as a simple double layer counterion, tending to reduce surface potential in proportion to $C_i z_i^2$ where C_i is its concentration and z_i its valency. In this it is apparently more effective than Na^+ at the same concen-tration, as anticipated from the z^2 factor. The other effect of Ca^{2+} is physio-logically antagonistic to the first. Increasing concentrations affect membrane structure in an unknown way, perhaps involving Ca^{2+} binding to a group such as phosphate, causing a progressive increase in membrane resistance. Electrostatic binding with anionic groups near the cell surface, involving competition with monovalent cations has been postulated to raise membrane resistance. (Luttgau and Niedergerke, 1958; Singer and Tasaki, 1968). Such duplicity of action may be a basic factor in the calcium paradox.

The evidence available therefore lends support to the idea that the primary common action of inducers of pinocytosis in free-living amoebae is a reduction in the negative potential of the cell surface.

B. Pinocytosis Induction in Tissue Cells

The confused and frequently contradictory literature on mammalian cell pinocytosis will not be discussed at length here. Unlike free-living cells, most mammalian cells require serum proteins in the medium and their presence greatly complicates interpretation of the results. Whereas Cohn and Parks (1967a,b) claim to have demonstrated preferential pinocytosis of polyanions by macrophages, Allison (1968) has been unable to confirm their findings. The binding of negative polyions to cell surfaces studied electrophoretically has been reported by Lipman (1968) but removal of cations from the surface does not seem to be excluded. Pulvertaft and Weiss (1963) have reported that reduction in ionic strength at constant osmotic pressure induces pinocytosis in polymorphs. If this is correct, it may imply that increased surface negative potential may be able to trigger off pinocytosis in these cells. However, the work of Ryser (1968) on the stimulation of albumin uptake by basic polypeptides in sarcoma S180 II cells is most easily interpreted as showing that adsorption of polycations promotes pinocytosis. Furthermore, it is quite clear that mere overall surface potential changes do not afford a satisfactory explanation of his results, since poly-D-lysine is more active than poly-L-lysine. His results certainly suggest that both molecular size and chemical specificity, as well as overall charge, are factors which a satisfactory theory must explain. It would be interesting to know whether free-living amoebae can also exhibit stereoisomeric specificity.

C. Change in Membrane Permeability

Having suggested a common thread in the action of inducing substances which may provide the environmental input in the pinocytosis of free-living amoebae, the next question is how the information passes to the cytoplasm. A vital clue to this step was provided by Brandt and Freeman (1967) and by Gingell and Palmer (see Gingell, 1967a). These investigators found that the initiation of pinocytosis was accompanied by a decrease in electrical resistance of the surface membrane, implying an increase in ionic permeability. Brandt and Freeman showed that 10 mM sodium chloride lowered membrane resistance ten or twenty-fold, while RNAase and lysozyme also lowered membrane resistance. Using a technique employing a single intracellular glass microelectrode, filled with 3M KCl (figure 9), Gingell and Palmer found in addition that removal of the stimulus was accompanied by a recovery of membrane resistance. Figure 10 is an experimental record of resistance changes occurring on adding sodium chloride and then washing it off with

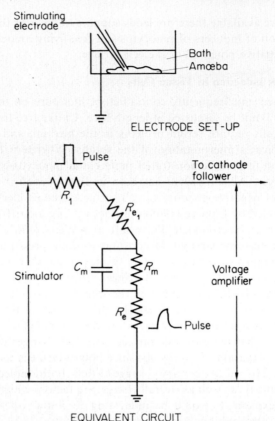

Figure 9. The upper diagram shows the electrode arrangement used in experiments on *Amoeba proteus* and the lower diagram gives an outline of the circuit. R_{e_1} is the resistance of the stimulating and recording electrode, R_e is the resistance of the indifferent extracellular electrode, and R_1 is the output resistance of the stimulator ($10^9\,\Omega$). Characteristic pulse profiles before and after the current pulses pass through the cell membrane are indicated.

culture medium. Similarly, when the cells are treated with 0·1 per cent RNAase at pH 7 the resistance falls within a few minutes to an immeasurably low value. After eluting the protein with 0·1 per cent NaCl the resistance returns to normal (figure 11).

As mentioned previously, Ca^{2+} in low concentration does not induce pinocytosis, a result which is apparently not in accord with simple surface potential theory which predicts that the effectiveness of a counter-ion should be related to the square of its valency. It is thus extremely interesting that Brandt and Freeman found that 0·1–0·5 mM Ca^{2+} reversed the effect of inducing agents and raised the threshold of induction. This is in complete

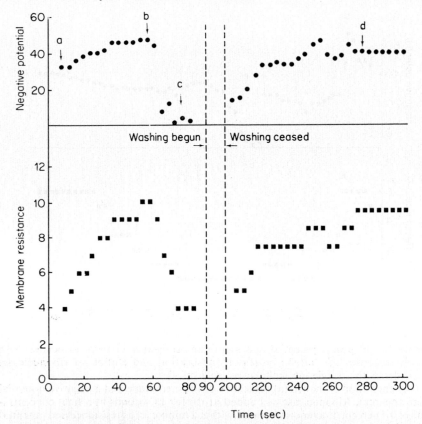

Figure 10. Electrical potential and cell membrane resistance changes preceding motile responses of *Amoeba proteus* through the action of sodium chloride. Recorded with a single intracellular microelectrode. (a) penetration of the microelectrode into a streaming cell: as it 'sealed in' the resistance and negative potential recorded increased to steady values. (b) rapid addition of NaCl to a final concentration of 8·5 mM caused a rapid fall in potential to zero and the resistance decreased. (c) locomotion had practically ceased after 20 secs. NaCl was washed away with Chalkley's medium and the potential rose steadily. Movement began 3 mins after washing began. Due to the recording technique resistance changes shown are only relative. Although potential is uncalibrated a value between −20 to −40 mV was frequently found in untreated cells.

accord with the action of Ca^{2+} on surface membranes of excitable and inexcitable cells, (Frankenhaeusen and Hodgkin, 1957; Bruce and Marshall, 1965; Loewenstein, 1966), where it is probably important in maintaining structural integrity (see Shah and Schulman, 1965; and see McClare, 1968).

It may be significant that reduction of membrane resistance and the onset of pinocytosis is invariably accompanied by cessation of locomotion.

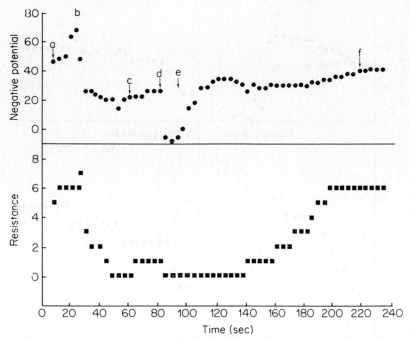

Figure 11. Electrical potential and membrane resistance changes associated with motile responses of *Amoeba proteus* to adsorption and elution of ribonuclease. Recorded with a single intracellular microelectrode.

On penetration of the cell (a) resistance of the membrane and a negative potential were apparent. Ribonuclease was added at (b) for 18 seconds at a final concentration of 0·5 percent. Locomotion ceased within a minute (c) as resistance and potential fell. Fluid in the electrode bath was made 0·1 M with NaCl during 10 seconds (d). Resistance fell to nearly zero and a positive potential was measured. At (e), 20 seconds later, washing with a continuous flow of Chalkley's medium started and continued throughout the recovery period. The potential soon became less positive and resistance increased. At (f) pseudopodal activity was seen. Electrode resistance remained constant during the experiment, but zero potential changed. The fall in resistance on addition of NaCl to less than that of the electrode outside the cell is possibly an artefact due to lowering of electrode resistance in the environment of high ionic strength within the low resistance amoeba.

Recovery of the membrane resistance occurs on removal of the pinocytotic-inducing substance and appears to be associated with the disappearance of pinocytotic channels and the onset of pseudopodal activity.

D. Differentiation of Naegleria

The remarkable reversible transformation between amoeboid and flagellate forms of *Naegleria gruberi* has been discussed by Willmer (1963). Transition to the flagellate phase is stimulated by reduction in ionic strength of the medium, which is not an osmotic effect, as well as by certain hormones. It is striking, as pointed out by Chapman-Andresen, that the salts which cause pinocytosis in free-living amoebae favour the amoeboid phase of *N. gruberi*. The monovalent cations potassium, sodium, ammonium, choline and tetraethylammonium all have a quantitatively similar action, favouring the amoeboid phase, while divalent cations magnesium, hexamethonium and decamethonium are more potent in this action.

Willmer also found that deoxycorticosterone and progesterone, which affect cation discrimination in higher animal cells, promoted flagellation of amoebae at low concentrations but had the opposite effect at higher concentrations, preventing flagellation. Progesterone also acts differently on amoebae and flagellates, which suggests that they are different physiologically as well as morphologically (Willmer, 1963).

Willmer has postulated that the change from amoeba to flagellate is accompanied by a change in surface properties, which is a cellular response to altered ionic balance between cell and environment due to salt and hormonal action on the membrane. The fact that a surface change occurs is substantiated by the finding (Gingell, 1966) that polylysine is far more lytic to flagellates than amoebae. In the light of recent evidence on hormone action on membranes discussed earlier (Sutherland and coworkers, 1967; Rasmussen and Tenehouse, 1968), it is likely that both salts and hormones affect the cell membrane, perhaps activating cyclic AMP, and produce changes in membrane function and cellular morphology.

The effects of monovalent cations, and the more pronounced effects of divalent cations support a hypothesis that they act by reducing the surface potential of the negative membrane, thereby initiating a change in membrane function paralleled by deoxycorticosterone and progesterone.

E. Response of Amphibian Egg Surface to Adsorbed Polycations

A further example of a possible surface potential mediated transducer mechanism is the contractile response of fertilized *Xenopus laevis* eggs and early cleavage stage cells to adsorbed polyelectrolytes and ionic detergents (figure 12). When 0·1 percent solutions of poly-L-lysine of molecular weight 2,600, 50,000 or 250,000, which is strongly cationic over a wide pH range, is applied locally near the cell surface membrane, a localized non-propagated contraction occurs within 1–2 minutes in normal Steinberg's medium. (Gingell, 1967a, 1970). Similar responses are obtained with poly-L-ornithine, poly-L-arginine histone sulphate and certain samples of RNAase, but not with sodium poly-L-glutamate (mol.wt. 68,000), sodium heparin or un-polymerized amino acids. The cationic detergent hexadecyl trimethyl-ammonium bromide at 0·5 mM and also the anionic detergent sodium

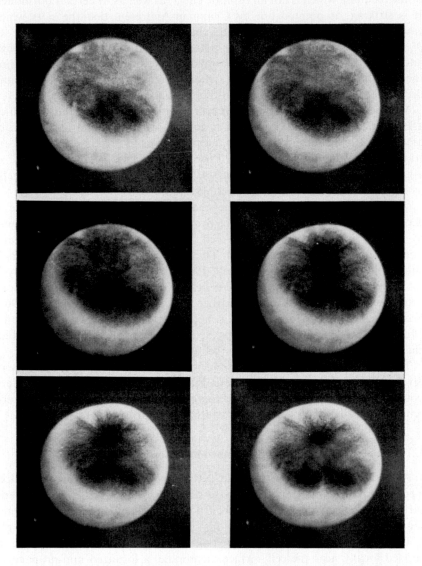

Figure 12. Successive stages of the contractile response of a fertilized *Xenopus laevis* egg to 1mM sodium dodecyl sulphate applied locally to the animal pole are shown. The sequence runs left to right and top to bottom. The first frame depicts the untreated egg, the second was taken immediately after treatment. Subsequent frames are separated by approximately 1 minute intervals. (Magnification approximately × 25)

dodecyl sulphate at 1·0 mM cause contraction, but the non-ionic detergents 'Triton X-100' and 'Pluronic F 68' are completely ineffective. Higher concentrations of ionic detergents readily lyse the membrane.

Electron microscopy of glutaraldehyde-fixed osmium-stained sections (Gingell 1967a, 1970) shows that polylysine-treated regions are thrown into extensive villus-like folds as a result of contraction in the subjacent cortex, beneath which there is a local accumulation of pigment granules (figure 13a,b,c). The membrane is coated, as described by Katchalsky and coworkers (1959) for red cells, with adsorbed polycation but there is no evidence that the membrane itself contracts, unlike the proposed action of polyelectrolytes on nuclear membranes (Korohoda and coworkers, 1968). Localized membrane lysis is sometimes seen in the electron microscope, and it is possible that polycations can produce micropunctures in the membrane (but see below). It is known that saponin has such an effect (Seeman, 1967).

Reversible relaxation of the contracted region is accelerated by bathing eggs in 1 per cent sodium polyglutamate of molecular weight 60,000 or sodium heparin at neutral pH. On replacement of this medium by Steinberg's medium, which contains 0·34 mM Ca^{2+} and 0·83 mM Mg^{2+}, recontraction of the fully relaxed region occurs even after a hiatus of thirty minutes.

Using a double intracellular microelectrode recording technique (Gingell and Palmer, 1968), the membrane resistance to applied rectangular current pulses during the application of polyelectrolytes was measured (figure 14). Prior to visible contraction the resistance falls from approximately 14 kΩ cm^2 to less than 2 kΩ cm^2 a few seconds after adding polyelectrolyte, and remains low during subsequent contraction. Following RNAase treatment a spontaneous recovery of resistance, followed by relaxation in about 8 minutes was observed repeatedly but other samples of RNAase have since been found to give no contractile response. The reason for this is not clear. While 1 per cent polyglutamate alone has no effect on control membrane resistance, it promotes slow relaxation and restoration in the resistance of surface membrane previously treated with polylysine, as shown in figure 15 where it can be seen that polylysine causes a fall in resistance from 16 kΩ to 3 kΩ. Calculation shows that even a single 50 Å radius hole would reduce the resistance almost to zero, so there is good evidence that the fall in resistance is not due to membrane rupture, but represents a generally modified membrane resistance.

In the absence of external divalent cations with or without 3 mM disodium EDTA (ethylenediaminetetraacetate) no contractile response can be evoked, though membrane resistance falls on adding polycation. This suggests that extracellular Ca^{2+} or Mg^{2+} is necessary for contraction. On addition of normal Steinberg's medium, which contains these cations, rapid contraction results without any change in membrane resistance.

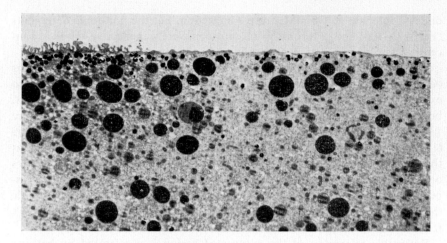

Figure 13a. View at low magnification of a region of the animal pole of a fertilized egg treated locally with polylysine. The extremity of the treated region is recognizable at the left-hand side by superficial protrusions and pigment granule accumulation in the cytoplasm. The right-hand side has not been treated and shows no signs of contraction. (Magnification × 1,000)

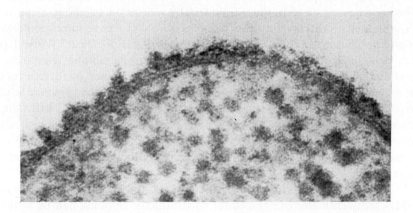

Figure 13b. High power view of the contracted cell surface after treatment with polylysine. Abundant material adsorbed to the outside of the plasma membrane can be seen. The plasma membrane does not appear to be asymmetric. (Magnification × 316,000)

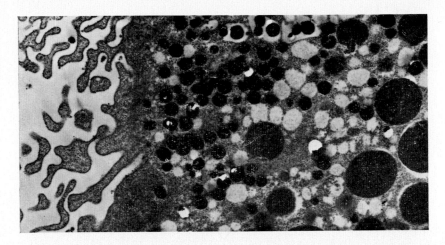

Figure 13c. Animal pole contracted after localized treatment with polylysine. The surface is thrown into microvillus-like protrusions, the region separating the pigment granules and the cell surface has become considerably thickened and all cellular inclusions are closely aggregated. The section is slightly tangential. (Magnification $\times$ 9,000)

These experiments demonstrate that strongly polycationic molecules act in an apparently non-specific way at the cell surface resulting in a fall in membrane resistance, followed by contraction in the cytoplasm just beneath the membrane. The requirement of divalent cations suggested that contraction might be initiated by cationic movement across the membrane to a cortical contractile zone. This hypothesis was tested by iontophoresis of K^+, Cl^-, Na^+, Mg^{2+} and Ca^{2+} immediately beneath the surface membrane (see Gingell, Garrod and Palmer, 1970). Only Ca^{2+} was seen to induce contraction. Spontaneous relaxation after a few minutes might possibly indicate the presence of a Ca^{2+} sequestering system in the cortex, and it is tempting to compare the cortical contractile apparatus with that of muscle. Relaxation induced by polyanions may be tentatively attributed to reversible binding to previously adsorbed polycations, a reaction which might lower the potential near the membrane surface. Sequestering of Ca^{2+} by polycations may play a role in inhibiting contraction, but extensive binding of Ca^{2+} in solution is unlikely, because membrane resistance is unaffected by the polyanion solutions used over long periods, whereas resistance falls progressively in Ca^{2+} free solutions.

The increase in permeability caused by the apparently non-specific adsorption of strongly charged molecules suggests that the charge carried rather than the configuration of the adsorbed molecule is the factor which

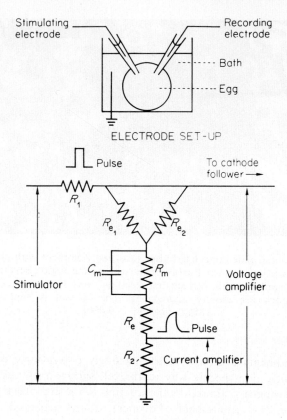

ELECTRODE SET-UP

EQUIVALENT CIRCUIT

Figure 14. The upper diagram shows the electrode arrangement used in experiments on *Xenopus laevis* eggs and the lower diagram gives an outline of the circuit. R_{e_1} is the resistance of the stimulating electrode, R_{e_2} the resistance of the recording electrode and R_e that of the indifferent electrode. R_m is the resistance of the cell membrane and C_m its capacitance. R_1 is the output resistance of the stimulator, $10^9 \ \Omega$. R_2 is a resistance of $10^5 \ \Omega$ across which current is measured. Characteristic pulse profiles before and after the current passes through the cell membrane are indicated.

determines membrane response. In other words the change in membrane electrostatic surface potential due to charge density change may well be the critical event.

Lack of overriding chemical specificity in the ligand–cell surface interaction responsible for initiating the permeability transformation is clear. While the action of ribonuclease is in question, histone sulphate, poly-L-lysine HBr, poly-L-ornithine HBr and hexadecyltrimethylammonium bromide

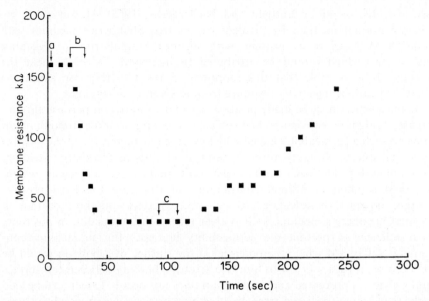

Figure 15. Membrane resistance changes associated with contraction and relaxation of *Xenopus laevis* egg in response to poly-L-lysine applied to the cell surface. Recorded with two intracellular microelectrodes. (a) resistance of the egg membrane. (b) on addition of poly-L-lysine locally the resistance rapidly fell from 160 KΩ to 30 KΩ and contraction began. (c) excess poly-L-lysine was washed out of the electrode bath and 0·1 per cent Na polyglutamate was added, without any immediate effect. Five minutes later the resistance was 170 KΩ and relaxation was advanced.

are active; all are highly ionized cationic molecules at around neutral pH. Non-ionic species and neutral amino acids are ineffective, as are polyanions. However the small anion sodium dodecyl sulphate is active, and therefore represents something of a problem.

If the charge carried is responsible for initiating the permeability change then it must be concluded that either an increase in positive or negative charge can be effective. On the other hand this detergent, and perhaps hexadecyltrimethylammonium bromide may interact hydrophobically with the surface by virtue of their short hydrocarbon chains. Lack of activity of large highly polymeric non-ionic detergents does not necessarily obviate this interpretation.

Unfortunately it is not possible to parallel a resistive effect of a rise in negative surface potential by the obvious method of reducing the concentration of external sodium which is the predominant cation. This is because reducing the ratio $Na^+:Ca^{2+}$ in the bulk phase increases the ratio Ca^{2+}/Na^+ near (possibly bound to) the surface, and results in increased membrane resistance see also discussion by Luttgau and Niedergerke, 1958). However, on re-placing normal medium by divalent cation free Steinberg's medium with 0·00058 M NaCl made isotonic with sorbitol a small rise in resistance still occurs which cannot be attributed to increased $Ca^{2+}:Na^+$ near the surface. It is possible that this represents a resistive response simply to increased surface negativity resulting from lowered ionic strength.

If the surface must be made positive in order to initiate a permeability in-crease, it might be anticipated that sufficiently raising the ionic strength might prevent such a permeability increase by lowering the magnitude of the positive potential due to the polycation. Treatment of eggs in Steinberg's medium containing 0·5 M NaCl with 1 per cent polylysine (molecular weight 195,000) resulted in normal contraction: in this case the local positive surface potential is calculated to be approximately one-third of that in normal Steinberg's medium, so it is clear that such a reduction in positivity is insufficient to prevent the permeability increase. But in medium con-taining NaCl>1M where the potential at the surface of a polyion would be reduced by at least a factor of four, electrostatic binding between cell surface and polyion is prevented and contraction does not occur. Thus it can be said only that increasing ionic strength does not prevent a permeability trans-formation due to adsorbed polycation below the point at which the reduction in potential prevents ionic binding of the polyion.

It must be concluded that although most of the evidence suggests that a non-specific increase in surface positivity is the common factor linking the action of polycations, the action of ionic detergents is still far from settled. Nevertheless, it is clear that subsequent changes in membrane organization result in an increase in Ca^{2+} permeability and probably also to larger mole-cules. It has not been completely ruled out that actual puncture of the membrane may occur. Cytoplasmic contraction, the overt cellular response, is normally initiated by Ca^{2+} ions interacting reversibly with a cortical contractile system, though alternative operation of the system with Ba^{2+} and Sr^{2+} but not Mg^{2+} shows that it lacks rigid specificity, and resembles the muscular actinomyosin mechanism. Although in the case of pinocytosis in *Amoeba* the nature of the ionic output from the transducer is unknown, the overall mechanism shares in common with that described for *Xenopus*, an increase in membrane permeability resulting in cytoplasmic motile response.

The discovery of Ca^{2+}-activated contractile region beneath the cell membrane has important implications in its own right. It is known to be responsible for contractile closure of wounds in the cell surface (Gingell, 1969) previously believed to be a function of the surface coat (Holtfreter, 1943). It is probably also involved in the initial stages of cytoplasmic division in the amphibian blastula as well as changes in free cell surface area which occur, for example, during invagination of the archenteron. In these circumstances, it seems likely that the contractile process is activated by local release of bound Ca^{2+}, possibly from elements which also have a sequestering function and induce relaxation.

F. Responses to d.c. Electric Fields

Galvantactic responses have long been recognized in free-living cells (see Jones and coworkers, 1966) and such behaviour may be instrumental in the movements of osteoblasts in response to currents set up in stressed bone (Bassett, 1965) and in regeneration (Becker, 1961). Curiously, there are few published reports of the behaviour of higher metazoan cells in uniform electric fields, the outcome of which would be crucial to any theory of electrical guidance.

Jahn (1966) has proposed a mechanism for galvanotasis based on capacitative current flow inside the cell. However, by analogy with the stimulus for pinocytosis in large free-living amoebae, it seems possible that a change in membrane properties might occur; at the anodal side the surface potential would become more positive, perhaps resulting in a permeability increase to certain ions. Furthermore, Ca^{2+} in the membrane at the anodal end would experience an electrical force tending to move it into the cytoplasm, which might be related to the observed anodally initiated cytoplasmic contraction. It would be enlightening to know whether permeability changes can occur under electric field conditions in cells which are able to respond galvanotactically.

IV. CELLULAR CONTACT INTERACTIONS

The attempt to define intercellular contact rigorously is fraught with difficulty; although uncomplicated at the light-microscopic level, the concept acquires a certain ambiguity at the molecular level. For the purpose of this discussion contact will be taken to mean a chemical or physical interaction resulting from close apposition of cell surface membranes.

There are a number of well documented cases where contact between cells modulates cellular behaviour. These can be loosely grouped together under the headings of cell movement, cell growth, developmental interactions and immunological responses, which one considered on the basis of the role played by the surface membrane, in terms of sensor, transducer and channel concepts (Wolpert and Gingell, 1969). It is not the purpose of this chapter to examine in detail the evidence for cellular contact in these processes, nor to discuss the degree of specificity required in each, but merely to point out that the close approach of cells could lead to changes in surface potential which might be instrumental in eliciting cellular responses. Perhaps the most interesting possibility for such intercellular interaction in relation to the channel concept is to be found in the functional coupling phenomenon, leading to the formation of a potential information channel between cells.

A. Transductive Genesis of a Channel

It has been argued that close apposition of cell membranes might cause a rise in electrostatic surface potentials, and consequently lead to changes in membrane permeability or function. Therefore, the greatest interest is attached to the work of Furshpan and Potter (1969) and Loewenstein and his colleagues who have shown that a variety of excitable and inexcitable cells in contact sometimes have localized regions of increased intercellular ionic permeability. This is the functional coupling phenomenon which is described by Loewenstein (1966). In most cases where the ultrastructural basis of functional coupling in vertebrate tissue has been investigated, tight junctions have been found between the cells (Furshpan and Potter, 1969). In invertebrate material it has been suggested that septate desmosomes are involved (Bullivant and Loewenstein, 1968).

Loewenstein (1967) has described the onset of functional coupling between like cells of disaggregated sponges of *Haliclona* and *Microciona* species. Cells of the same species spontaneously aggregated in artificial sea water and electrical coupling was established between 1 and 40 minutes after apparent contact. No communication was evident between cells of the same species pressed together in artificial sea water with lowered divalent cation concentrations even after some hours, but on restoring divalent cations coupling became detectable.

Coupling does not depend on cell orientation, showing that any part of the cell surface is potentially capable of permeability transformation. The fact that cells of different species are not mutually adherent and do not functionally couple, suggests that the factors necessary for adhesion are also necessary for the genesis of functional coupling.

Loewenstein (1968) has put forward a hypothesis, based on the known sensitivity of surface and junctional membranes to divalent cation concentration, to explain the onset of functional coupling, suggesting that the future junctional region is sealed off and divalent cations are somehow removed, rendering the junctional membrane leaky.

There seem to be two main difficulties with this suggestion. First, it does not explain how or why divalent cations are removed. Secondly, if all the surface is potentially capable of forming a permeable junction and is pumping out Ca^{2+} (Loewenstein, 1967), the exit of Ca^{2+} must be stopped in the region where coupling occurs, unless Ca^{2+} in this compartment is relatively rapidly removed. It would seem that the feature lacking in this analysis is consideration of the localized change in membrane properties brought about at contact, and it is suggested that a transductive surface potential change may be the initial event.

V. SUMMARY

It is suggested that the cell membrane surface electrostatic potential is capable of regulating membrane function in non-excitable cells. Environmental factors which modulate surface potential are considered as input stimuli into a transducer mechanism located in the membrane. Output from the transducer may be represented by changes in transmembrane ionic fluxes, or configurational changes at the cytoplasmic side of the membrane. Either may be capable of transmitting information about environmental events to the cytoplasm and lead to characteristic cellular response to the stimulus.

The most significant points in the theoretical section are that negative cell membrane surface potential is reduced towards zero by an increase in ionic strength of the medium and by adsorption of cationic molecules, whereas the potential is predictably increased by the close approach of cell surfaces. Possible effects of magnetic and electric fields are also considered. Changes in surface potential brought about by these means may thus lead to cellular responses.

A number of biological phenomena including pinocytosis in free-living amoebae, a contractile response at the surface of amphibian eggs, transformation of *Naegleria gruberi*, electric field responses and intercellular contact interaction are discussed in relation to this hypothesis.

Although the mechanism proposed affords a fairly satisfactory explanation of the initial events involved in pinocytosis of free-living cells, as well as the induced contractile response of amphibian eggs, the situation is not so clear in tissue cells where some degree of molecular specificity may be necessary in addition to potential changes at the cell surface. There is no direct experimental evidence that surface potential changes are important in the development of intercellular communication. It could however provide a means whereby the simple approach of two cells can spontaneously reciprocally modulate each other's membrane properties, which may well lead to a change in membrane function.

Surface potential changes which have been discussed provide no basis for specificity *per se*. Specificity might reside in the cytoplasmic responsive system, or at the level of membrane-input interaction. This has been discussed by Wolpert and Gingell (1969) and is not considered here.

It remains to be seen whether the concept of membrane surface potential as a central part of a transducer mechanism has general applicability in cellular interactions.

N

Acknowledgement

I would like to thank Dr. A. E. Warner for her criticism of the manuscript.

REFERENCES

Abercrombie, M. (1961) *Exptl. Cell Res.* Suppl. **8**, 188
Abercrombie, M. (1967) *Nat. Cancer Inst. Monogr.*, **26**, 249
Allison, A. C. (1968) Personal communication
Bangham, A. D. and D. Papahadjopoulos (1966) *Biochim. Biophys. Acta.*, **126**, 181
Bangham, A. D., M. M. Standish and N. Miller (1965a) *Nature*, **208**, 1295
Bangham, A. D., M. M. Standish and J. C. Watkins (1965b) *J. Mol. Biol.*, **13**, 138
Barnothy, J. M. (1964) *Biological Effects of Magnetic Fields*, Plenum Press, New York
Barr, L. (1965) *J. Theoret. Biol.*, **9**, 351
Bassett, C. A. L. (1965) *Sci. Amer.*, **213**, 18
Becker, R. O. (1961) *J. Bone Joint Surg.*, **43A**, 643
Brandt, P. W. (1958) *Exptl. Cell Res.*, **15**, 300
Brandt, P. W. and A. R. Freeman (1967) *Science*, **155**, 582
Brandt, P. W. and G. D. Pappas (1960) *J. Biophys. Biochem. Cytol.*, **8**, 675
Bruce, D. L. and J. M. Marshall (1965) *J. Gen. Physiol.*, **49**, 151
Bullivant, S. and W. R. Loewenstein (1968) *J. Cell Biol.*, **37**, 621
Bygrave, F. L. (1967) *Nature*, **214**, 667
Changeux, J. P., J. Thiery, Y. Tung and C. Kittel (1967) *Proc. Natl. Acad. Sci. U.S.*, **57**, 335
Chapman-Andresen, C. (1958) *Compt. Rend. Trav. Lab. Carlsberg*, **31**, 77
Chapman-Andresen, C. (1962) *Compt. Rend. Trav. Lab. Carlsberg.*, **33**, 73
Chapman, D., V. B. Kamat, J. de Grier and S. A. Penkett (1968) *J. Mol. Biol.*, **31**, 101
Cohn, Z. A. and E. Parks (1967a) *J. Exp. Med.*, **125**, 213
Cohn, Z. A. and E. Parks (1967b) *J. Exp. Med.*, **125**, 457
Cook, G. M. W. (1968) *Biol. Rev.*, **43**, 363
Dean, P. M. and E. K. Matthews (1968) *Nature*, **219**, 389
Dresser, D. W. and N. A. Mitchison (1968) *Advan. Immunol.*, **8**, 129
Frankenhaeuser, B. (1960) *J. Physiol.*, **152**, 159
Frankenhaeuser, B. and A. L. Hodgkin (1957) *J. Physiol.*, **137**, 218
Furshpan, E. J. and D. D. Potter (1969) *Current Topics in Developmental Biology*, Vol. 3, Academic Press, New York
Gingell, D. (1966) Unpublished observations
Gingell, D. (1967a) Ph.D. Dissertation, University of London
Gingell, D. (1967b) *J. Theoret. Biol.*, **17**, 451
Gingell, D. (1968) *J. Theoret. Biol.*, **19**, 340
Gingell, D. (1970) *J. Embryol. Exp. Morph.*, **23**, 583
Gingell, D. and J. F. Palmer (1968) *Nature*, **217**, 98
Gingell, D., D. R. Garrod and J. F. Palmer (1970) In *Calcium and Cellular Function*, J. A. Churchill, London
Goldman, D. E. (1943) *J. Gen. Physiol.*, **27**, 37
Grobstein, C. (1961) *Exptl. Cell Res. Suppl.* **8**, 234
Gustafson T. and L. Wolpert (1967) *Biol. Rev.*, **42**, 442
Hardy, D. A. and N. R. Ling (1969) *Nature*, **221**, 545
Hasselbach, W. (1964) *Prog. Biophys. Biophys. Chem.*; **14**, 169
Hill, T. L. (1967) *Proc. Natl. Acad. Sci. U.S.*, **58**, 111
Hodgkin, A. L. and W. K. Chandler (1965) *J. Gen. Physiol.*, **48**, 27
Hodgkin, A. L. and A. F. Huxley (1952) *J. Physiol. (London)*, **117**, 500
Holtfreter, J. (1943) *J. Exp. Zool.*, **93**, 251
Jahn, T. L. (1966) *J. Cell Physiol.*, **68**, 135
Jones, A. R., T. L. Jahn and J. R. Fonesca (1966) *J. Cell Physiol.*, **68**, 127
Katchalsky, A., D. Danon and A. Nevo (1959) *Biochim. Biophys. Acta.*, **33**, 120

Kholodov, Y. A. (1966) *National Aeronautics and Space Administration Technical Translation*, NASA TTF—465
Korn, E. D. (1966) *Science*, **153**, 1491
Korohoda, W., J. A. Forrester, K. G. Moreman and E. J. Ambrose (1968) *Nature*, **217**, 615
Kroeger, H. and M. Lezzi (1966) *Ann. Rev. Enzymol.*, **11**, 1
Lesseps, R. J. (1967) *J. Cell Biol.*, **34**, 173
Lipman, M. (1968) In R. Fleishmajer and R. Billingham (Eds.), *Proceedings of a Symposium on Epithelial Mesenchymal Interactions*, Williams and Wilkins, Baltimore, p. 208
Loewenstein, W. R. (1966) *Ann. N.Y. Acad. Sci.*, **137**, 441
Loewenstein, W. R. (1967) *Develop. Biol.*, **15**, 503
Loewenstein, W. R. (1968) *Perspectives Biol. Med.*, **11**, 262
Lubin, M. (1967) *Nature*, **213**, 451
Luttgau, H. C. and Niedergerke, R. (1958) *J. Physiol.*, **143**, 486
MacGillivray, A. D. (1969) *J. Theoret. Biol.*, **23**, 205
McClare, C. W. F. (1968) *Nature*, **216**, 766
Moscona, (1968) *Develop. Biol.*, **18**, 250
Mueller, P. and D. O. Rudin (1968) *Nature*, **217**, 583
Nachmias, V. T. (1968) *Exptl. Cell Res.*, **51**, 347
Nelson, D. S. and S. V. Boyden (1967) *Brit. Med. Bull.*, **23**, 15
Papahadjopoulos, D. and A. D. Bangham (1966) *Biochim. Biophys. Acta.*, **126**, 185
Parsegian, A. (1969) *Biophys. J. Soc. Abstracts*, **9**, A41
Passow, H. (1964) *Réunion Internationale de Biophysique*, Paris
Passow, H. (1965) *Proceedings of the International Union of Physical Sciences, Tokyo*, **1V**, 555
Podolsky, R. J. (1968) *Symp. Soc. Exp. Biol.*, **22**, 87
Polissar, M. J. (1954) In F. H. Johnson, H. Eyring and M. J. Polissar (Eds.), *The Kinetic Basis of Molecular Biology*, John Wiley, New York
Pulvertaft, R. S. V. and L. Weiss (1963) *J. Pathol. Bacteriol.*, **85**, 473
Rambourg, A. and C. P. Leblond (1967) *J. Cell Biol.*, **32**, 27
Rammler, D. H. and A. Zaffaroni (1967) *Ann. N.Y. Acad. Sci.*, **141**, 13
Rasmussen, H. and A. Tenenhouse (1968) *Proc. Natl. Acad. Sci. U.S.*, **59**, 1364
Ryser, H. J. P. (1968) *Science*, **159**, 390
Schumaker, V. N. (1958) *Exptl. Cell Res.*, **15**, 314
Seeman, P. (1967) *J. Cell Biol.*, **32**, 55
Shah, D. O. and J. H. Schulman (1965) *J. Lipid Res.*, **6**, 341
Singer, and I. Tasaki (1968) In '*Biological Membranes, Fact and Function*', Academic Press, New York
Steinberg, M. S. (1964) In *Cellular Membranes in Development*, Academic Press, New York
Stoker, M. (1967) In *Current Topics in Developmental Biology*, **2**, 107
Subak-Sharpe, H., R. Burk and J. Pitts (1966) *Heredity*, **21**, 342
Sutherland, E. W., G. A. Robison and R. W. Butcher (1968) *Circulation*, **37**, 279
Verwey, E. J. W. and J. Th. G. Overbeek (1948) *Theory of the Stability of Lyophobic Colloids*, Elsevier, Amsterdam
Wall, F. T. and J. Berkovitz (1957) *J. Chem. Phys.*, **26**, 114
Weiss, P. (1949) *Yale J. Biol. Med.*, **19**, 235
Willmer, E. N. (1963) *Symp. Soc. Exp. Biol.*, **17**, 215
Wolpert, L. and D. Gingell (1969) In *CIBA Symposium on Homeostatic Regulators*, J. A. Churchill
Wolpert, L. and D. Gingell (1968) *Symp. Soc. Exp. Biol.*, **22**
Wyatt, H. V. (1964) *J. Theoret. Biol.*, **6**, 441

APPENDIX

Figure 3 shows the potential profile between two surfaces bearing equal fixed charges σ of the same sign on planes I, II ($x=0$, $2d$) which are situated at a distance $x = -t$ from an impenetrable region. The degree of penetrability of the superficial material of thickness $-t$ is α.

We define the conditions

$$x = -t \quad \psi = \psi_t$$

$$\left.\begin{array}{l} x = 0 \\ x = 2d \end{array}\right\} \psi = \psi_0$$

$$x = d \quad \psi = \psi_d$$

Overall electrical neutrality requires that the fixed charge σ is equal to the counter-ion charge ρ. Since the system is symmetrical about d

$$\sigma = -\int_{-t}^{-0} \rho dx - \int_{+0}^{d} \rho dx \tag{1}$$

where $+0$ and -0 conveniently indicate the direction from which zero is approached. Poisson's equation in one dimension can be written,

$$\frac{d^2\psi}{dx^2} = -\frac{4\pi\rho}{D} \tag{2}$$

where D is the dielectric constant of the aqueous phase in bulk. Substituting ρ from (2) into (1)

$$\sigma = -\frac{D}{4\pi}\left\{ \int_{-t}^{-0} -\frac{d^2\psi}{dx^2}dx - \int_{+0}^{d} -\frac{d^2\psi}{dx^2}dx \right\}$$

$$= \frac{D}{4\pi}\left\{ \left(\frac{d\psi}{dx}\right)_{-0} - \left(\frac{d\psi}{dx}\right)_{-t} + \left(\frac{d\psi}{dx}\right)_{d} - \left(\frac{d\psi}{dx}\right)_{+0} \right\}$$

Midway between the planes I and II symmetry requires that

$$\left(\frac{d\psi}{dx}\right)_{d} = 0$$

By Gauss's law it is also necessary that at the uncharged boundary $x = -t$

$$\left(\frac{d\psi}{dx}\right)_{-t} = 0$$

Therefore, the expression for σ reduces to

$$\sigma = \frac{D}{4\pi}\left\{ \left(\frac{d\psi}{dx}\right)_{-0} - \left(\frac{d\psi}{dx}\right)_{+0} \right\} \tag{3}$$

The fundamental differential equation (Verwey and Overbeek, 1948) in one dimension

$$\frac{d^2\psi}{dx^2} = \frac{8\pi nze}{D}\sinh(ze\psi/kT)$$

where $k=$Boltzmann's constant; $z=$counterion valency; $e=$charge on the election; $n=$number of ions per c.c. in bulk solution, reduces for small values of ψ to

$$\frac{d^2\psi}{dx^2} = \frac{8\pi ne^2z^2\psi}{DkT} = \kappa^2\psi \qquad (4)$$

Where κ is the Debye-Hückel parameter, modified in the region $-0 \geqslant x \geqslant -t$ by a factor $(1-\alpha)$ to take into account the incomplete penetrability of the surface layer (Haydon and Tayler, 1960).

Particular solution of the simple differential equation (4) (Gingell, 1967b) leads to expression for $\frac{d\psi}{dx}$ in the regions $-0 \geqslant x \geqslant -t$ and $0 \geqslant x \geqslant +0$.

Substitution for $\left(\frac{d\psi}{dx}\right)_{+0}$ and $\left(\frac{d\psi}{dx}\right)_{-0}$ thus obtained in equation (3) putting $\beta=(1-\alpha)^{\frac{1}{2}}$, gives

$$\sigma = -\frac{D\kappa\psi_0}{4\pi}(\beta\tanh \kappa\beta t+\tanh \kappa d) \qquad (5)$$

ψ_0 can be expressed in terms of ψ_c, thus,

$$\psi_t = -\frac{4\pi\sigma}{D\kappa}\text{sech } \kappa\beta t(\beta\tanh \kappa\beta t+\tanh \kappa d)^{-1} \qquad (6)$$

Equations (5) and (6) are the required explicit relationships.

Index